# Acoustic Imaging with Electronic Circuits

**Advances in Electronics and Electron Physics**

*Edited by*
L. MARTON
*Smithsonian Institution, Washington, D.C.*

*Associate Editor*
CLAIRE MARTON

SUPPLEMENTS

1. Electroluminescence and Related Effects, 1963 HENRY F. IVEY
2. Optical Masers, 1964 GEORGE BIRNBAUM
3. Narrow Angle Electron Guns and Cathode Ray Tubes, 1968 HILARY MOSS
4. Electron Beam and Laser Beam Technology, 1968 L. MARTON AND A. B. EL-KAREH, EDS.
5. Linear Ferrite Devices for Microwave Applications, 1968 WILHELM H. VON AULOCK AND CLIFFORD E. FAY
6. Electron Probe Microanalysis, 1969 A. J. TOUSIMIS AND L. MARTON, EDS.
7. Quadrupoles in Electron Lens Design, 1969 P. W. HAWKES
8. Charge Transfer Devices, 1975 CARLO H. SÉQUIN AND MICHAEL F. TOMPSETT
9. Sequency Theory: Foundations and Applications, 1977 HENNING F. HARMUTH
10. Computer Techniques for Image Processing in Electron Microscopy, 1978 W. O. SAXTON
11. Acoustic Imaging with Electronic Circuits, 1979 HENNING F. HARMUTH
12. Image Transmission Techniques, 1979 WILLIAM K. PRATT

# Acoustic Imaging with Electronic Circuits

HENNING F. HARMUTH

DEPARTMENT OF ELECTRICAL ENGINEERING
THE CATHOLIC UNIVERSITY OF AMERICA
WASHINGTON, D.C.

1979

ACADEMIC PRESS New York San Francisco London
*A Subsidiary of Harcourt Brace Jovanovich, Publishers*

ACADEMIC PRESS, INC.
111 Fifth Avenue, New York, New York 10003

*United Kingdom Edition published by*
ACADEMIC PRESS, INC. (LONDON) LTD.
24/28 Oval Road, London NW1 7DX

LIBRARY OF CONGRESS CATALOG CARD NUMBER: 63–12814

ISBN 0–12–014571–5

PRINTED IN THE UNITED STATES OF AMERICA

*79 80 81 82 83 84 9 8 7 6 5 4 3 2 1*

*To Heinz M. Schlicke*

*Interference Control Co., Milwaukee, Wisconsin*

*for fifteen years of help*

# Contents*



* Equations are numbered consecutively within each of Sections 1.1 to 9.5. Reference to an equation in a different section is made by writing the number of the section in front of the number of the equation, e.g., Eq. (2.2-9) for Eq. (9) in Section 2.2.

Illustrations and tables are numbered consecutively within each section, with the number of the section given first, e.g., Fig. 5.4-2, Table 3.1-1.

References are characterized by the name of the (first) author, the year of publication, and a lower cased latin letter if more than one reference by the same (first) author is listed for the year.

## Foreword

Early in 1977 we published as Supplement 9 to *Advances in Electronics and Electron Physics* a monograph entitled "Sequency Theory: Foundations and Applications" by Henning F. Harmuth. It shows his versatility that, a relatively short time later, we can present another monograph as a further supplement to this series.

The subject of acoustic imaging is a rather new one. While Dr. Harmuth restricts himself in this volume to the problem of the use of electrical circuits in producing acoustic images, the ample bibliography allows the student of the wider problem to dig deeper. The interest aroused by the earlier monograph makes it most likely that the present one will find a warm reception.

L. Marton
C. Marton

# Preface

Nature uses acoustic waves to a much lesser extent than electromagnetic waves for the transmission of information. Our ears are sensitive up to a frequency of about 16 kHz, which means we can receive about 32,000 samples of a sound wave per second. On the other hand, a TV picture, with the standard of 625 lines and 25 frames per second, delivers $625^2 \times 25 = 9{,}765{,}625$ samples per second to our eyes; the number is still higher if the pictures are in color rather than in black and white. This shows that our eyes receive information at a rate of at least two or three orders of magnitude higher than our ears.

The large size of the sensor arrays required for the reception of acoustic waves seems to be the explanation for this different exploitation of acoustic and electromagnetic waves in nature. The sensor array for acoustic waves consists typically of two ears, while the sensor array for light waves in the human eye contains more than a hundred million sensors.

The scientific and practical development of image generation by acoustic rather than electromagnetic waves was doubtlessly handicapped by the lack of an example provided by nature. When the possibility and the potential of image generation by acoustic waves was finally recognized, the development was along the lines previously explored for electromagnetic waves. The acoustic lens was based on the optical lens, sonar followed radar, and acoustical holography was an offspring of optical holography.

Acoustic imaging with electronic circuits broke with the traditional pattern. This method was invented and developed without any example in optics or "electromagnetics." Indeed, the method could only be applied to optical imaging if the response time of the electronic circuits could be reduced by a factor of about $10^{-8}$.

The transition from the first theoretical concept to the first experimental equipment was very fast, less than five years. This had much to do with the high state of development of our electronic technology. Equally important was the quick recognition of the potential of real time, reliable acoustic imaging for submarines. A. Cecelsky, a former submarine officer of the U.S. Navy, initiated the first development program within the Office of Naval Research; he was succeeded by M. A. Blizard, who continued the strong support until the first equipment was built. Credit for the successful experimental work is also due to J. F. Ballon, R. D. Matulka, and D. D. Pizinger, all officers of the U.S. Navy assigned to the Office of Naval Research. C. McKinney of the Applied Research Laboratories of the University of Texas at Austin was of great help during the tests at the Lake Travis Test Station near Austin; the author is greatly indebted to him.

# Acoustic Imaging with Electronic Circuits

# 1 Introduction

## 1.1 Imaging with Sound Waves

There are currently four known methods for the generation of images by means of electromagnetic or acoustic waves. The oldest method uses lenses, which can be used for both electromagnetic and acoustic waves. Historically the second method is the echo principle used in radar and sonar; in the form of synthetic aperture radar and sonar, it produces images with an impressive resolution. The third method is holography; it was introduced theoretically by Gabor (1949, 1951) and implemented when the laser became available (Collier *et al.*, 1971; Kock, 1973). The fourth method uses two-dimensional, spatial electric filters; it was developed theoretically and experimentally during the last few years (Harmuth, 1976, 1977; Harmuth *et al.*, 1974). The first moving images by this method were obtained by J. Dierks of the Applied Research Laboratories of the University of Texas at the Lake Travis Test Station near Austin, Texas.

Refer to Fig. 1.1-1a to see what process is performed to generate an image by means of either electromagnetic or acoustic waves. Three wavefronts propagate from the points P1, P2, and P3 of the object plane to a

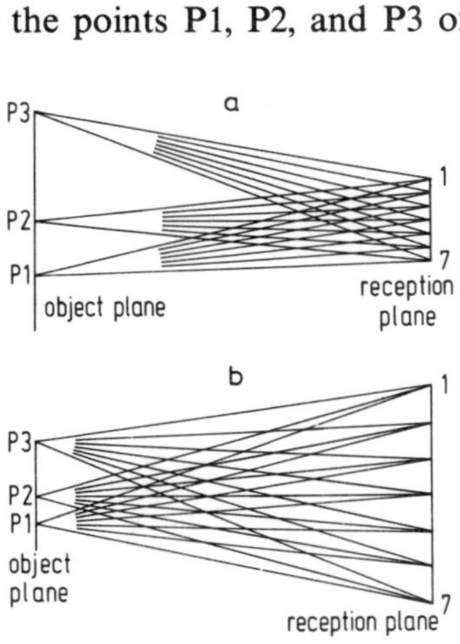


FIG. 1.1-1. Wavefronts generated at three points P1, P2, and P3 in the object plane and received at many points in the reception plane. (a) Reception plane smaller than object plane (telescope). (b) Reception plane larger than object plane (microscope).

reception plane. At every point of the reception plane one receives the sum of three wavefronts. The process of *imaging* means that this sum is decomposed into its components, so that one regains separately the amplitudes of the waves coming from P1, P2, and P3. Mathematically speaking, a linear transformation of the individual wavefronts originating in the object plane produces their sum in the reception plane; the inverse of this transformation reproduces the original wavefronts, which we call the image of the object plane. The various methods of image generation differ primarily in the way they produce the inverse transformation.

In Fig. 1.1-1a the reception plane is smaller than the object plane, while the opposite is true in Fig. 1.1-1b. We use the term *telescope* if the reception plane is smaller, and the term *microscope* if it is larger than the object plane.

The object planes in Fig. 1.1-1 need no further explanation, but it is not evident which physical structure can represent a reception plane. Figure 1.1-2

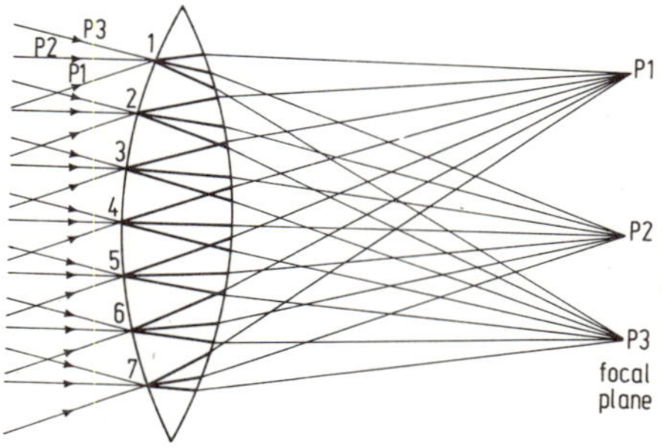


FIG. 1.1-2. Principle of the image generation by a lens.

shows a reception "plane" implemented by a lens. The points 1 to 7 on the reception planes in Fig. 1.1-1 are now the points 1 to 7 on the surface of the lens. The wavefronts from the three points P1, P2, and P3—which are assumed to be at infinite distance in order to have planar rather than spherical wavefronts at the lens—arrive as a sum at the points 1 to 7 but are separated by the lens into three image points in the focal plane.

We have used the term reception plane for the surface of the lens even though it is curved. Later, it will be seen that both the object plane and the reception plane may be curved, but this generalization will be tacitly ignored until needed.

The convex lens in Fig. 1.1-2 produces an image if the propagation velocity of the wavefront is slower in the lens than outside the lens. This is generally so for electromagnetic waves if the lens is made of glass and if it is surrounded by air. An acoustic wave propagates generally faster in a liquid than in a gas, and still faster in a solid. An acoustic lens in air is thus generally concave rather than convex, while an acoustic lens in water—the most important medium for acoustic imaging—may be either concave or convex, depending on the velocity of sound in the material of the lens.

The reception planes in Fig. 1.1-1 and the lens in Fig. 1.1-2 show seven receiving points, and the theory will be developed for a large but finite number of points. This is in contrast to most theoretical work on optical imaging, which assumes nondenumerably many points in order to be able to use differential calculus. This approach works in optics, even though a lens contains a finite number of glass molecules, the eye contains a finite number of light-sensitive cones, and a photographic film contains a finite number of light-sensitive molecules. The finite numbers are so large that the mathematical idealization is sufficiently accurate. This is no longer so in acoustic imaging with electronic circuits. The reception plane consists, in this case, of an array of hydrophones, and their number is conspicuously different from nondenumerably infinite.

Holographic acoustic images have not yet been produced satisfactorily,[1] but sonar images have been. The sonar images differ from our usual concept of an image, which is based on our experience with images formed by our eyes. Figure 1.1-3 shows the difference between the two kinds of images. The

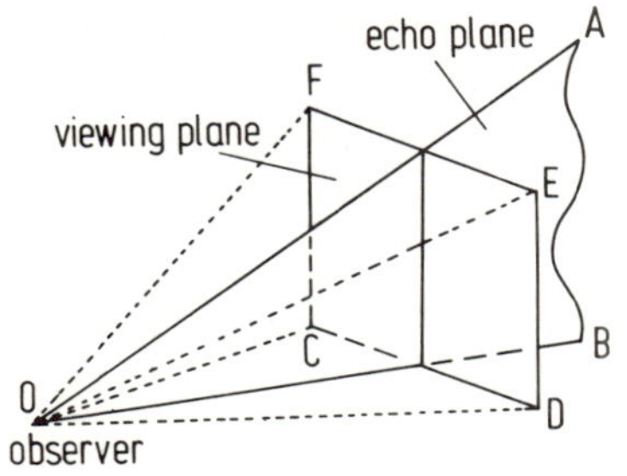


FIG. 1.1-3. The echo plane OAB of which the sonar principle produces images, and the viewing plane CDEF of which the eye and the lens produce images.

sonar principle produces an image of the *echo plane*, while the usual visual images, such as photographs or TV pictures, show an image of the *viewing*

[1] Imaging via acoustical holography using the boundary between a liquid and a gas or a crystal and a gas works theoretically, but its practical implementation proved difficult (Boutin and Mueller, 1967; Farrah *et al.*, 1970; Fritzler *et al.*, 1969; Goetz, 1970; Hildebrand and Brenden, 1972; Korpel and Desmares, 1969; Korpel and Whitman, 1969; Marom *et al.*, 1971; Metherel *et al.*, 1969; Mueller and Sheridan, 1966; Mueller and Keating, 1969; Mueller *et al.*, 1969; Mueller, 1971). The emphasis shifted to the implementation of the principle by electronic circuits; holography in this form has long been known in electrical engineering as synchronous demodulation (Booth and Sutton, 1974; Thorn *et al.*, 1974).

*plane*, which is perpendicular to the echo plane. Sonar images for medical diagnosis have been produced successfully (von Ramm and Thurstone, 1975; Kisslo and von Ramm, 1975).

## 1.2 Acoustic Lens versus Electronic Processing

Figure 1.1-2 shows an optical or acoustic lens that receives three planar wavefronts from the points P1, P2, and P3 at infinity. The lens concentrates the wavefronts in the three points P1, P2, and P3 in the focal plane. The propagation time of any part of a wavefront from its origin at infinity to its image point in the focal plane is the same, regardless of where the wavefront strikes the surface of the lens. The wavefront from point P1 at infinity requires more time to reach the point 1 on the surface of the lens than the points 2, 3, ..., 7. However, the propagation time from the surface of the lens to point P1 in the focal plane is shortest for the section of the wavefront striking the lens at point 1 and becomes increasingly larger for sections of the wavefront striking at the points 2, 3, ..., 7. The lens and the space between the lens and the focal plane act like many delay lines. There is a delay line for every point on the surface of the lens and for every angle of incidence. The heavy lines inside the lens in Fig. 1.1-2 show three "delay lines" originating at each one of the seven points on the surface of the lens.

Let us translate the action of the lens into electric circuits. In order to remain practical, we will use sound waves rather than light waves for illustration. The seven points on the surface of the lens in Fig. 1.1-2 are replaced by seven microphones or hydrophones in Fig. 1.2-1, which transform

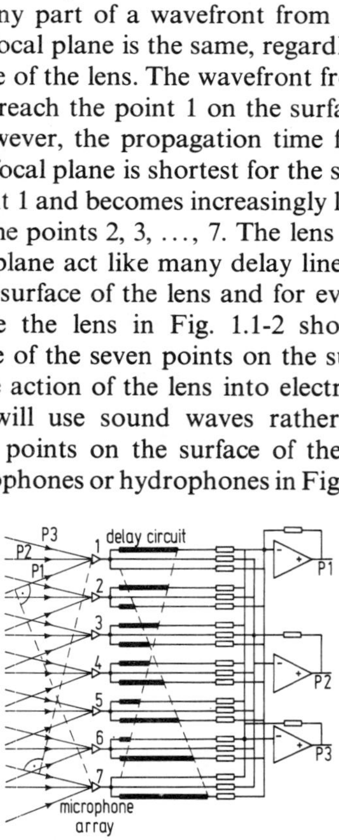


FIG. 1.2-1. Electric equivalent of an acoustic lens according to Fig. 1.1-2, using microphones, electric delay circuits, and summing amplifiers.

the acoustic oscillations into electric oscillations. Let us assume that the seven electromagnetic waves produced travel through the delay circuits shown with the same phase velocity as the acoustic wavefront traveled before reaching the microphones; the other parts of the electric circuit produce negligible delays. The traveling time of a wave from P1 at infinity, first as acoustic wave then as electromagnetic waves, to the output terminal P1 of the topmost summing amplifier, will thus be the same, regardless of which microphone received the sound wave.

The dashed lines in Fig. 1.2-1 are perpendicular to the line of propagation of the respective sound waves. One can readily see how these lines are used to determine the electric length of the delay circuits; the delay produced by these circuits is proportionate to their shown lengths.

Figure 1.2-1 shows seven microphones but only three points P1, P2, and P3 that are resolved. One will expect that an array of seven microphones can resolve seven points without ambiguity. This is shown in Fig. 1.2-2.

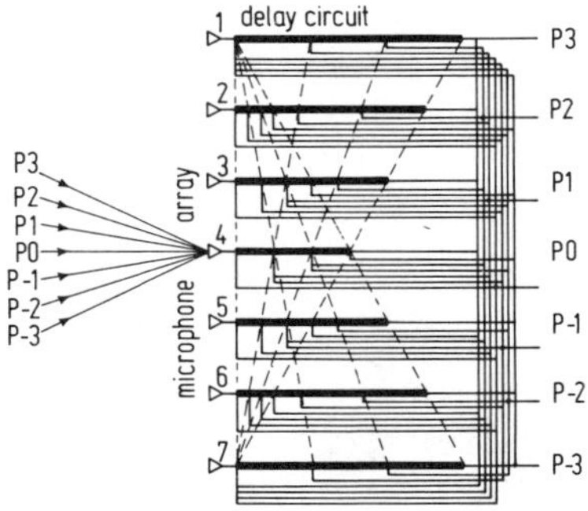


FIG. 1.2-2. Improved circuit of Fig. 1.2-1 yielding seven resolved points.

There are again seven microphones as in Fig. 1.2-1, but delay circuits with taps are connected to the microphones instead of a multitude of delay circuits. The location of the taps is determined by lines perpendicular to the lines of propagation from the points P-3, ... P3 to the microphone array. The resistors and summing amplifiers shown in Fig. 1.2-1 have been omitted in Fig. 1.2-2.

The image produced by the lens in Fig. 1.1-2 is a mirror image, since the sequence of the points P1, P2, and P3 in the focal plane is the reversed sequence of the points at infinity. The same holds true for Fig. 1.2-1. No such reversal occurs in Fig. 1.2-2. It is quite obvious that the location of the output terminals P1 to P3 in Fig. 1.2-1 or P-3 to P3 in Fig. 1.2-2 can be chosen at will. This possibility not only avoids side reversal but also provides a means to eliminate some of the distortions of the simple lens for wide viewing angles.

The imaging process in Fig. 1.1-2 is explained for one dimension only, but we extend it habitually to two dimensions by rotating the cut of the lens around the line from point 4 on the lens surface to P2 in the focal plane. This implies the use of polar coordinates. The choice of polar coordinates is based on technology. It is easier to compute and grind a lens in polar than in cartesian coordinates. The image is usually required to have a rectangular shape; e.g., modern photographs and TV screens are rectangular rather than circular, which favors cartesian coordinates.

The extension of electronic circuits from one to two dimensions can be done in polar coordinates too, but technology favors cartesian coordinates. Let us assume the circuit of Fig. 1.2-2 is mounted on a printed circuit card, without the microphones, but with the resistors and summing amplifiers that were left out in Fig. 1.2-2. The transition to two dimensions is accomplished by stacking seven cards vertically and another seven cards horizontally as shown in Fig. 1.2-3. The input terminals of the cards are the terminals 1 to 7

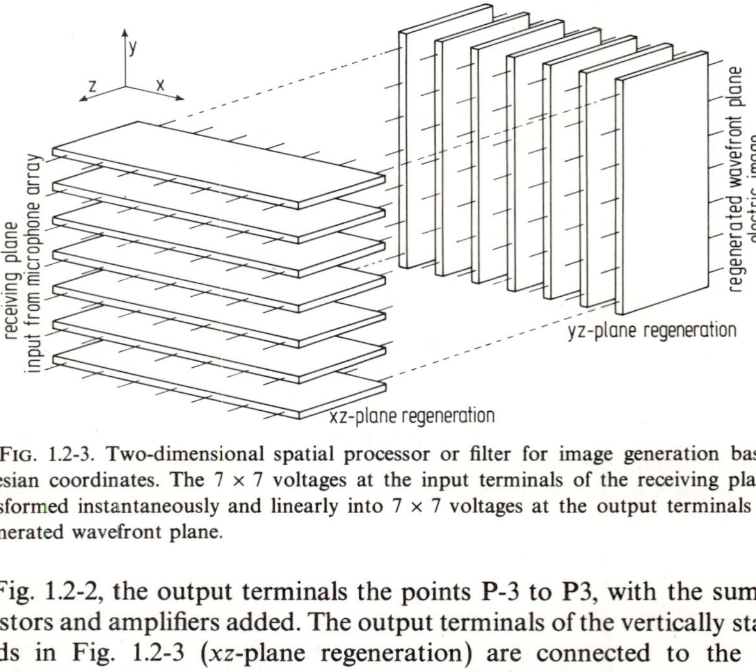


FIG. 1.2-3. Two-dimensional spatial processor or filter for image generation based on cartesian coordinates. The 7 × 7 voltages at the input terminals of the receiving plane are transformed instantaneously and linearly into 7 × 7 voltages at the output terminals of the regenerated wavefront plane.

in Fig. 1.2-2, the output terminals the points P-3 to P3, with the summing resistors and amplifiers added. The output terminals of the vertically stacked cards in Fig. 1.2-3 (*xz*-plane regeneration) are connected to the input terminals of the horizontally stacked cards (*yz*-plane regeneration) as indicated by the four dashed lines; the other connections are not shown, in order to avoid obscuring the picture.

A quadratic array of 7 × 7 microphones is connected to the input terminals of the vertical stack of cards in Fig. 1.2-3. The received acoustic wavefront is transformed by the microphones into a two-dimensional set or array of voltages, and the processor of Fig. 1.2-3 transforms this set of voltages into another set of voltages, which is the *electric image*. The next task is to transform the electric image into a visible optical image. Any electrooptical converter will do this. The best known converter is the television tube, but a light-emitting diode array could be used too.

Let us deviate for a moment and consider the implementation of the processor in Fig. 1.2-3 in polar coordinates. Figure 1.2-4 shows the arrange-

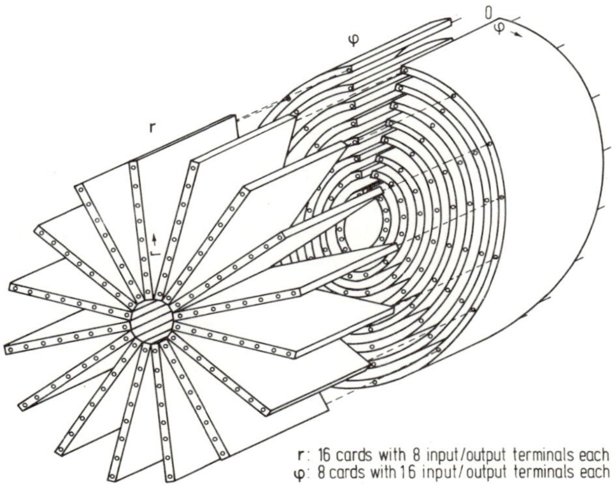


FIG. 1.2-4. Principle of circuit performing a two-dimensional spatial transform in polar coordinates. The cards denoted **r** contain the circuit of Fig. 1.2-2 but with 8 rather than 7 input and output terminals, while the cards denoted **φ** contain the same circuit but with 16 terminals.

ment of printed circuit cards according to Fig. 1.2-2 for polar coordinates. This illustration assumes an array of $8 \times 16 = 128$ microphones rather than $7 \times 7 = 49$ microphones as in Fig. 1.2-3. The microphones have to be mounted according to the pattern of polar rather than cartesian coordinates. The practical implementation of Fig. 1.2-4 appears much more complicated than that of Fig. 1.2-3, but this is not so. Both illustrations show only how the printed circuit cards have to be connected to the microphones, to each other, and to the electrooptical converter. The cards can all be fabricated in the usual way and mounted in the conventional printed circuit card hangers; only the wiring between the cards has to be done according to Figs. 1.2-3 and 1.2-4. The mechanical construction of the microphone array and of the electrooptical display differs. Figure 1.2-5 shows the arrangement of an essentially equal number of microphones in a polar array of $8 \times 16 = 128$ microphones and in a cartesian array of $10 \times 10 = 100$

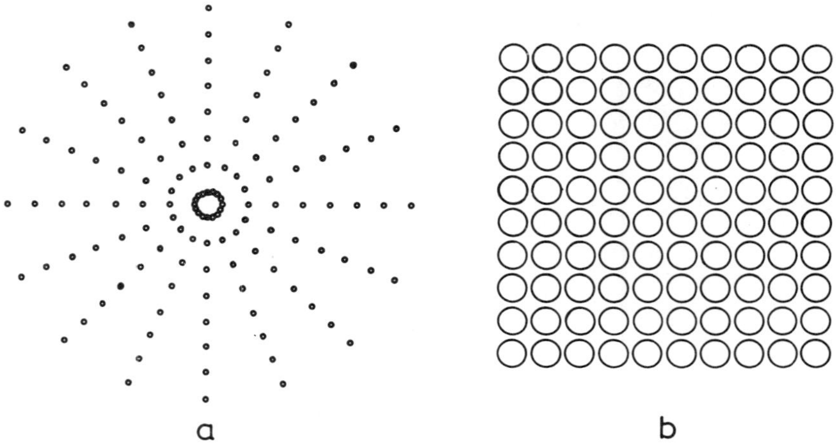


FIG. 1.2-5. Arrays with $8 \times 16 = 128$ microphones in polar coordinates (a), and with $10 \times 10 = 100$ microphones in cartesian coordinates (b).

microphones. The electrooptical converter should be constructed analogously. For a light-emitting diode display, one must mount the diodes like the microphones in Fig. 1.2-5. A cathode ray tube, as used in television, is scanning according to cartesian coordinates, but polar coordinate scanning is often used for radar displays.

The microphones in Fig. 1.2-5b cover the area of the array more uniformly than those in Fig. 1.2-5a. This translates practically into a uniform resolution of images based on cartesian coordinates, while polar coordinates favor a resolution that is better in the center than at the edges. The difference does not show up in optics, since optics is based, as pointed out before, on nondenumerably many points on the surface of the lens and in the focal plane of Fig. 1.1-2, which would imply nondenumerably many microphones in Fig. 1.2-5 and nondenumerably many terminals in Figs. 1.2-3 and 1.2-4. In principle, one cannot say whether uniform or nonuniform resolution is preferable. For photographs or TV pictures, we want generally uniform resolution. The eye, on the other hand, has conspicuously better resolution in the center than farther away from the center. The reason is obviously that we can move the eyes to get the best resolution where we want it, but we cannot change the resolution of a photograph or TV picture once it has been produced.

The construction of a microphone or hydrophone array according to Fig. 1.2-5b is preferable to that in Fig. 1.2-5a, since the microphones or hydrophones in the center of Fig. 1.2-5a must be very small. The use of cartesian coordinates is thus favored by the receptor array as well as by our preference for rectangular pictures. The electronic processor can easily be built for cartesian coordinates too; thus this is the preferred system of coordinates.

The choice is not so clear-cut if a lens is used. In principle, one could use two cylindrical lenses positioned perpendicular to each other like the printed circuit cards in Fig. 1.2-3 and base everything on cartesian coordinates. However, lenses for imaging are usually circular, and the mismatch of a lens favoring polar coordinates and an image in cartesian coordinates causes a loss. This loss can readily be inferred for the popular photographic format of 24 × 36 $mm^2$ from Fig. 1.2-6a. A circular lens produces an image

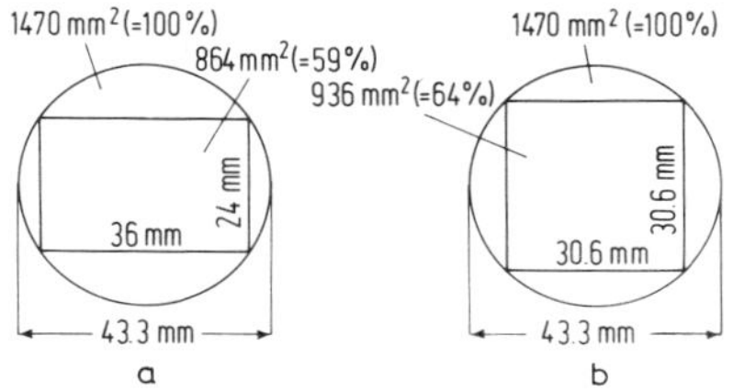


FIG. 1.2-6. A rectangular image with a ratio width/height = 3/2 uses 59% of the circular image (a), while a square image uses 64% of the circular image (b).

with a diameter of 43.3 mm that has nowhere less resolution than in the corners of the rectangular area. The area of the circle is 1470 $mm^2$, while that of the rectangle is 864 $mm^2$ or 59% of the area of the circle. Hence, only 59% of the image produced by the lens is used. Figure 1.2-6b shows that a square image format will increase the used image area to 936 $mm^2$ or 64% of the area of the circle. The loss of area is no longer important enough in photography to produce circular instead of rectangular images, but telescopes and microscopes continue to use the circular format. The mismatch of coordinate systems is a problem for acoustic lenses, since their diameter is not very large compared with the wavelength. A diameter of 10 mm for an optical lens means 20,000 wavelengths for blue light[1] and 16,700 wavelengths for red light. A sinusoidal sound wave in water has a wavelength of about 1.5 cm for a frequency of 100 kHz, and about 1.5 mm for a frequency of 1 MHz. A lens diameter of 100 wavelengths implies lenses of 1.5 m and 15 cm diameter, which is about as large as one can realistically make them. The large diameter of the optical lens in terms of wavelength implies that the resolution of the photographic camera is limited by the film rather than the lens, and this is the reason why we can afford to use only 59% or 64% of the image points produced by the lens according

[1] The wavelength of blue light is about 0.45 $\mu$m, that of red light about 0.65 $\mu$m.

to Fig. 1.2-6. The number of points resolved by the film is typically in the order of a few percent of the points resolved by the lens. This is not so for telescopes and microscopes used for scientific observations, and it is not so for acoustic lenses; the diameter of the lens limits the number of resolvable points. If one uses only 64% of the points resolved by the lens due to a mismatch of coordinate systems according to Fig. 1.2-6b, one must increase the area of the lens by $100 \times 0.36/0.64 = 56\%$.

Let us turn to the basic equipment required for the generation of acoustic images. Since the primary medium for the operation of this equipment is water, which includes the human body if acoustic imaging is used instead of X rays for medical diagnosis, we will use the terms hydrophone and sound projector rather than microphone and loudspeaker. Figure 1.2-7 shows a sound projector that insonifies the object plane. The

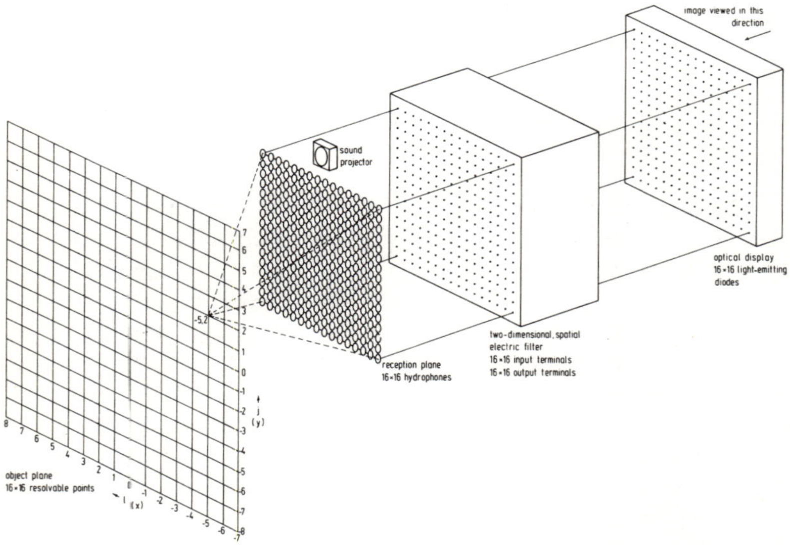


FIG. 1.2-7. The principle of acoustic imaging by means of a two-dimensional, spatial electric filter.

sound produced may be a continuous sinusoidal wave, a pulsed sinusoidal wave, or a pulse without any sinusoidal carrier.[1] The sound wave is scattered in the object plane, which one may take to be the bottom of the sea. The many partial wavefronts produced by the individual scattering points sum to

[1] If pulses are used and one wants to produce moving rather than still images, one must radiate at least 25 pulses per second.

a wavefront that is received by the 16 × 16 hydrophones in the reception plane. The 16 × 16 sound pressures measured at any moment by the hydrophones are connected through a linear transformation to the sound pressures in the object plane. There are, of course, many more than 16 × 16 scattering points in the object plane. However, since we measure only 16 × 16 sound pressures and know that there is a linear relationship between the sound pressures in the object plane and in the reception plane, we can calculate only the sound pressures at a set of 16 × 16 points in the object plane. In principle, we can choose these points, but the previous discussion showed that a rectangular display calls for operation in cartesian coordinates, and the 16 × 16 points in the object plane in Fig. 1.2-7 are thus chosen according to cartesian coordinates. Hence, we act as if the received wavefront were produced by scattering at the 16 × 16 points of the grid of the object plane in Fig. 1.2-7. This is, of course, not what really happens. The sound pressures obtained in this way for the 16 × 16 points of the grid represent actually some average of all the scattering points in a certain area around the 16 × 16 grid points.

It all sounds very different from what we are used to in optics, but this is so only because we use differential calculus and nondenumerably many points in the theory of optics. The brightness of an image "point" produced by a lens always represents in reality some average brightness of a certain area in the object plane. This "certain area" is small relative to the area of the object plane, but it can be absolutely very large; indeed, the absolute size of the "certain area" increases toward infinity as the distance between lens and object plane increases toward infinity.

The determination of the sound pressures at the 16 × 16 grid points of the object plane from the 16 × 16 sound pressures measured by the hydrophones is done by the two-dimensional, spatial electric filter in Fig. 1.2-7. For the time being, this filter consists of 2 × 16 printed circuit cards with 16 input terminals each, analogous to the 2 × 7 cards with 7 input terminals each shown in Fig. 1.2-3. The input voltages to the filter represent the sound pressures measured by the hydrophones; the output voltages represent the sound pressures at the grid points in the object plane.

The output voltages of the electric filter are fed to an optical display consisting of light-emitting diodes (Fig. 1.2-7). The brightness of the diodes represents the sound pressures at the grid points of the object plane. Practical equipment uses a sampling circuit and a television display, which means a cathode ray tube, instead of the light-emitting diodes.

To summarize: acoustic imaging with electronic circuits uses sound waves to obtain information about the object plane, electric voltages to process this information, and light waves to present the processed information to our eyes. If we had an organ like the eye that was sensitive to two-dimensional, spatial sets of electric voltages, we could save the electrooptical conversion.

If acoustic technology were more versatile, we could avoid the conversions to and from electric voltages, but this is not even possible when one uses acoustic lenses.

For acoustic imaging with a lens, one would put an acoustic lens in front of the hydrophone array in Fig. 1.2-7 and eliminate the two-dimensional, spatial electric filter. The hydrophones would be directly connected to the optical display. This type of imaging is described in papers by Beaver *et al.* (1976), Jones and Gilmour (1974, 1976), Maginness *et al.* (1976), Meindl (1976a,b), and Meindl *et al.* (1976).

Refer to Fig. 1.2-8 for a more detailed discussion of the difference

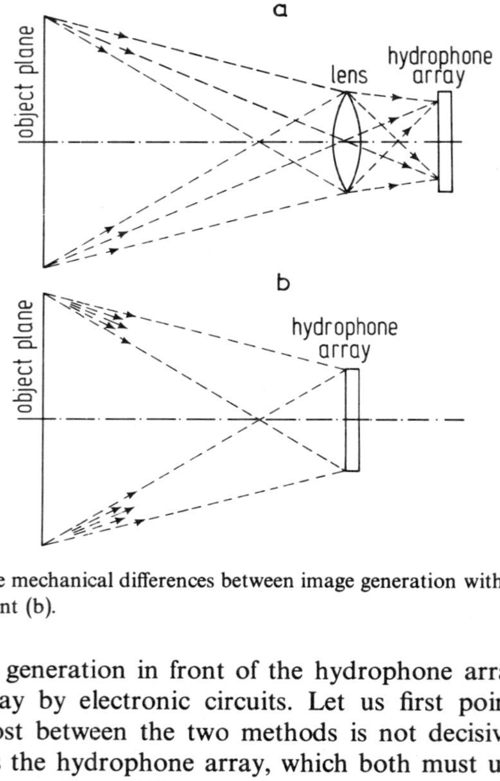


FIG. 1.2-8. The mechanical differences between image generation with a lens (a) and with electronic equipment (b).

between image generation in front of the hydrophone array by a lens or behind the array by electronic circuits. Let us first point out that the difference in cost between the two methods is not decisive,[1] because the costliest item is the hydrophone array, which both must use, and because there are more important differences.

Consider the imaging equipment developed by Jones and Gilmour (1974,

[1] This statement is true only for image generation with its implied high resolution. There are uses requiring a much poorer resolution for which imaging with a lens is considerably cheaper than with electronic circuits.

1976). It uses an acoustic lens system with a diameter of 25 cm and a distance of about 75 cm from the surface of the first lens to the hydrophone array. The volume of the enclosed space is about 50 $dm^3$ or liters. Since the space is filled either with water or with the material of the lenses, the weight of the equipment in front of the hydrophone array is upward of 50 kg. If used underwater, the weight would not matter; but it would matter in applications, such as medical diagnosis, where the equipment is surrounded by air. The operating frequency of the equipment is 3 MHz, which implies a wavelength of 0.5 mm. We cannot now build electronic imaging circuits for a frequency of 3 MHz, since the required operational amplifiers are not yet available at acceptable cost for such a high frequency.

The range of an acoustic imaging system operating in water at 3 MHz is a few meters. This is sufficient for medical diagnosis and for viewing through very turbid water. In clean water, we can see just as far with light, and there is thus no need to use acoustic waves. If we want to achieve a viewing range of 100 m and more, we must decrease the frequency drastically. Consider the reduction from 3 MHz to 300 kHz. The diameter of the lens would increase from 25 cm to 2.5 m, and the length of the lens system from 75 cm to 7.5 m. The volume would not increase by a factor of 10 but by a factor of $10^3$, that is from 50 $dm^3$ to 50 $m^3$. The weight would go from 50 kg to 50 tons, but this would not be important since a long-range viewing system would only be used underwater. However, the inertia and the drag of such a large size acoustic lens system would be hard to justify for practical use.

The fields of application for acoustic lenses and electronic circuits are thus well distinguishable. The size of the lens system becomes more acceptable as the frequency increases, while the implementation of the electronic circuits becomes easier as the frequency decreases. At this time, the acoustic lens is definitely preferable well above 1 MHz, while electronic circuits are preferable well below[1] 1 MHz. Progress in technology is going to increase this frequency of 1 MHz, since electronic circuits are advancing rapidly, while there is no such change in geometric optics or rather in the geometry of imaging.

Figure 1.2-9 shows a hydrophone array with the dimensions $1.25 \times 1.25$ $m^2$. It is evident that maneuvering an array of this size is difficult and that one would want to avoid compounding the problem by adding several lenses of about the same size at a distance of several times 1.25 m. The array

[1] The fields of application are not so clearly separated for acoustic imaging in air. Since the velocity of sound in air is about one-fifth the velocity in water, one is inclined to replace 1 MHz by 200 kHz. However, electronic circuits readily operate at 200 kHz and higher frequencies, and a decision between lens or electronic circuits is thus more difficult to make.

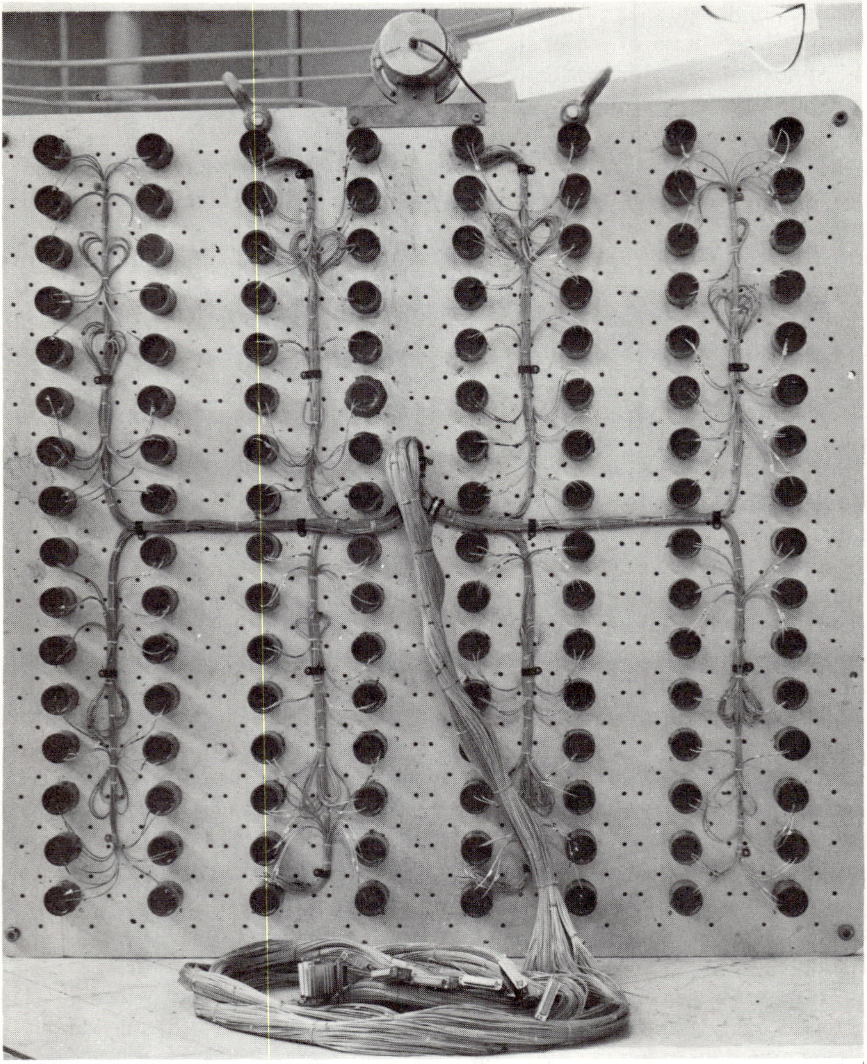

FIG. 1.2-9. Array with 16 × 16 hydrophones during assembly. A sound projector is shown on top.

contains 16 × 16 hydrophones, even though it appears to contain only half as many. A sound projector is mounted on top, in correspondence to the array and projector shown in Fig. 1.2-7.

The equipment in Fig. 1.2-7 processes simultaneously sound pressures at many hydrophones or voltages at many terminals. This makes the equipment very powerful but also expensive. Two different routes have been pursued to trade performance for cost. Equipment using the echo principle produces two-dimensional images in the echo plane by using only one row of hydrophones instead of the two-dimensional array (von Ramm and Thurstone, 1975). Another way is to forgo the generation of moving images and produce only still images; the parallel processing can, in this case, be replaced by serial processing (Marom *et al.*, 1971; Booth and Sutton, 1974).

Our current electronic technology is basically one-dimensional, while images are inherently two-, three-, or four-dimensional ($xy$, $xyz$ or $xyt$, $xyzt$). The one-dimensional operation of most electronic equipment is usually expressed by saying that it operates serially rather than parallel. The serial operation of digital computers is a great handicap for their application to image processing. If we look for long-term guidance on how to develop acoustic imaging equipment, we must look beyond our current one-dimensional electronic technology. How does nature solve the problem of image generation? Due to the rapid time variation of light waves, the eye uses a lens in front of the "cone array," rather than a cone array in front with all the processing done in the brain. However, the cones feed parallel to the brain, and this was apparently important enough to put the brain right behind the eyes, even though it would be much better protected if located close to the heart and the lungs. Such a location would have required a long bundle of about 100 million nerves, which is as hard to implement as a bundle of 100 million wires. One can readily see that the wires coming out of the hydrophone array in Fig. 1.2-9 pose a big problem if one advances from 16 × 16 hydrophones to substantially more. The long-term engineering solution is to put the processing equipment right behind the array, even though one would prefer to put the hydrophone array into the water and the processing equipment into a dry place.[1] The analogy to the location of the brain is striking. Electronic equipment working parallel in two dimensions can be built now, but it is still too bulky to be mounted directly onto a hydrophone array. However, there is no sign that the rapid development of electronics will slow down in the foreseeable future. Hence, we will emphasize parallel operation in both theory and circuit design.

[1] A more promising long-term solution is provided by the synthetic aperture techniques discussed in Chapter 9.

## 1.3 Survey of Applications

Acoustic imaging with electronic circuits has reached a stage where the scientific results have to be transferred to development and practical use. This is always a difficult task since the potential users do not know what can be done, and no engineering development is started until the potential users demand it. Hence, in addition to the development of theory and experimental equipment, one must develop useful applications to make acoustic imaging with electronic circuits more than a scientific feat.

There are currently two large recognized areas of application. The one is in seeing underwater, the other in medical diagnosis as an alternative to X rays.

The best short-term applications of seeing underwater are military ones. Finding and classifying submarines, detecting skin divers (or "swimmers" in naval parlance), finding mines and other obstacles in shipping channels, locating sunken ships and objects that fell into the sea, helping a submarine navigate close to ice or other inhomogeneities where it is harder to detect than in homgeneous water, etc.

Submariners are very reluctant to radiate sound waves because of the risk of detection. This situation is very similar to the reluctance to radiate radio waves in times past. The development of spread spectrum techniques over the last twenty years made it possible to reduce the spectral power density of radio signals below that of the natural background noise. This makes it very difficult to detect such radiation without knowledge of the spreading code.[1] The spread spectrum techniques can be used in principle for acoustic imaging too. Instead of using a pure sinusoidal wave or a signal with narrow relative bandwidth for insonification, one may use one with large bandwidth and low spectral power density. Furthermore, the signal may be changed according to a code. At present, we cannot build such a system, since we are still struggling to get better resolution, better dynamic range, etc. Once these problems are overcome, we will turn to the implementation of spread spectrum techniques and synthetic aperture techniques as surely as we did in radio communications and radar.

Over the long run, the civilian applications of acoustic imaging are much more important than the military ones. The developing technology of deep-sea mining calls for equipment that provides a longer viewing distance than the few meters achieved by TV cameras with searchlights. Offshore oil drilling advances steadily farther out and into deeper water. The pipelines

[1] See the special issue on spread spectrum communications of *IEEE Trans. Commun.* **COM-25**, No. 8 (1977).

as well as the drilling towers require maintenance underwater, and this water is frequently so polluted that it is essentially opaque to light waves. Sound waves will penetrate such polluted water.

Once military and commercial applications have brought down the cost of acoustic imaging, one can easily see archaeologists using it to find and recover sunken Phoenician and Greek ships (treasure hunters will prefer Spanish galleons). Those who find this statement overly optimistic may reflect on the change in diving from an expensive commercial operation to the sport of skin diving over the last forty years. The electronic circuits used for acoustic imaging are bulky and expensive, but they are not bulkier or more expensive than the first electronic computers. The computer was reduced to the pocket calculator in thirty years, and there is no reason why a comparable reduction should not be achievable for acoustic imaging equipment.

Let us turn to medical diagnosis. We are currently using X rays to look inside the human body. Sound waves will penetrate the body just as well, and they are currently used to a limited extent for diagnosis. Acoustic images avoid the radiation hazards that are a necessary by-product of X-ray images. Furthermore, X-ray images show differences in the scattering and absorption of X rays, while acoustic images show them for sound waves. Hence, there is no direct exchange of X-ray and acoustic images. This implies that there should be cases where the acoustic image shows more than the X-ray image, and vice versa. Hence, besides the elimination of the radiation hazard, there should be cases in which acoustic images permit a better diagnosis.

If sound can be used to look into the human body, it can be used to look into other objects. In principle, one could foresee acoustic microscopes that produce images of the interior rather than of the surface of many objects. However, the higher velocity of sound in solids and the resolution required put these applications beyond the reach of our current technology.

The generation of acoustic images in air, on the other hand, is easier than in water due to the lower velocity of sound. The difficulty here is that acoustic imaging in air has to compete with optical imaging with light and infrared waves. One could imagine Secret Service agents operating in the dark with infrared-scopes, being outclassed by their opponents who use acoustic imaging equipment as well as infrared-scopes, but this appears to be a very limited application. There have been other suggestions, such as looking through tropical foliage from a helicopter to see what is hidden below or looking through fog and clouds, but none of these potential applications is as convincing as seeing underwater or medical diagnosis.

# 2 Image Generation by Linear Transformations

## 2.1 Transformation in the Medium and Inverse Transformation by Tapped Delay Circuits

Let a wave with the time variation $u(t)$ originate in a point of the object plane. $u(t)$ may be considered to represent the sound pressure, but the investigation applies to any other acoustic or electromagnetic quantity represented by $u(t)$. The wave produces output voltages at all hydrophones in the reception plane, as indicated for the point $l = -5, j = +2$ in Fig. 1.2-7. Let us first calculate the output voltages for one dimension, assuming a plane wave coming from a point at infinity. Figure 2.1-1 shows a line array with $2^n = 16$ hydrophones equally spaced with distance $d$. A hydrophone $k$ has the distance $(k - 2^n/2 + 1)d = id$ from the center of the array. Note that the center is somewhat awkward to define for an even number $2^n$ of hydrophones. An odd number of hydrophones would avoid this difficulty, but powers of 2 lead to simple circuits.

Let the plane wave $W(\beta)$ arrive from the direction $\beta$ and let it have the time variation $u(t)$. It shall produce the voltage $v(2^n/2 - 1, t) = v(k, t)$ at the terminals of the hydrophone $k = 2^n/2 - 1 = 7$ in the center of the array in Fig. 2.1-1. This voltage is the arbitrary reference for the output voltage of all other hydrophones. One may readily recognize that the output voltage $v(k, t) = v(i, t)$ of the general hydrophone $i$ is defined by the relation

$$\begin{aligned} v(i, t) &= v\{[t + i(d/v) \sin \beta]/T\} \\ i &= k - 2^n/2 + 1 \end{aligned} \qquad (1)$$

where $v$ is the propagation velocity of the wavefront, and $T$ the normalization time.

A change from $u(t)$ to $Kv(i, t)$ is the transformation of the sound wave by the medium between object plane and reception plane; $K$ is a constant that accounts for $v(i, t)$ being an output voltage of a hydrophone rather than a sound pressure in front of the hydrophone. The inverse transformation will produce the image of the point radiating the sound wave $u(t)$. To perform this inverse transformation, we need a circuit with $2^n$ input terminals connected to the hydrophones that produce $2^n$ output voltages. The $2^n$ input voltages $v(i, t)$ of Eq. (1) should produce an output voltage at only one terminal, or at least very nearly so.

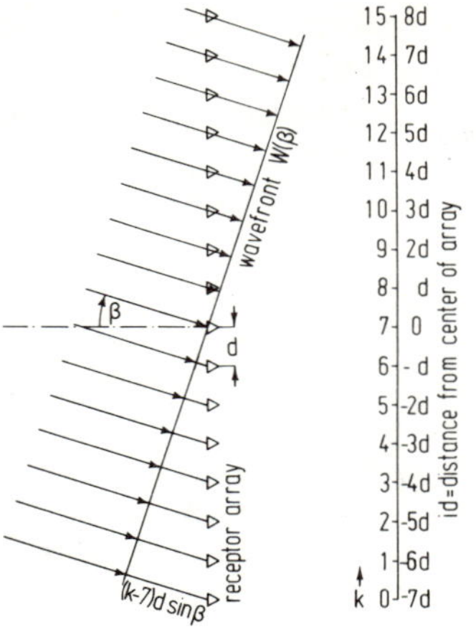


FIG. 2.1-1. Calculation of the output voltages of the receptors of a line array produced by a plane wavefront coming from the point $\beta$ at infinity.

There are currently three known methods to perform this inverse transformation with electronic circuits. The first is based on time-invariant delay circuits. It is the most understandable one, but also the most frustrating one for practical use. The second is based on the Fourier transform. It only works if the time variation of the voltages $v(i, t)$ in Eq. (1) is sinusoidal or very nearly so. Despite this drawback, this method is the most practical one at the present. The third method is based on sampled delay circuits. It is an attractive alternative to the Fourier transform since it lends itself to integrated circuit construction. However, there are doubts whether this method remains attractive when one wants to implement more sophisticated features of image generation.

Let us assume we have a delay circuit that delays the voltage $v(i, t)$ of Eq. (1) by the time $t_\gamma$:

$$t_\gamma = i(d/v) \sin \gamma \tag{2}$$

where $\gamma$ is the *observation angle.*[1] This is a nominal angle for which the circuit is designed to receive a wavefront, while it actually comes from direction $\beta$. An array of $2^n$ hydrophones permits one to design the circuit for $2^n$ observation angles. Hence, $\gamma$ can have $2^n$ different values.

[1] The negative delay times are no problem, since one can add a constant delay $t_0$ that makes $t_0 + t_\gamma$ always positive. The same result can be achieved by using the output voltage of the hydrophone $k = 15$ in Fig. 2.1-1 for $\gamma > 0$ and that of the hydrophone $k = 0$ for $\gamma < 0$ as reference.

The distinction between observation angle $\gamma$ and *angle of incidence* $\beta$ is usually not made in optics. The reason is that the mathematical methods used approximate the imaging process as if the image of a continuum of points rather than a finite number of points could be produced. For an explanation, refer to Fig. 2.1-2 which shows an acoustic lens producing an

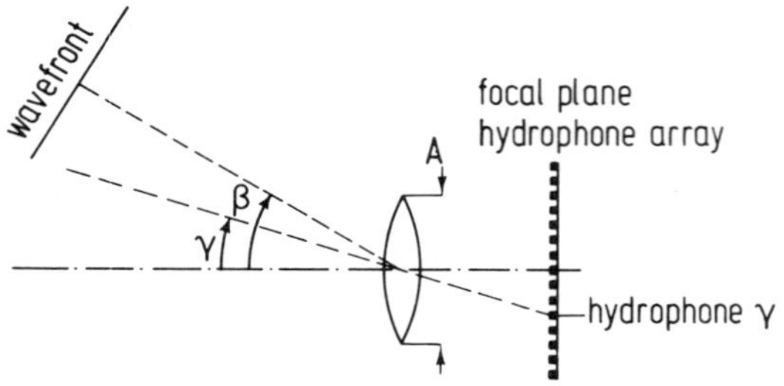


FIG. 2.1-2. Definition of the line aperture $A$, the observation angle $\gamma$, and the angle of incidence $\beta$.

image on a hydrophone array.[1] Obviously, only as many image points can be generated as there are hydrophones. There is one observation angle for each hydrophone, but there are many more possible angles of incidence $\beta$. The same applies to optics, but we have to replace the hydrophones by the light-sensitive cones and bars of the eye or by the light-sensitive molecules of a photographic film, etc. The number of cones, bars, and molecules is so great that we can treat them successfully as nondenumerably many points and use the mathematical methods for the continuum. Acoustic imaging requires more realistic mathematical methods based on a finite number of hydrophones, at least until we can build arrays with millions of hydrophones.

Let us introduce $t_\gamma$ from Eq. (2) into Eq. (1):

$$v(i, t - t_\gamma) = v\{[t + i(d/v)(\sin\beta - \sin\gamma)]/T\} \tag{3}$$

For $\beta = \gamma$ all voltages for $i = 0, \pm 1, \pm 2, \ldots$ are synchronous and can be added algebraically by a summing circuit; the time variation of $u(t)$ or $v(i, t)$ is of no importance. The output voltage representing the observation angle $\gamma$ is the image of the wavefront with angle of incidence $\beta = \gamma$ or of the point $\beta = \gamma$ at infinity.

The next step of sophistication is to ask for the output voltage if $\beta$ and $\gamma$ are not equal. Furthermore, one may ask for the output voltages representing the angles of observation other than $\gamma = \beta$. Both questions turn

[1] An acoustic lens used for image generation may be convex or concave, depending on the relative sound velocities in the lens and the medium around the lens. We always show a convex lens, which makes it easier to think in terms of optics.

out to be two sides of the same problem. In order to solve it, we must specify the time variation $u(t)$ of the wave or of the output voltage $v(i, t)$. We chose a sinusoidal time variation because it is required for imaging by means of the Fourier transform, which is of primary practical interest here. However, one should be very cautious about using sinusoidal waves for acoustic imaging. In order to avoid reverberations, we must use pulsed waves rather than continuous sinusoidal waves. If the pulses are of the order of 1 ms or larger and we use a sinusoidal carrier wave with a frequency of 100 kHz, we gain the advantage of needing a much smaller array than without the carrier. The carrier introduces a much higher attenuation, but the trade of more attenuation for a smaller array is well worth it. On the other hand, if the pulse duration is reduced to 1 $\mu$sec, we need a carrier frequency of upward of 10 MHz. The achievable reduction of the array size is no longer of interest, but the vastly increased attenuation is. An example[1] of sonar imaging equipment that avoids a sinusoidal carrier for this reason is the equipment of von Ramm and Thurstone (1975).

Let us replace the general function $v(i, t - t_\gamma)$ of Eq. (3) by a sinusoidal function:

$$v(i, t - t_\gamma) = V \sin\{2\pi[vt + id(\sin \beta - \sin \gamma)]/\lambda\} \tag{4}$$

where $\lambda = vT$ is the wavelength.

The sum of all voltages $v(i, t - t_\gamma)$ for $i = -2^n/2 + 1, \ldots, 2^n/2$ is denoted by $v(\gamma, \beta, t)$:

$$\begin{aligned} v(\gamma, \beta, t) &= \sum_{i=-2^n/2+1}^{2^n/2} V \sin[2\pi(vt + id\delta)/\lambda] \\ &= 2^n V \Bigg[\sin(2\pi vt/\lambda) \sum_{i=-2^n/2+1}^{2^n/2} 2^{-n} \cos(2\pi d\delta i/\lambda) \\ &\quad + \cos(2\pi vt/\lambda) \sum_{i=-2^n/2+1}^{2^n/2} 2^{-n} \sin(2\pi d\delta i/\lambda)\Bigg] \end{aligned} \tag{5}$$

$$\begin{aligned} \delta &= \sin \beta - \sin \gamma = 2 \sin \tfrac{1}{2}(\beta - \gamma) \cos \tfrac{1}{2}(\beta + \gamma) \\ &\doteq (\beta - \gamma) \cos \tfrac{1}{2}(\beta + \gamma) \qquad \text{for} \quad \beta - \gamma \ll 1 \\ &\doteq \beta - \gamma \qquad \text{for} \quad \beta \ll 1 \quad \text{and} \quad \gamma \ll 1 \end{aligned}$$

[1] Another example of the use of acoustic pulses without a sinusoidal carrier is provided by the South American oilbirds (*Steatornis caripensis*) and the Southeast Asian swiftlets (genus *Collocalia*), who use clicks in the range of human hearing for navigation in the dark (Ross, 1965, particularly the oscillogram on p. 290). The better known navigation system of bats uses chirps, similar to the signals of chirp radar, at ultrasonic frequencies. The click sonar of the birds was brought to the author's attention by H. H. Schreiber of Grumman Aerospace Corporation.

where $\beta - \gamma$ is the angle of incidence which refers to the observation angle $\gamma$ rather than to the array axis.

Each term in the last two sums of Eq. (5) represents the area of a rectangle with the base equal to $2^{-n}$ and the height equal to either $\cos(2\pi d\delta i/\lambda)$ or $\sin(2\pi d\delta i/\lambda)$. The sums themselves represent the area under step functions with steps of width $2^{-n}$. For large values of $2^n$ one may represent these sums by integrals:[1]

$$\begin{aligned}\sum_{i=-2^n/2+1}^{2^n/2} 2^{-n}\cos(2\pi d\delta i/\lambda) &\doteq \int_{-1/2}^{1/2} \cos[2\pi(2^n d\delta/\lambda)(2^{-n}i)]\,d(2^{-n}i)\\ &\doteq \frac{\sin \pi A\delta/\lambda}{\pi A\delta/\lambda}\\ \sum_{i=-2^n/2+1}^{2^n/2} 2^{-n}\sin(2\pi d\delta i/\lambda) &\doteq \int_{-1/2}^{1/2} \sin[2\pi(2^n d\delta/\lambda)(2^{-n}i)]\,d(2^{-n}i) = 0\\ A &= 2^n d = \text{aperture}\end{aligned} \tag{6}$$

Note that the aperture $A$ is not defined by the area of the hydrophone array or of the lens in Fig. 2.1-2, as is usual in optics, but by the length $2^n d$ of the array.

The substitution of Eq. (6) into Eq. (5) yields:

$$\begin{aligned}v(\gamma, \beta, t) &= V\frac{A}{d}\frac{\sin[\pi A(\sin\beta - \sin\gamma)/\lambda]}{\pi A(\sin\beta - \sin\gamma)/\lambda}\sin(2\pi vt/\lambda)\\ &\doteq V\frac{A}{d}\frac{\sin[\pi A(\beta-\gamma)/\lambda]}{\pi A(\beta-\gamma)/\lambda}\sin(2\pi vt/\lambda) \qquad \text{for} \quad \beta, \gamma \ll 1\end{aligned} \tag{7}$$

We are interested here in the principle of imaging. Hence, the simplified function

$$v(\gamma, \beta) = \frac{\sin[\pi A(\sin\beta - \sin\gamma)/\lambda]}{\pi A(\sin\beta - \sin\gamma)/\lambda} \doteq \frac{\sin[\pi A(\beta-\gamma)/\lambda]}{\pi A(\beta-\gamma)/\lambda} \tag{8}$$

holding for small values of $\beta$ and $\gamma$ is plotted in Fig. 2.1-3 for $\gamma = 0, \pm\lambda/A, \pm 2\lambda/A$. These curves give the variation of the normalized amplitude $v(\gamma, \beta)$ of the voltage $v(\gamma, \beta, t)$ as a function of $\beta$ for these particular choices of the observation angle $\gamma$. These curves are usually referred to as beam pattern or diffraction pattern.

The principle of a circuit producing $v(\gamma, \beta, t)$ is shown by Fig. 2.1-4. Tapped delay circuits are connected to the output terminals of the hydro-

[1] The exact summation is possible too, but is of no practical interest since the approximation is sufficiently accurate (see Urick, 1967, p. 42).

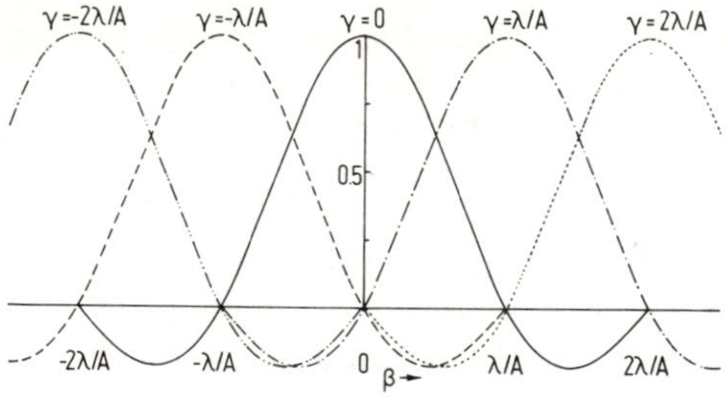


FIG. 2.1-3. The normalized amplitudes $v(\gamma, \beta)$ according to Eq. (8) for $\beta \ll 1$, $\gamma \ll 1$. For larger values of $\beta$ and $\gamma$ one has to substitute $\sin\beta$ and $\sin\gamma$ for $\beta$ and $\gamma$; the values $\pm\lambda/A$, $\pm 2\lambda/A, \ldots$ of the scale are replaced by $\pm\sin(\lambda/A)$, $\pm\sin(2\lambda/A), \ldots$.

phones. The delay times at the taps are $t_0 + (k - 2^4/2 + 1)(d/v)\sin\gamma$. These delay times are shown in detail for the taps $k = 4$, $\gamma = -5\lambda/A$ and $k = 11$, $\gamma = 5\lambda/A$. The voltages for a fixed value of $\gamma$ and all values $k = i + 7 = 0, \ldots, 15$ are summed to yield $v(\gamma, \beta, t)$ of Eq. (7). For the values $\gamma = -7\lambda/A, \ldots, +8\lambda/A$ shown in Fig. 2.1-4, one obtains the 16 voltages $v(-7\lambda/A, \beta, t)$ to $v(+8\lambda/A, \beta, t)$. Note that this circuit works for nonsinusoidal voltages too. The *image voltages* $v(\gamma, \beta, t)$ of Eq. (7) are in the general case replaced by the sum

$$v(\gamma, \beta, t) = \sum_{i=-2^n/2+1}^{2^n/2} v\{[t + i(d/v)(\sin\beta - \sin\gamma)]/T\} \tag{9}$$

according to Eq. (3).

Let a sound source be located at infinity. The amplitude of the voltage

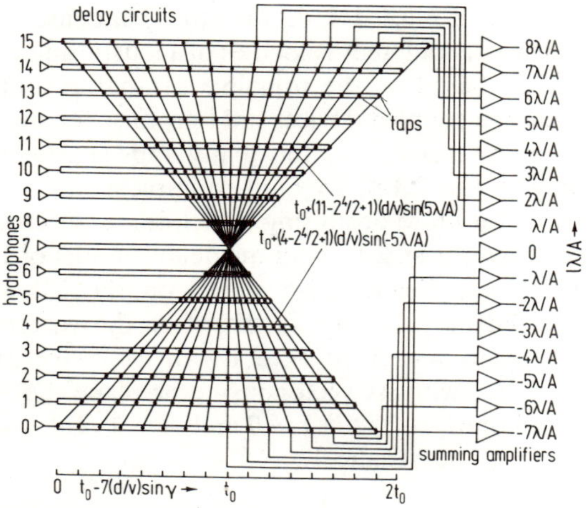


FIG. 2.1-4. Producing the images of 16 points by means of delay circuits.

$v(0, \beta, t)$ obtained by summing the voltages at the taps along the line $\gamma = 0$ in Fig. 2.1-4 will vary like the curve $\gamma = 0$ in Fig. 2.1-3 as a function of $\beta$; the voltage $v(\lambda/A, \beta, t)$ obtained by summing the voltages at the taps along the line $\gamma = \lambda/A$ in Fig. 2.1-4 will vary like the curve $\gamma = \lambda/A$ in Fig. 2.1-3, and so on. Let two sound sources located in the directions $\sin \beta = 0$ and $\sin \beta = \lambda/A$ at infinity generate the sinusoidal waves $U_0 \sin(2\pi vt/\lambda + \alpha_0)$ and $U_{\lambda/A} \sin(2\pi vt/\lambda + \alpha_{\lambda/A})$. They will produce the normalized amplitudes $v(0, 0) = 1$ and $(U_{\lambda/A}/U_0)v(\lambda/A, \lambda/A) = U_{\lambda/A}/U_0$, while all the other normalized amplitudes $v(\gamma, 0)$ and $v(\gamma, \lambda/A)$, with $\gamma = -7\lambda/A, \ldots, -\lambda/A, +2\lambda/A, \ldots, +8\lambda/A$, are zero. Hence, the two sound sources will be resolved without mutual interference. The ratio $\lambda/A = \sin \varepsilon$ defines the classical resolution angle $\varepsilon$ for sinusoidal waves.

The principle of the delay circuits discussed here is the same principle as used by the lens for image generation. The lens and the space between the lens and the focal plane provide paths with different delays, so that a plane wave coming from infinity requires the same time to reach the image point in the focal plane, regardless of where the plane wave reaches the surface of the lens. The summing of all the waves with various delays is done by the eye or a photographic film. Hence, the lens with the associated detector can perform the linear operations of delay and summation. Linearity of the operations is a necessity if we want to produce the images of many points simultaneously. But there are many more linear operations than delay and summation. For instance, multiplication is a very useful operation that the usual lens does not perform. Time-variable linear operations are a whole class of operations beyond the capabilities of the lens. Electronic circuits can readily perform multiplications and time-variable operations. The time-invariant delay circuits of Fig. 2.1-4, on the other hand, are not practical for acoustics due to the vastly different velocities of acoustic and electromagnetic waves. Delay circuits based on the principle of analog shift registers and implemented by charge coupled devices hold promise for the future. At this time, they are too expensive, and the many taps at different locations, required according to Fig. 2.1-4, are hard to implement.

Before we search for more promising principles for electrical circuits, let us generalize the array of Fig. 2.1-1 and the delay circuits of Fig. 2.1-4 from one to two dimensions. Figure 2.1-5 shows on the left a planar wave $W(\beta_x, \beta_y)$ represented by either its wavefront or the vector normal to it. It is decomposed into a wavefront $W(\beta_x)$ with its vector in the $xz$-plane and a wavefront $W(\beta_y)$ with its vector in the $yz$-plane. Let the wavefront arrive at the hydrophone $k = 2^n/2 - 1 = 7$ and $m = 2^n/2 - 1 = 7$. The wavefront still has to travel the distance

$$(k - 2^n/2 + 1)d \sin \beta_x \tag{10}$$

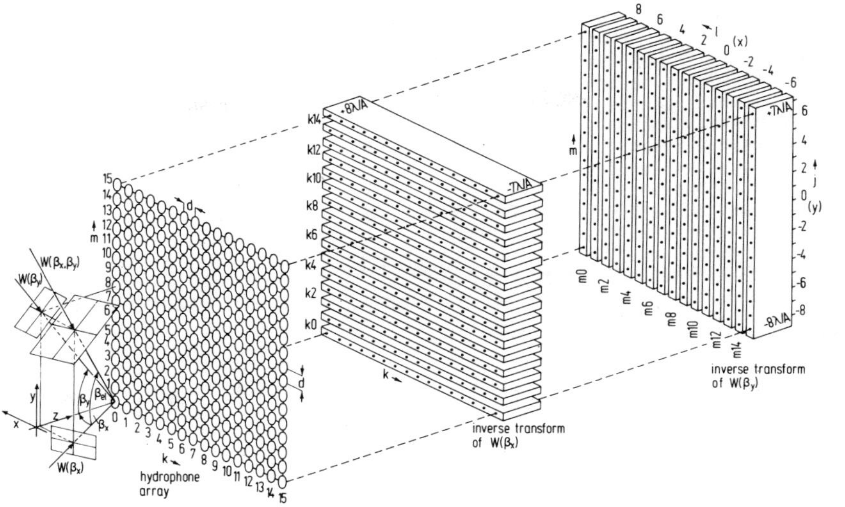


FIG. 2.1-5. Generating the inverse transform in two dimensions using cartesian coordinates.

to reach the hydrophone $k$ in the row $m = 2^n/2 - 1 = 7$. This is the same distance as in Eq. (1), except that $\beta$ is replaced by $\beta_x$. In addition, the wave has to travel the distance

$$(m - 2^n/2 + 1)d \sin \beta_y \tag{11}$$

to reach the hydrophone $m$. The generalization of the voltage $v(i, t)$ of Eq. (1) to two dimensions is thus the voltage $v(i, h, t)$:

$$\begin{gathered} v(i, h, t) = v\{[t + i(d/v) \sin \beta_x + h(d/v) \sin \beta_y]/T\} \\ i = k - 2^n/2 + 1, \qquad h = m - 2^n/2 + 1 \end{gathered} \tag{12}$$

The angle $\beta_x$ is usually called azimuth, but $\beta_y$ is *not* the elevation. The elevation is denoted $\beta_{el}$ in Fig. 2.1-5. Azimuth and elevation refer to spherical coordinates, while we are using cartesian coordinates.

The delay time $t_\gamma$ of Eq. (2) is replaced by $t_{\gamma, xy}$

$$t_{\gamma, xy} = i(d/v) \sin \gamma_x + h(d/v) \sin \gamma_y \tag{13}$$

where $\gamma_x$ is the observation angle in the $xz$-plane and $\gamma_y$ the observation angle in the $yz$-plane.

The sinusoidal function $v(i, t - t_\gamma)$ of Eq. (4) is generalized to $v(i, h, t - t_{\gamma, xy})$:

$$\begin{aligned} v(i, h, t - t_{\gamma, xy}) = V \sin\{2\pi[vt &+ id(\sin \beta_x - \sin \gamma_x) \\ &+ hd(\sin \beta_y - \sin \gamma_y)]/\lambda\} \end{aligned} \tag{14}$$

The sum voltage $v(\gamma, \beta, t)$ in Eq. (5) is replaced by $v(\gamma_x, \gamma_y, \beta_x, \beta_y, t)$ and the summation is over $h$ as well as $i$:

$$v(\gamma_x, \gamma_y, \beta_x, \beta_y, t) = \sum_{i=2^n/2+1}^{2^n/2} \sum_{h=2^n/2+1}^{2^n/2} V \sin[2\pi(vt + id\delta_x + hd\delta_y)/\lambda]$$

$$\begin{aligned} i &= k - 2^n/2 + 1, \qquad h = m - 2^n/2 + 1 \\ \delta_x &= \sin \beta_x - \sin \gamma_x, \qquad \delta_y = \sin \beta_y - \sin \gamma_y \end{aligned} \tag{15}$$

Using the approximation of Eq. (6), which shows that the sum of terms $\sin(2\pi id\delta_x/\lambda)$ and $\sin(2\pi hd\delta_y/\lambda)$ equals zero, one obtains the following double sum:

$$\begin{aligned} v(\gamma_x, \gamma_y, \beta_x, \beta_y, t) &= 2^{2n}V \sin(2\pi vt/\lambda) \sum_{i=-2^n/2+1}^{2^n/2} 2^{-n} \cos(2\pi id\delta_x/\lambda) \\ &\quad \times \sum_{h=2^n/2+1}^{2^n/2} 2^{-n} \cos(2\pi hd\delta_y/\lambda) \\ &\doteqdot V\left(\frac{A}{d}\right)^2 \frac{\sin[\pi A(\sin \beta_x - \sin \gamma_x)/\lambda]}{\pi A(\sin \beta_x - \sin \gamma_x)/\lambda} \\ &\quad \times \frac{\sin[\pi A(\sin \beta_y - \sin \gamma_y)/\lambda]}{\pi A(\sin \beta_y - \sin \gamma_y)/\lambda} \sin(2\pi vt/\lambda) \end{aligned} \tag{16}$$

$$A = 2^n d, \qquad i = k - 2^n/2 + 1, \qquad h = m - 2^n/2 + 1$$

The normalized amplitude of $v(\gamma_x, \gamma_y, \beta_x, \beta_y, t)$ is denoted $v(\gamma_x, \gamma_y, \beta_x, \beta_y)$:

$$v(\gamma_x, \gamma_y, \beta_x, \beta_y) \doteqdot \frac{\sin[\pi A(\sin \beta_x - \sin \gamma_x)/\lambda]}{A(\sin \beta_x - \sin \gamma_x)/\lambda} \frac{\sin[\pi A(\sin \beta_y - \sin \gamma_y)/\lambda]}{A(\sin \beta_y - \sin \gamma_y)/\lambda} \tag{17}$$

For small values of $\beta_x$, $\beta_y$, $\gamma_x$, and $\gamma_y$ one obtains the functions of Fig. 2.1-3 in two dimensions. A computer plot of the function for $\gamma_x = \gamma_y = 0$ and $\sin \beta_x \doteqdot \beta_x$, $\sin \beta_y \doteqdot \beta_y$ is shown in Fig. 2.1-6.

A practical circuit to obtain $v(\gamma_x, \gamma_y, \beta_x, \beta_y, t)$ may be derived from Eq. (15). The received $2^n \times 2^n$ voltages are first delayed and summed over $i$ or $k$, then over $h$ or $m$. This can be done, in principle, by producing printed circuit cards with the circuit of Fig. 2.1-4, stacking $2^n$ such cards vertically and another $2^n$ such cards horizontally as shown in Fig. 2.1-5, and interconnecting them as indicated there by the dashed lines. The voltages $v(\gamma_x, \gamma_y, \beta_x, \beta_y, t)$ are obtained at the output terminals of the second stack of cards. These terminals are labeled horizontally from $l = -7$ to $l = +8$, corresponding to the values $\gamma_x = -7\lambda/A$ to $\gamma_x = +8\lambda/A$; vertically they are labeled from $j = -8$ to $j = +7$, corresponding to the values $\gamma_y = -8\lambda/A$

to $\gamma_y = +7\lambda/A$. The different labeling of the vertically and horizontally stacked cards is caused by the even number of cards in each stack and one's desire to equalize the number of output terminals in each quadrant $l < 0$, $j < 0$ to $l > 0$, $j > 0$. The different labeling causes no practical problem, since the cards of Fig. 2.1-4 only have to be flipped around so that the

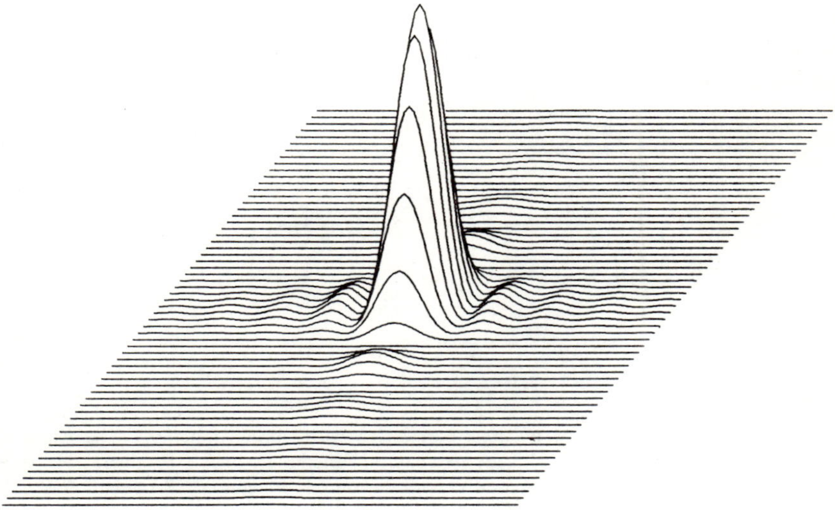

FIG. 2.1-6. The diffraction or beam pattern $v(\gamma_x, \gamma_y, \beta_x, \beta_y)$ according to Eq. (17) for $\gamma_x = \gamma_y = 0$ and small values of $\beta_x$, $\beta_y$.

input terminal 15 becomes the new terminal 0. The output terminals must then be labeled from $-8\lambda/A$ at the bottom to $+7\lambda/A$ at the top, which corresponds to the labeling of the vertically stacked cards in Fig. 2.1-5.

## 2.2 Inverse Transformation by Means of the Fourier Transform

We know from Fourier optics that the operation of a lens cannot only be described in terms of delays and summations, but also in terms of a two-dimensional, spatial Fourier transform. Let us see whether this description does not lead to more practical circuits than tapped delay circuits. The need to treat the number of hydrophones as finite, rather than as nondenumerably, makes the mathematical investigation very different from the one used in Fourier optics. Hence, we will discuss this approach first in a nonmathematical, qualitative way.

It is well known that correlation with sample functions is a good way to detect a signal. Figure 2.2-1 shows the principle. There are two possible signals denoted "expected function" on top of Figs. 2.2-1a and b. These signals are sampled sinusoidal pulses with either one or two cycles. Note that "sampled function" means that only the voltages shown by heavy lines are received. The dashed envelope of these samples is plotted only to emphasize that the samples represent two sinusoidal functions.

In order to decide which of the two expected functions has been received, we multiply the expected function in Figs. 2.2-1a and 2.2-1b with the sample functions 1 and 2. These multiplications are represented by the products 1 and 2.

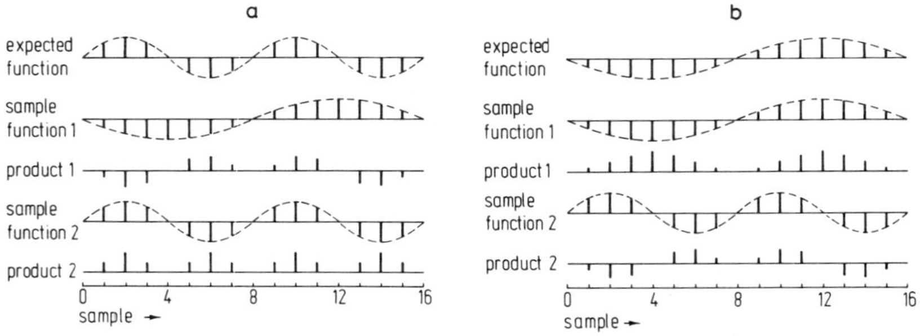


FIG. 2.2-1. Signal detection by cross-correlation with sample functions. The expected functions on top are multiplied with the two sample functions 1 and 2. Product 1 has only positive samples if the expected function equals the sample function 1. Product 2 has only positive samples if the expected function equals the sample function 2.

Let the products be summed. The sum of product 2 in Fig. 2.2-1b and the sum of product 1 in Fig. 2.2-1a equal zero since for any positive sample there is a negative one with equal magnitude. The sum of product 1 in Fig. 2.2-1b and the sum of product 2 in Fig. 2.2-1a, on the other hand, are not zero. Hence, "sum of product 1 unequal zero" means that the sinusoidal function with one cycle was received, while "sum of product 2 unequal zero" means that the sinusoidal function with two cycles was received. This is the principle of signal detection by cross-correlation with sample functions.

There are two points one must observe for the application of this principle to image generation. (a) The signals or functions do not have to be continuous but can be represented by samples. (b) The signals or functions do not have to be *time* signals or *time* functions; the abscissas in Fig. 2.2-1 are labeled sample and not time for this reason.

Consider now 16 hydrophones in a straight line with equal distances as shown in Fig. 2.2-2. A planar wave with sinusoidal time variation arrives at an angle $\varepsilon$ with the array axis. The pressure of the wave at a certain time is indicated by the 16 sinusoidal functions. The pressure at the

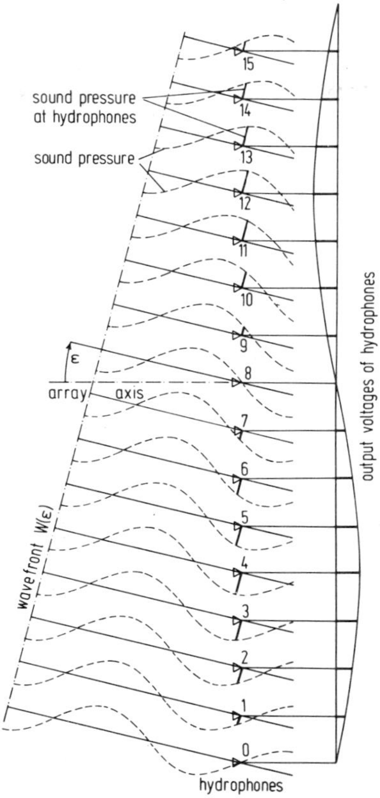


FIG. 2.2-2. Planar wave with sinusoidal time variation arriving at an angle ε with the axis of a line array of hydrophones.

hydrophones is represented by the samples of these functions shown by heavy lines. The pressure at the hydrophone 0 equals zero; it decreases for the hydrophones 1, 2, 3, 4, increases for the hydrophones 5, 6, 7, and becomes zero for hydrophone 8. The pressure then becomes positive for the hydrophones 9 to 15.

The hydrophones produce output voltages proportional to the pressure. These voltages are represented by the samples 0 to 15 on the right of Fig. 2.2-2. One can readily see that these output voltages are nothing else than the "expected function" in Fig. 2.2-1b.

Let us turn to Fig. 2.2-3. It shows the same hydrophone array as before, but the sinusoidal wave now arrives at an angle 2ε with the array axis. Again, the pressure at a certain time is represented by 16 sinusoidal functions, and the pressure at the hydrophones by the samples of these

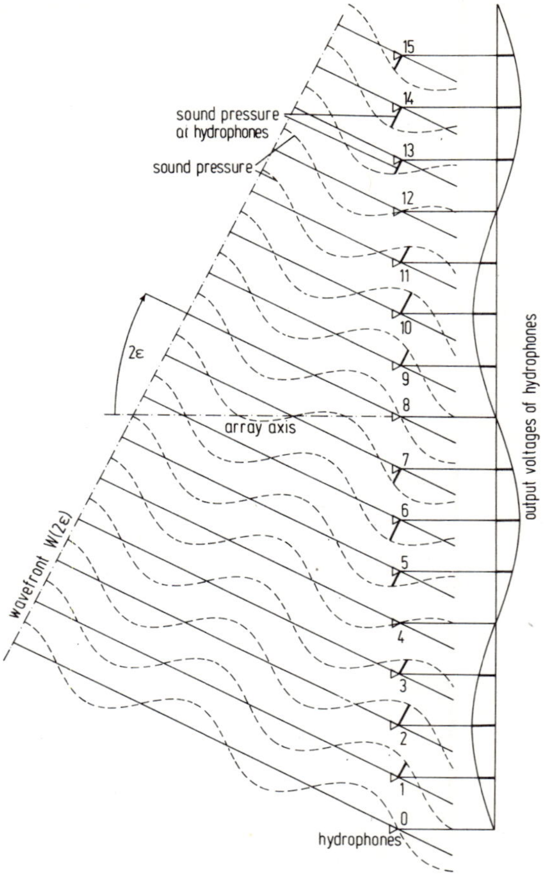


FIG. 2.2-3. Planar wave with sinusoidal time variation arriving at an angle $2\varepsilon$ with the axis of a line array of hydrophones.

functions shown by heavy lines. The pressure is zero at the hydrophones 0, 4, 8, and 12; it is positive at the hydrophones 1, 2, 3, 9, 10, 11, and it is negative at the remaining hydrophones. The output voltages are represented by the samples on the right, and they are equal to the samples of the "expected function" in Fig. 2.2-1a.

Let us now elaborate this principle quantitatively. We start with the voltage $v(i, t)$ of Eq. (2.1-1), but we must specialize its time variation to that of a sinusoidal function:

$$\begin{aligned} v(i, t) &= V \sin[2\pi(vt + id \sin \beta)/\lambda] \\ \lambda &= vT, \qquad i = k - 2^n/2 + 1, \qquad k = 0 \ldots 2^n - 1 \end{aligned} \tag{1}$$

Instead of delaying $v(i, t)$ by $t_\gamma$ according to Eq. (2.1-2) before summation, we multiply $v(i, t)$ by the following time-invariant factors:

$$\tfrac{1}{2}\sqrt{2}, \qquad \sin[2\pi i(d/\lambda)\sin\gamma], \qquad \cos[2\pi i(d/\lambda)\sin\gamma] \tag{2}$$

The sum over $i$ of the three products yields $v(0, \beta, t)$, $v(\gamma, \beta, s, t)$, and $v(\gamma, \beta, c, t)$:

$$\begin{aligned} v(0, \beta, t) &= \frac{1}{2}\sqrt{2} \sum_{i=-2^n/2+1}^{2^n/2} V \sin[2\pi(vt + id\sin\beta)/\lambda] \\ &= \frac{1}{2}\sqrt{2}\, 2^n V \Big\{ \sin(2\pi vt/\lambda) \sum_{i=-2^n/2+1}^{2^n/2} 2^{-n} \cos[2\pi i(d/\lambda)\sin\beta] \\ &\quad + \cos(2\pi vt)/\lambda) \sum_{i=-2^n/2+1}^{2^n/2} 2^{-n} \sin[2\pi i(d/\lambda)\sin\beta] \Big\} \end{aligned} \tag{3}$$

$$\begin{aligned} v(\gamma, \beta, s, t) &= \sum_{i=-2^n/2+1}^{2^n/2} V \sin[2\pi(vt + id\sin\beta)/\lambda] \sin[2\pi i(d/\lambda)\sin\gamma] \\ &= 2^n V \Big\{ \sin(2\pi vt/\lambda) \sum_{i=-2^n/2+1}^{2^n/2} 2^{-n} \cos[2\pi i(d/\lambda)\sin\beta] \\ &\quad \times \sin[2\pi i(d/\lambda)\sin\gamma] \\ &\quad + \cos(2\pi vt/\lambda) \sum_{i=-2^n/2+1}^{2^n/2} 2^{-n} \sin[2\pi i(d/\lambda)\sin\beta] \\ &\quad \times \sin[2\pi i(d/\lambda)\sin\gamma] \Big\} \end{aligned} \tag{4}$$

$$\begin{aligned} v(\gamma, \beta, c, t) &= \sum_{i=-2^n/2+1}^{2^n/2} V \sin[2\pi(vt + id\sin\beta)/\lambda] \cos[2\pi i(d/\lambda)\sin\gamma] \\ &= 2^n V \Big\{ \sin(2\pi vt/\lambda) \sum_{i=-2^n/2+1}^{2^n/2} 2^{-n} \cos[2\pi i(d/\lambda)\sin\beta] \\ &\quad \times \cos[2\pi i(d/\lambda)\sin\gamma] \\ &\quad + \cos(2\pi vt/\lambda) \sum_{i=-2^n/2+1}^{2^n/2} 2^{-n} \sin[2\pi i(d/\lambda)\sin\beta] \\ &\quad \times \cos[2\pi i(d/\lambda)\sin\gamma] \Big\} \end{aligned} \tag{5}$$

$$i = k - 2^n/2 + 1$$

One may readily recognize that the functions $v(0, \beta, t)$, $v(\gamma, \beta, s, t)$, and $v(\gamma, \beta, c, t)$ are the discrete Fourier transform of the space-time function

$v(i, t)$ of Eq. (1) for the discrete spatial variable $i$, if $(d/\lambda) \sin \gamma$ satisfies the following condition:

$$\begin{aligned} 2^n(d/\lambda) \sin \gamma = (A/\lambda) \sin \gamma &= l && \text{for} \quad \gamma > 0 \\ &= -l && \text{for} \quad \gamma < 0 \\ &= 0 && \text{for} \quad \gamma = 0 \end{aligned} \tag{6}$$

$$l = 1, 2, \ldots \qquad 2^n d = A$$

These are the same values $\sin \gamma = 0, \pm\lambda/A, \pm 2\lambda/A, \ldots$ for which Fig. 2.1-3 was plotted.

For large values of $2^n$, one may approximate the sums in Eqs. (3)–(5) by integrals:

$$\begin{aligned} v(0, \beta, t) &\doteq \frac{1}{2}\sqrt{2}\, 2^n V \Big\{ \sin(2\pi vt/\lambda) \int_{-1/2}^{1/2} \cos[2\pi(2^{-n}i)(2^n d/\lambda) \sin \beta] d(2^{-n}i) \\ &\quad + \cos(2\pi vt/\lambda) \int_{-1/2}^{1/2} \sin[2\pi(2^{-n}i)(2^n d/\lambda) \sin \beta] d(2^{-n}i) \Big\} \\ &\doteq \frac{1}{2}\sqrt{2}\, V(A/d) \frac{\sin[\pi(A/\lambda) \sin \beta]}{\pi(A/\lambda) \sin \beta} \sin(2\pi vt/\lambda) \end{aligned} \tag{7}$$

$$\begin{aligned} v(\gamma, \beta, s, t) \doteq \frac{VA}{2d} \Bigg( & \frac{\sin[\pi(A/\lambda)(\sin \beta - \sin \gamma)]}{\pi(A/\lambda)(\sin \beta - \sin \gamma)} \\ & - \frac{\sin[\pi(A/\lambda)(\sin \beta + \sin \gamma)]}{\pi(A/\lambda)(\sin \beta + \sin \gamma)} \Bigg) \cos(2\pi vt/\lambda) \end{aligned} \tag{8}$$

$$\begin{aligned} v(\gamma, \beta, c, t) \doteq \frac{VA}{2d} \Bigg( & \frac{\sin[\pi(A/\lambda)(\sin \beta - \sin \gamma)]}{\pi(A/\lambda)(\sin \beta - \sin \gamma)} \\ & + \frac{\sin[\pi(A/\lambda)(\sin \beta + \sin \gamma)]}{\pi(A/\lambda)(\sin \beta + \sin \gamma)} \Bigg) \sin(2\pi vt/\lambda) \end{aligned} \tag{9}$$

$$2^n d = A$$

The voltage $v(\gamma, \beta, s, t)$ is integrated over $t$ to change the time variation from $\cos(2\pi vt/\lambda)$ to $\sin(2\pi vt/\lambda)$:

$$v^{\dagger}(\gamma, \beta, s, t) = -(2\pi v/\lambda) \int v(\gamma, \beta, s, t)\, dt \tag{10}$$

The letter $l$ or $-l$ is substituted for $(A/\lambda)\sin\gamma$ from Eq. (6), and the following sums and differences are formed:

$$v(l, \beta, t) = \tfrac{1}{2}[v(\gamma, \beta, c, t) + v\dagger(\gamma, \beta, s, t)]$$

$$= \frac{VA}{2d} \frac{\sin \pi[(A/\lambda)\sin\beta - l]}{\pi[(A/\lambda)\sin\beta - l]} \sin(2\pi v t/\lambda) \qquad (11)$$

$$v(-l, \beta, t) = \tfrac{1}{2}[v(\gamma, \beta, c, t) - v\dagger(\gamma, \beta, s, t)]$$

$$= -v(l, \beta, t) + v(\gamma, \beta, c, t)$$

$$= \frac{VA}{2d} \frac{\sin \pi[(A/\lambda)\sin\beta + l]}{\pi[(A/\lambda)\sin\beta + l]} \sin(2\pi v t/\lambda) \qquad (12)$$

The voltages $v(0, \beta, t)$, $v(l, \beta, t)$, and $v(-l, \beta, t)$ in Eqs. (7), (11), and (12) are identical with $v(\gamma, \beta, t)$ in Eq. (2.1-7) if one substitutes 0, $+l$, and $-l$ for $(A/\lambda)\sin\gamma$. Hence, one may replace the delay circuits in Fig. 2.1-4 by circuits that perform a discrete, spatial Fourier transform on the $2^n$ sinusoidal input voltages $v(i, t)$ of Eq. (1) according to Eqs. (3)–(5), a phase shift according to Eq. (10), and summations according to Eqs. (11) and (12). Note that the delay circuits work for any time variation of the input voltage, and the same holds true for the spatial Fourier transform. However, the phase shift according to Eq. (10) restricts the beam forming with the help of the Fourier transform to input voltages with sinusoidal time variation. One may, of course, replace the phase shifting through integration in Eq. (10) by a phase shifting method that operates over a wider frequency band, if one is interested in such a wider band. There is usually little incentive to do so in acoustic imaging, since a sound projector readily produces a wave with sinusoidal time variation, and phase shifting by integration requires simple, low-cost circuits. The huge number of beams one has to form makes the cost of the circuits the primary criterion.

The practical design of imaging circuits, or beam formers, based on the Fourier transform has grown into a specialty of its own. Hence, we postpone its discussion until Chapter 3 and turn to the third method of image generation with electronic circuits.

## 2.3 Inverse Transformation by Means of Sampled Storage Circuits

The delay circuits in Fig. 2.1-4 may be implemented by shift registers. The input signal must be transformed from analog to digital for the use of digital shift registers, but one could also use analog shift registers implemented, e.g., by charge coupled devices. The output voltages of the summing amplifiers in Fig. 2.1-4 would be step functions if analog shift registers were used, and binary numbers if digital shift registers were used. In both cases one obtains sampled functions rather than continuous ones.

The use of sampled functions leads to the implementation of the inverse transform by sampling filters. The principle will be explained with the help of Fig. 2.3-1. A plane wave producing the voltage $f(t)$ at the center receptor

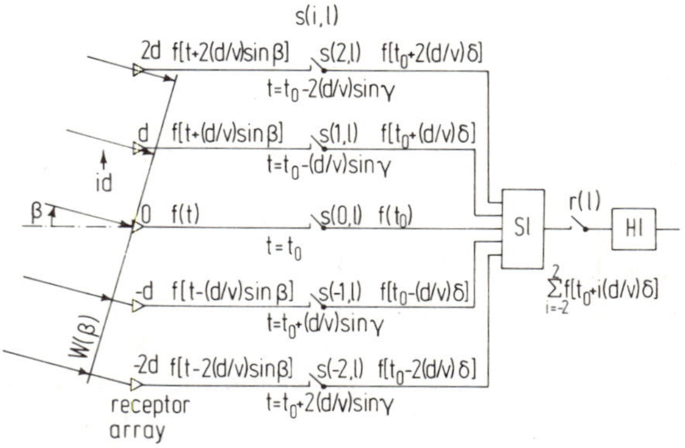


FOG. 2.3-1. Principle of the inverse transformation by means of sampled storage circuits. $\delta = \sin\beta - \sin\gamma$; H$l$, analog hold circuit; S$l$, analog sum-and-hold circuit.

$i = 0$ produces the voltage $f[t + i(d/v)\sin\beta]$ at the receptor $i$. Let the switch $s(i, l)$ be closed momentarily at the time $t_0 - i(d/v)\sin\gamma$. The sampled voltage $f[t_0 + i(d/v)(\sin\beta - \sin\gamma)] = f[t_0 + i(d/v)\delta]$ is obtained and fed to the analog sum-and-hold circuit S$l$. After samples have been summed from all switches s($i$, $l$), $i = -2, \ldots, +2$, one obtains the sum

$$\sum_{i=-2}^{+2} f[t_0 + i(d/v)\delta] \tag{1}$$

This is the same sum as that of Eq. (2.1-5), except that $V\sin(2\pi vt/\lambda)$ has been replaced by $f(vt_0/\lambda)$ and the limits of the sum are slightly different. Hence, a simple sum-and-hold circuit according to Fig. 2.3-1 can be substituted for the delay and summing circuits of Fig. 2.1-4 if one is satisfied with a sampled output voltage.

The sum-and-hold circuit in Fig. 2.3-1 is connected via the readout switch r($l$) to a hold circuit H$l$. This switch transfers a completed sum $\sum f[t_0 + i(d/v)\delta]$ from the summing circuit S$l$ to the hold circuit H$l$ and permits a new cycle of sampling and summing to start.

The circuit of Fig. 2.3-1 is shown in Fig. 2.3-2 for 16 input terminals $k = 0, \ldots, 15$ or $i = -7, \ldots, +8$. The 16 input terminals imply 16 possible observation angles $\gamma$ determined by the relation $\sin\gamma = l\lambda/A$, $l = -7, \ldots, +8$,

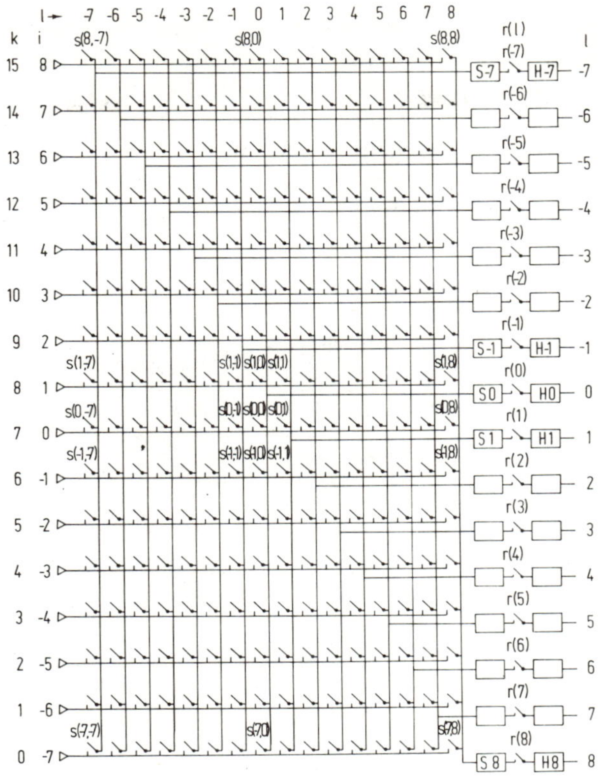


FIG. 2.3-2. Circuit for the resolution of 16 points using 16 receptors and sampled storage circuits. The location of the hydrophone relative to the center $i = 0$ is $id$. The observation angles are $\sin \gamma = l\lambda/A$. The selection switches are denoted s($i$, $l$) and the readout switches $r(l)$. S$l$, sum-and-hold circuit; H$l$, hold circuit. The output voltage at any terminal $l$ in the notation of Fig. 2.3-1 is given by the sum $\sum_{i=-7}^{8} f[t_0 - i(d/v)(\sin \beta - l\lambda/A)]$.

according to Eq. (2.2-6). Hence, there are $16^2$ switches s($i$, $l$), $i = k - (2^n/2 - 1) = k - (16/2 - 1) = -7, \ldots, +8$ and $l = -7, \ldots, +8$. Furthermore, there are 16 sum-and-hold circuits S$l$, readout switches r($l$) and hold circuits H$l$.

The timing diagram for the switches s($i$, $l$) and r($l$) of Fig. 2.3-2 is shown in Fig. 2.3-3. The number $i$ is plotted vertically on the left, the number $l$ denotes the slanting lines. The intersection of a line $i$ with a line $l$, marked by a dot, indicates the closing time of switch s($i$, $l$). The time is plotted horizontally; $t_0$ is an arbitrary reference time. Note that all switches s(0, $-7$) to s(0, 8) are closed at the same time $t = t_0$; this time is represented by the intersection of the 15 lines $i = 0$, $l = -7, \ldots, +8$.

The extension of the sampled storage principle from one to two dimensions is rather difficult. The 16 input terminals in Fig. 2.3-2 call for $16^2$ switches. A two-dimensional array with $16^2$ input terminals would require $16^4$ switches. The timing circuits associated with those switches and the wiring would make the method undesirable. For practical equipment one must be able to replace one two-dimensional transform by two one-dimensional transforms. Figure 2.3-4 shows a possible way to accomplish this.

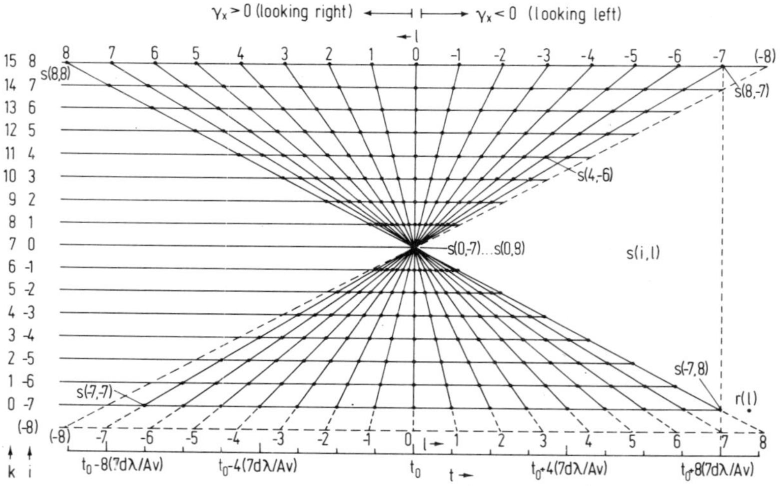


FIG. 2.3-3. Closing times of the selection switches s($i$, $l$) and the readout switches r($l$). The observation angles are $\sin \gamma = l\lambda/A$.

The $16^2$ receptors feed into the two-dimensional sampling filter consisting of a stack of 16 cards with the circuit of Fig. 2.3-2. The cards are denoted $km$ with $m = 0, \ldots, 15$. The output terminals of these cards feed to a storage and distribution circuit with 16 cards $ml$, $l = -7, \ldots, +8$. The circuit of these cards is shown in Fig. 2.3-5. The cards $ml$ sum the output voltages of the cards $km$ for all values $m = 0, \ldots, 15$ and a certain value of $l$. Only four of the $16^2$ interconnecting wires are shown in Fig. 2.3-4 to avoid obscuring the picture. Let the switches on all cards $km$, $m = 0, \ldots, 15$, of the two-dimensional sampling filter be operated according to the time diagram of Fig. 2.3-3. Each card $km$ will then produce the same 16 output voltages, and the summation in one of the cards of the storage and distribution circuit will yield the voltage of one output terminal of a card $km$ multiplied by 16. Let now the switch $t(0)$ in Fig. 2.3-5 be closed. Each card $ml$ in Fig. 2.3-4 will then feed a voltage to the input

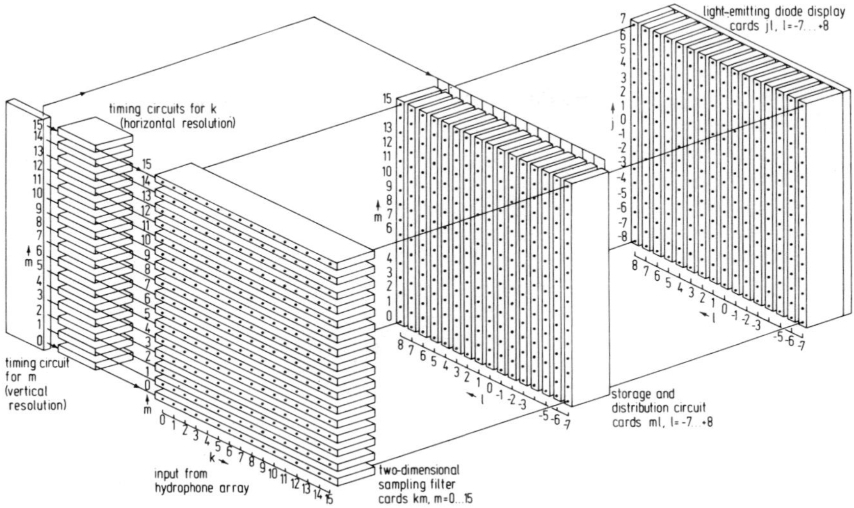


FIG. 2.3-4. Image generation by means of a two-dimensional sampling filter.

terminal $j = 0$ of the cards $jl$, $l = -7, \ldots, +8$, of the light-emitting diode display. The 16 light-emitting diodes of the row $j = 0$ will show the *line image* for the cartesian elevation angle $\gamma_y = 0$, or $(A/\lambda) \sin \gamma_y = j = 0$, and all azimuth angles $(A/\lambda) \sin \gamma_x = l = -7, \ldots, +8$.

The center column with the heading $j = 0$ in Fig. 2.3-6 shows the time diagram for the generation of the line image for the cartesian elevation angle

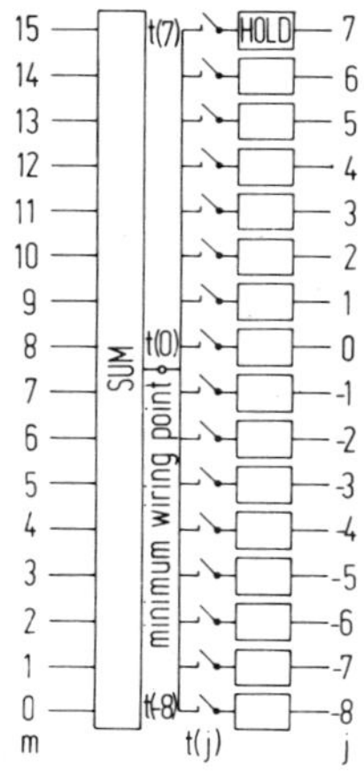


FIG. 2.3-5. Storage and distribution circuit for sampling in two-dimensional cartesian coordinates. SUM, summer; HOLD, hold circuit.

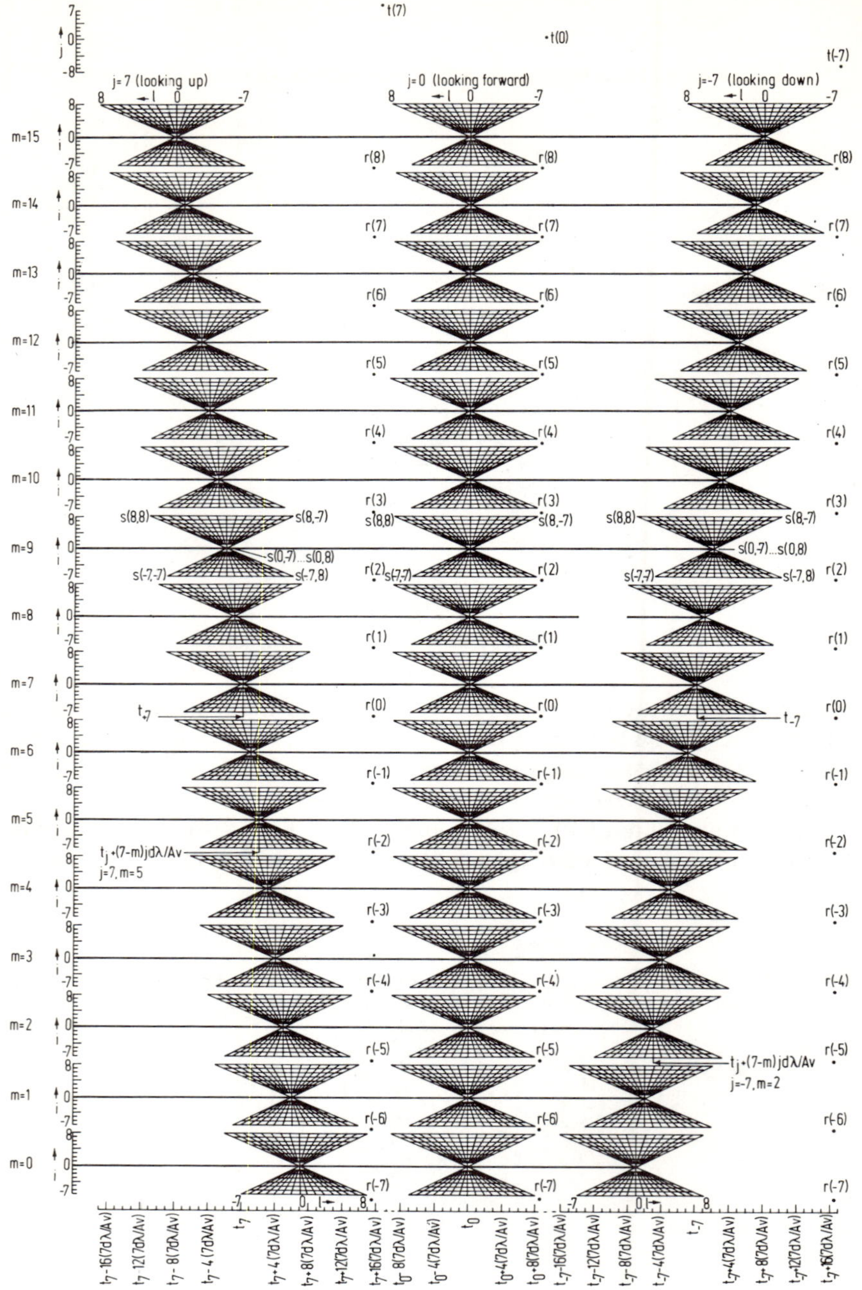


FIG. 2.3-6. Time diagram for the two-dimensional sampling filter of Fig. 2.3-4 for the horizontal angles $(A/\lambda)\sin\gamma_x = i$, $i = -7, \ldots, +8$, and the vertical angles $(A/\lambda)\sin\gamma_y = j$, $j = -7, 0, +7$. The times $t_j + (7 - m)jd\lambda/Av$ give the operating times for the switches $s(-7,0), \ldots, s(8,0)$; the number 7 must be replaced by $2^n/2 - 1$ for arrays with $2^n$ input terminals.

$\gamma_y = 0$. The 16 timing diagrams for $m = 0, \ldots, 15$ are all identical and are smaller scale versions of the time diagram of Fig. 2.3-3, except that the operating time of the switch $t(0)$ of Fig. 2.3-5 is shown on top of Fig. 2.3-6 in addition to the operating times of the switches s($i$, $l$) and r($l$).

If we want to generate a line image for a cartesian elevation angle $(A/\lambda) \sin \gamma_y = j = -8, \ldots, +7$ one must operate the switches of the cards $km$ with certain delays for $m = 0, \ldots, 15$. The resulting time diagrams are shown in Fig. 2.3-7 for $(A/\lambda) \sin \gamma_y = j = +7$ and $(A/\lambda) \sin \gamma_y = j = -7$. The

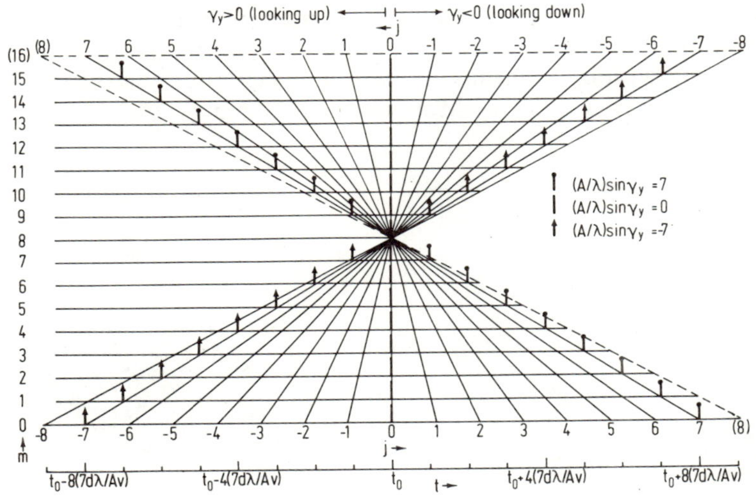


FIG. 2.3-7. Time diagram for the synchronizing pulses from the "timing circuit for $m$" to the "timing circuit for $k$" in Fig. 2.3-4. The synchronizing pulses for $j = -7, 0, +7$ are shown explicitly.

relative delays of the "center points" of the time diagrams for $m = 0, \ldots, 15$ are the same as the relative delays for $l = +7$ or $l = -7$ in Fig. 2.3-3; the "center points" show the operating times of the switches s(0, $-7$), ..., s(0, 8). Note that the operating times in Fig. 2.3-7 are not indicated as in Fig. 2.3-3 by dots at the intersections of the horizontal lines $i$ and the slanting lines $l$, but by the intersections only.

Let us discuss the practical implementation of the timing circuits for the switches s($i$, $l$), r($l$), and $t(j)$. For each card $km$, $m = 0, \ldots, 15$, in Fig. 2.3-4 one needs a timing circuit that operates the switches s($i$, $l$) and $r(l)$ according to Fig. 2.3-3. These 16 *timing circuits for k* are shown in Fig. 2.3-4 on the left. It is assumed that they are driven by clock pulses provided by the *timing circuit for m* in Fig. 2.3-4. The time diagram for the pulses at the terminals $m = 0, \ldots, 15$ of this circuit is shown in Fig. 2.3-7. Let us assume that the line image for the cartesian elevation angle $(A/\lambda) \sin \gamma_y = 0$ is to be produced. The pulses at the terminals $m = 0, \ldots, 15$ have to be simultaneous as shown by the pulses for $j = 0$, $m = 0, \ldots, 15$ in Fig. 2.3-7. An elevation angle $(A/\lambda) \sin \gamma_y = +7$ calls for a timing of the pulses as shown for $j = +7$, $m = 0, \ldots, 15$. Similarly, an elevation angle $(A/\lambda) \sin \gamma_y = -7$ requires the timing shown for $j = -7$, $m = 0, \ldots, 15$. Note that the timing diagrams of Figs. 2.3-3 and 2.3-7 differ in two points only: the dots representing the closing times of switches in Fig. 2.3-3 are replaced by pulses in Fig. 2.3-7, occurring at the terminals of the timing circuit for $m$ in Fig. 2.3-4; and the lines $l = -8$ or 8 in Fig. 2.3-3 are interchanged with the lines $j = 8$ or $-8$ in Fig. 2.3-7.

Image generation by means of sampled storage circuits was investigated in more detail in a thesis by Mohamed (1976), but no circuits have been built so far.

# 3 Implementation of the Fourier Transform and Beam Formers by Analog Circuits

## 3.1 Fast Fourier Transform Circuit with 16 Input Terminals

The spatial Fourier transform in one or two dimensions has found applications for beam forming, image processing, correlation, etc. The transform is usually performed either by computer or by means of Fourier optics. The computer is too slow when one wants to perform Fourier transforms of thousands to millions of samples per second. Fourier optics produces transforms with millions of samples in about $10^{-13}$ sec, but requires mechanical accuracies on the order of the wavelength of the light used. If electronic circuits are used in connection with Fourier optics, their response time will increase the actual time required per transform from $10^{-13}$ to $10^{-8}$ sec and more. We will discuss here an electronic analog circuit that produces a Fourier transform in about $10^{-6}$ sec. The largest equipment built with it so far—a two-dimensional acoustic beam former—transforms $16^2 = 256$ simultaneous samples (Harmuth *et al.*, 1974; Harmuth, 1976, 1977). The circuit design will be carried here to 64 samples for a one-dimensional transform, or $64^2 = 4096$ samples for a two-dimensional transform, but the design method will be sufficiently lucid to permit an extension to more samples.

Consider the Fourier transform of the sampled function $F(k)$ on top of Fig. 3.1-1 with the 16 samples A, ..., P. Since there are a finite number of samples, the term Fourier transform actually means the sampled form of the Fourier series. The discrete Fourier transform of the 16 samples of $F(k)$ requires 16 functions of the Fourier series in sampled form. The samples representing these 16 functions are shown in Fig. 3.1-1. The dashed lines help one recognize the more familiar continuous functions. The only surprise in Fig. 3.1-1 is the function $\cos 16\pi x$ where one might expect to find $\sin 16\pi x$. One could change this by taking the samples not at $k = 0, 1, \ldots, 15$ but at $k = 0.5, 1.5, \ldots, 15.5$.

The orthogonality of the sampled functions may readily be verified. For instance, the sum of the product of $\sin 2\pi x = \sin 2\pi k/16$ and $\sin 6\pi x = \sin 6\pi k/16$ yields:

$$\begin{aligned}\sum_{k=0}^{15} \sin(2\pi k/16)\sin(6\pi k/16) &= (0 + pr + q^2 - pr - 1 - pr + q^2 + pr + 0 \\ &\quad + pr + q^2 - pr - 1 - pr + q^2 + pr) \\ &= 2(2q^2 - 1) = 2(2\sin^2 \pi/4 - 1) = 0 \qquad (1)\end{aligned}$$

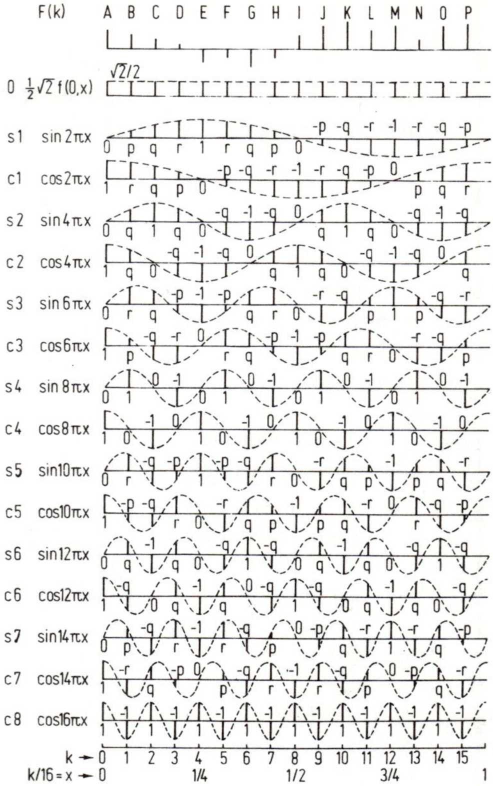


FIG. 3.1-1. The first 16 functions of the Fourier series represented by samples. $p = \sin \pi/8$, $q = \sin \pi/4$, $r = \sin 3\pi/8$.

The coefficients $a(j)$ of the Fourier transform are obtained by evaluating the following sums:

$$
\begin{aligned}
a(0) &= \sum_{k=0}^{15} F(k) f(0, k/16) \\
a_s(l) = a(2l-1) &= \sqrt{2} \sum_{k=0}^{15} F(k) \sin 2l\pi k/16 \\
a_c(l) = a(2l) &= \sqrt{2} \sum_{k=0}^{15} F(k) \cos 2l\pi k/16 \qquad (2) \\
a_c(8) = a(15) &= \sqrt{2} \sum_{k=0}^{15} F(k) \cos 16\pi k/16 \\
l = 1, 2, \ldots, 7, &\qquad x = k/16
\end{aligned}
$$

A practical circuit to produce the coefficient $a_s(1) = a(1)$, multiplied by $\sqrt{2}/2$, is shown in Fig. 3.1-2. The resistors with the values $R/0$, $R/p$, ..., $R$ produce the multiplications $F(k) \sin 2\pi k/16$ for $k = 0, \ldots, 15$, and the operational amplifier OP produces the summation. Negative values of $\sin 2\pi k/16$ are implemented by connecting the respective resistors to the inverting input terminal (−) of the operational amplifier.

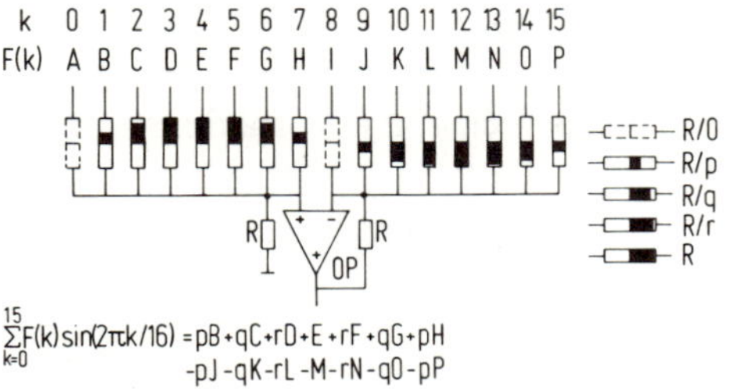


FIG. 3.1-2. Circuit based on the usual Fourier transform producing the Fourier coefficient $a_s(1) = a(1)$ of the sampled function $F(k)$ in Fig. 3.1-1. OP, operational amplifier; $p = \sin \pi/8$, $q = \sin \pi/4$, $r = \sin 3\pi/8$.

Sixteen circuits according to Fig. 3.1-2 with input resistors according to the values of the samples in Fig. 3.1-1 produce all 16 Fourier coefficients $a(0)$ to $a(15)$. However, a practical difficulty is encountered. Each operational amplifier is nominally connected to 16 input resistors; similarly, each driving source supplying the voltages that represent the samples A, ..., P is loaded by 16 resistors. This problem of loading increases with the number of samples of the Fourier transform. One could overcome it by using in Fig. 3.1-2 four amplifiers that sum four voltages each, and a fifth amplifier that sums the output voltages of these four amplifiers. Going this route, one derives circuits whose number of components increases essentially proportionate to $M^2$, where $M$ is the number of samples. One is immediately reminded that the computing time for the ordinary Fourier transform increases like $M^2$, but that the fast Fourier transform keeps this increase to $M \log M$, and one will suspect that a circuit exists for which the number of components increases nominally like $M \log M$. The principle of such circuits was derived several years ago (Harmuth *et al.*, 1974). More detailed descriptions were published by Gaubatz (1976, 1977). We will explain the circuit design here without using more mathematics than is contained in Eqs. (1) and (2).

Table 3.1-1 shows $k$, $F(k)$, and the values of the samples of the 16 functions of the Fourier series of Fig. 3.1-1. One can recognize a number of symmetries. For instance, in the row s1 the samples B and H of $F(k)$ are multiplied by $p$, and the samples J and P by $-p$. This suggests to produce the sums and differences A + I to H − P listed in Table 3.1-2. The Fourier coefficient $a_s(1)$ produced from row s1 has the following form:

$$a_s(1) = 0(\mathrm{A} - \mathrm{I}) + p(\mathrm{B} - \mathrm{J}) + q(\mathrm{C} - \mathrm{K}) + r(\mathrm{D} - \mathrm{L}) + 1(\mathrm{E} - \mathrm{M}) + r(\mathrm{F} - \mathrm{N}) + q(\mathrm{G} - \mathrm{O}) + p(\mathrm{H} - \mathrm{P})$$

One may still find symmetries in Table 3.1-2. For instance, the differences B − J and H − P in row s1 are both multiplied by $p$. Hence, we form the sums and differences (A + I) + (E + M) to (D − L) − (F − N) listed in Table 3.1-3. The Fourier coefficient $a_s(1)$ produced from row s1 now has the following form:

$$a_s(1) = 1(\mathrm{E} - \mathrm{M}) + p[(\mathrm{B} - \mathrm{J}) + (\mathrm{H} - \mathrm{P})] + q[(\mathrm{C} - \mathrm{K}) + (\mathrm{G} - \mathrm{O})] + r[(\mathrm{D} - \mathrm{L}) + (\mathrm{F} - \mathrm{N})]$$

The numbers 0, 8, 4, 12, ..., 16, 17, ... and 32, 33, ... in Table 3.1-3 refer to the terminals of the circuit in Fig. 3.1-3, which implements the Fourier transform according to the multiplications, summations, and

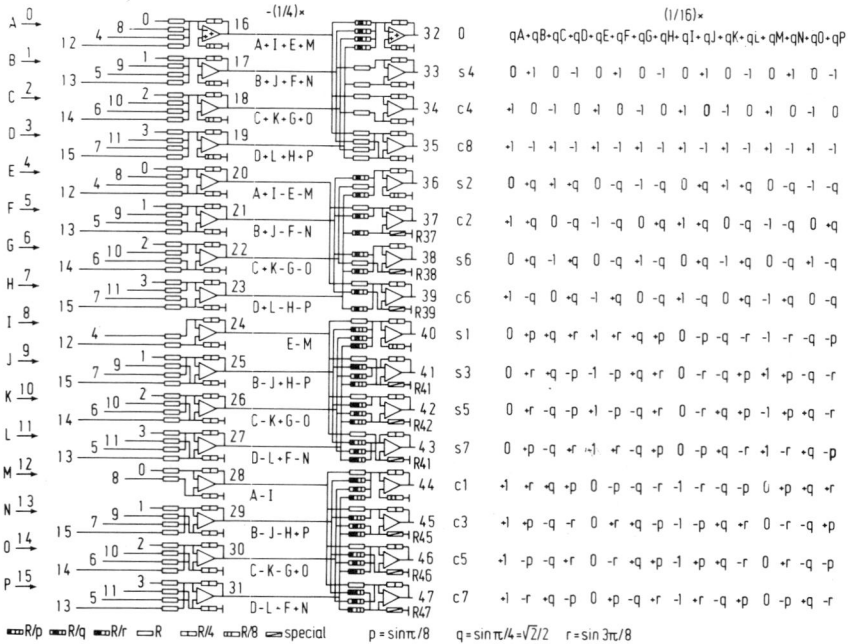

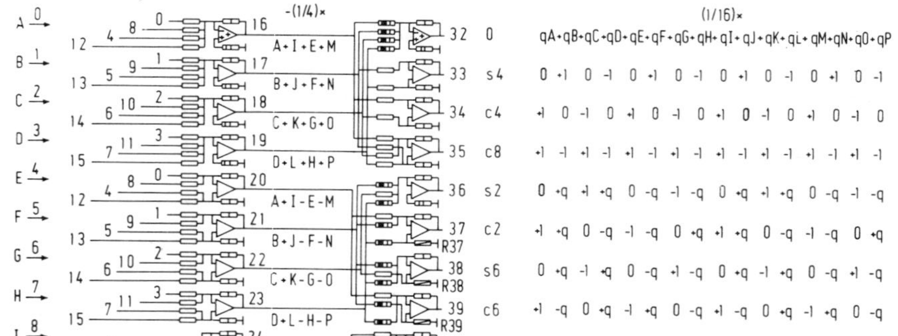


FIG. 3.1-3. Circuit performing a Fourier transform of 16 input voltages A, ..., P using a variation of the fast Fourier transform that produces the sum of four voltages at a time.

## TABLE 3.1-1

THE VALUES OF THE SAMPLES OF THE 16 FUNCTIONS OF THE FOURIER SERIES IN FIGURE 3.1-1[a]

| $k$ | 0 | 1 | 2 | 3 | 4 | 5 | 6 | 7 | 8 | 9 | 10 | 11 | 12 | 13 | 14 | 15 |
|---|---|---|---|---|---|---|---|---|---|---|---|---|---|---|---|---|
| $F(k)$ | $A$ | B | C | D | E | F | G | H | I | J | K | L | M | N | O | P |
| 0 | $q$ | $q$ | $q$ | $q$ | $q$ | $q$ | $q$ | $q$ | $q$ | $q$ | $q$ | $q$ | $q$ | $q$ | $q$ | $q$ |
| s1 | 0 | $p$ | $q$ | $r$ | 1 | $r$ | $q$ | $p$ | 0 | $-p$ | $-q$ | $-r$ | $-1$ | $-r$ | $-q$ | $-p$ |
| c1 | 1 | $r$ | $q$ | $p$ | 0 | $-p$ | $-q$ | $-r$ | $-1$ | $-r$ | $-q$ | $-p$ | 0 | $p$ | $q$ | $r$ |
| s2 | 0 | $q$ | 1 | $q$ | 0 | $-q$ | $-1$ | $-q$ | 0 | $q$ | 1 | $q$ | 0 | $-q$ | $-1$ | $-q$ |
| c2 | 1 | $q$ | 0 | $-q$ | $-1$ | $-q$ | 0 | $q$ | 1 | $q$ | 0 | $-q$ | $-1$ | $-q$ | 0 | $q$ |
| s3 | 0 | $r$ | $q$ | $-p$ | $-1$ | $-p$ | $q$ | $r$ | 0 | $-r$ | $-q$ | $p$ | 1 | $p$ | $-q$ | $-r$ |
| c3 | 1 | $p$ | $-q$ | $-r$ | 0 | $r$ | $q$ | $-p$ | $-1$ | $-p$ | $q$ | $r$ | 0 | $-r$ | $-q$ | $p$ |
| s4 | 0 | 1 | 0 | $-1$ | 0 | 1 | 0 | $-1$ | 0 | 1 | 0 | $-1$ | 0 | 1 | 0 | $-1$ |
| c4 | 1 | 0 | $-1$ | 0 | 1 | 0 | $-1$ | 0 | 1 | 0 | $-1$ | 0 | 1 | 0 | $-1$ | 0 |
| s5 | 0 | $r$ | $-q$ | $-p$ | 1 | $-p$ | $-q$ | $r$ | 0 | $-r$ | $q$ | $p$ | $-1$ | $p$ | $q$ | $-r$ |
| c5 | 1 | $-p$ | $-q$ | $r$ | 0 | $-r$ | $q$ | $p$ | $-1$ | $p$ | $q$ | $-r$ | 0 | $r$ | $-q$ | $-p$ |
| s6 | 0 | $q$ | $-1$ | $q$ | 0 | $-q$ | 1 | $-q$ | 0 | $q$ | $-1$ | $q$ | 0 | $-q$ | 1 | $-q$ |
| c6 | 1 | $-q$ | 0 | $q$ | $-1$ | $q$ | 0 | $-q$ | 1 | $-q$ | 0 | $q$ | $-1$ | $q$ | 0 | $-q$ |
| s7 | 0 | $p$ | $-q$ | $r$ | $-1$ | $r$ | $-q$ | $p$ | 0 | $-p$ | $q$ | $-r$ | 1 | $-r$ | $q$ | $-p$ |
| c7 | 1 | $-r$ | $q$ | $-p$ | 0 | $p$ | $-q$ | $r$ | $-1$ | $r$ | $-q$ | $p$ | 0 | $-p$ | $q$ | $-r$ |
| c8 | 1 | $-1$ | 1 | $-1$ | 1 | $-1$ | 1 | $-1$ | 1 | $-1$ | 1 | $-1$ | 1 | $-1$ | 1 | $-1$ |

[a] $p = \sin \pi/8$, $q = \sin \pi/4 = \sqrt{2}/2$, $r = \sin 3\pi/8$.

## TABLE 3.1-2

SUMS AND DIFFERENCES OF THE SAMPLES A, ..., P OF THE FUNCTION $F(k)$, AND THE VALUES 0, $\pm p$, $\pm q$, $\pm r$, $\pm 1$ OF THE SAMPLES OF THE 16 FUNCTIONS OF THE FOURIER SERIES BY WHICH THEY ARE MULTIPLIED

| | A + I | B + J | C + K | D + L | E + M | F + N | G + O | H + P | A − I | B − J | C − K | D − L | E − M | F − N | G − O | H − P |
|---|---|---|---|---|---|---|---|---|---|---|---|---|---|---|---|---|
| 0 | $q$ | $q$ | $q$ | $q$ | $q$ | $q$ | $q$ | $q$ | | | | | | | | |
| s1 | | | | | | | | | 0 | $p$ | $q$ | $r$ | 1 | $r$ | $q$ | $p$ |
| c1 | | | | | | | | | 1 | $r$ | $q$ | $p$ | 0 | $-p$ | $-q$ | $-r$ |
| s2 | 0 | $q$ | 1 | $q$ | 0 | $-q$ | $-1$ | $-q$ | | | | | | | | |
| c2 | 1 | $q$ | 0 | $-q$ | $-1$ | $-q$ | 0 | $q$ | | | | | | | | |
| s3 | | | | | | | | | 0 | $r$ | $q$ | $-p$ | $-1$ | $-p$ | $q$ | $r$ |
| c3 | | | | | | | | | 1 | $p$ | $-q$ | $-r$ | 0 | $r$ | $q$ | $-p$ |
| s4 | 0 | 1 | 0 | $-1$ | 0 | 1 | 0 | $-1$ | | | | | | | | |
| c4 | 1 | 0 | $-1$ | 0 | 1 | 0 | $-1$ | 0 | | | | | | | | |
| s5 | | | | | | | | | 0 | $r$ | $-q$ | $-p$ | 1 | $-p$ | $-q$ | $r$ |
| c5 | | | | | | | | | 1 | $-p$ | $-q$ | $r$ | 0 | $-r$ | $q$ | $p$ |
| s6 | 0 | $q$ | $-1$ | $q$ | 0 | $-q$ | 1 | $-q$ | | | | | | | | |
| c6 | 1 | $-q$ | 0 | $q$ | $-1$ | $q$ | 0 | $-q$ | | | | | | | | |
| s7 | | | | | | | | | 0 | $p$ | $-q$ | $r$ | $-1$ | $r$ | $-q$ | $p$ |
| c7 | | | | | | | | | 1 | $-r$ | $q$ | $-p$ | 0 | $p$ | $-q$ | $r$ |
| c8 | 1 | $-1$ | 1 | $-1$ | 1 | $-1$ | 1 | $-1$ | | | | | | | | |

TABLE 3.1-3

SUMS AND DIFFERENCES OF FOUR SAMPLES A, ..., P OF THE FUNCTION $F(k)$, AND THE VALUES 0, $\pm p$, $\pm q$, $\pm r$, $\pm 1$ OF THE SAMPLES OF THE 16 FUNCTIONS OF THE FOURIER SERIES BY WHICH THEY ARE MULTIPLIED

| | | 16 | 17 | 18 | 19 | 20 | 21 | 22 | 23 | 24 | 25 | 26 | 27 | 28 | 29 | 30 | 31 |
|---|---|---|---|---|---|---|---|---|---|---|---|---|---|---|---|---|---|
| | | 0 8 4 12<br>(A + I) + (E + M) | 1 9 5 13<br>(B + J) + (F + N) | 2 10 6 14<br>(C + K) + (G + O) | 3 11 7 15<br>(D + L) + (H + P) | 0 8 4 12<br>(A + I) − (E + M) | 1 9 5 13<br>(B + J) − (F + N) | 2 10 6 14<br>(C + K) − (G + O) | 3 11 7 15<br>(D + L) − (H + P) | 0 8 4 12<br>E − M | 1 9 7 15<br>(B − J) + (H − P) | 2 10 6 14<br>(C − K) + (G − O) | 3 11 5 13<br>(D − L) + (F − N) | 0 8 4 12<br>A − I | 1 9 7 15<br>(B − J) − (H − P) | 2 10 6 14<br>(C − K) − (G − O) | 3 11 5 13<br>(D − L) − (F − N) |
| 32 | 0 | $q$ | $q$ | $q$ | $q$ | | | | | | | | | | | | |
| 40 | s1 | | | | | | | | | 1 | $p$ | $q$ | $r$ | | | | |
| 44 | c1 | | | | | | | | | | | | | 1 | $r$ | $q$ | $p$ |
| 36 | s2 | | | | | 0 | $q$ | 1 | $q$ | | | | | | | | |
| 37 | c2 | | | | | 1 | $q$ | 0 | $-q$ | | | | | | | | |
| 41 | s3 | | | | | | | | | $-1$ | $r$ | $q$ | $-p$ | | | | |
| 45 | c3 | | | | | | | | | | | | | 1 | $p$ | $-q$ | $-r$ |
| 33 | s4 | 0 | 1 | 0 | $-1$ | | | | | | | | | | | | |
| 34 | c4 | 1 | 0 | $-1$ | 0 | | | | | | | | | | | | |
| 42 | s5 | | | | | | | | | 1 | $r$ | $-q$ | $-p$ | | | | |
| 46 | c5 | | | | | | | | | | | | | 1 | $-p$ | $-q$ | $r$ |
| 38 | s6 | | | | | 0 | $q$ | $-1$ | $q$ | | | | | | | | |
| 39 | c6 | | | | | 1 | $-q$ | 0 | $q$ | | | | | | | | |
| 43 | s7 | | | | | | | | | $-1$ | $p$ | $-q$ | $r$ | | | | |
| 47 | c7 | | | | | | | | | | | | | 1 | $-r$ | $q$ | $-p$ |
| 35 | c8 | 1 | $-1$ | 1 | $-1$ | | | | | | | | | | | | |

subtractions in Table 3.1-3. On the left are shown the input terminals 0, ..., 15 with the voltages A, ..., P. The terminals 0, 8, 4, 12 with the voltages A, I, E, M are connected to the amplifier on top left, and the sum $-\frac{1}{4}(A + I + E + M)$ is produced at the output terminal 16. This corresponds to the numbers and letters shown in the first column—headed by "16"—in Table 3.1-3. The sums produced by the other amplifiers on the left in Fig. 3.1-3 may readily be related to the letters and numbers in Table 3.1-3 by this example.

The voltages at the terminals 16, 17, 18, 19 are multiplied by $q$ and summed by the topmost amplifier on the right in Fig. 3.1-3; the sum $+\frac{1}{16}(qA + qB + \cdots + qP)$ is produced at the terminal 32. The first row in Table 3.1-3—denoted by the number 32 of the output terminal—represents this process. Correspondingly, the second row in Table 3.1-3 shows that terminal 40 delivers the voltage $+\frac{1}{16}\{1(E - M) + p[(B - J) + (H - P)] + q[(C - K) + (G - O)] + r[(D - L) + (F - N)]\}$, etc.

The factors $\frac{1}{4}$ and $\frac{1}{16}$ are used in order to maintain the same voltage level at all terminals. Without these scaling factors one would introduce unnecessary errors due to noise and drift.

The determination of the value of the resistors denoted "special" in Fig. 3.1-3 is widely discussed in the literature on operational amplifiers (Kostanty, 1972). The core of the matter is to have the same resistance to ground from the inverting and the noninverting input terminal of the amplifier. The simplest process is to connect equal resistors to the inverting and noninverting terminals, as shown in Fig. 3.1-4a, and to ground the

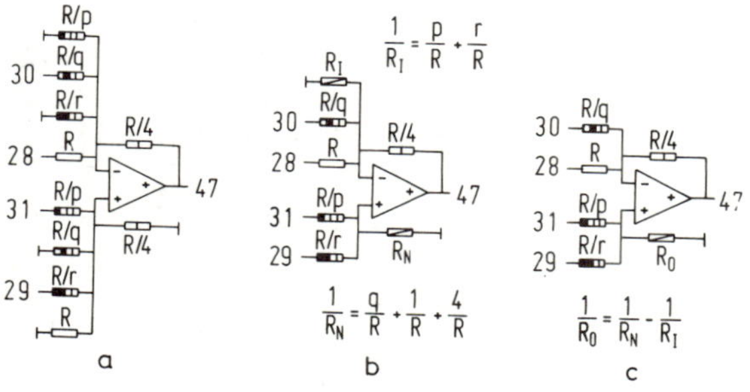


FIG. 3.1-4. The use of operational amplifiers for summation and subtraction.

resistors that are not needed for the summation of the input voltages. Figure 3.1-4a implements the summing circuit of Fig. 3.1-3 with the output terminal 47. This method appears to waste resistors, but this is so only if one uses individual resistors. If resistor networks are used, one may prefer to ground resistors in order to reduce the number of different networks.

The number of individual resistors in Fig. 3.1-4a can be reduced by first calculating the resulting resistances of the resistors connected directly to ground from the inverting and the noninverting terminal; these resistances are denoted $R_I$ and $R_N$ in Fig. 3.1-4b. One may then eliminate $R_I$ and replace $R_N$ by a resistance $R_0$ that follows from the relations

$$\frac{1}{R_0} = \frac{1}{R_N} - \frac{1}{R_I} \qquad \text{or} \qquad R_0 = \frac{R_N R_I}{R_I - R_N}$$

The last step is only possible if $R_I$ is larger than $R_N$. Figure 3.1-4c shows the circuit obtained in this way from Fig. 3.1-4b; the resistance $R_0$ is the value required for the resistor R47 in Fig. 3.1-3.

The principle of the 16-input fast Fourier transform circuit is generalized in the following two sections to 32, 64, and 12 input terminals, and the further extension to larger powers of 2 as well as to numbers of the form $2^k 3^m 5^n$—or generally numbers that can be represented by products of small prime numbers—will become evident. The main problem is that the size of the tables corresponding to Table 3.1-1 increases quadratically with the number of input terminals. Hence, a prime task is to simplify the derivation of the circuits.

## 3.2 Fast Fourier Transform Circuits with 32 and 64 Input Terminals

The values of the samples in Fig. 3.1-1 were obtained by computing the values of $\sin 2\pi lk/16$ and $\cos 2\pi lk/16$ for $k = 0, 1, \ldots, 15$ and $l = 1, 2, \ldots, 8$. This process is too cumbersome for more than 16 functions, since 32 functions require already the computation of $32^2 = 1024$ samples. A simpler way is to divide a circle into 32 equal sectors as shown in Fig. 3.2-1. The samples of the first sinusoidal function ($l = 1$) are obtained by plotting each value $0, p, q, r, \ldots$ as shown in Fig. 3.2-2 for $\sin 2\pi x = \sin 2\pi k/16 =$ s1. The samples of the function $\cos 2\pi x$ are obtained by starting in Fig. 3.2-1 at 1 rather than at 0. The samples of the functions for $l = 2$ are obtained by plotting every second value in Fig. 3.2-1; that means $0, q, s, \ldots$ for $\sin 4\pi x$, and $1, v, s, \ldots$ for $\cos 4\pi x$. For $l = 3$ one plots every third value, etc. In this fashion one can quite rapidly produce the $32^2$ samples shown in Fig. 3.2-2 and Table 3.2-1.

In analogy to the derivation of Table 3.1-3 from Table 3.1-1, one can now derive Table 3.2-2 from Table 3.2-1. It is easy to recognize the rule by which the sums of four voltages on top of Tables 3.1-3 and 3.2-2 are written. Table 3.1-3 lists the functions 0, s1, c1, ..., c8 in a sequence, while Table 3.2-2 lists on the left the functions 0, s2, c2, ..., c16 with even values of $l$, and the functions s1, c1, ..., c15 with odd values of $l$ on the right.

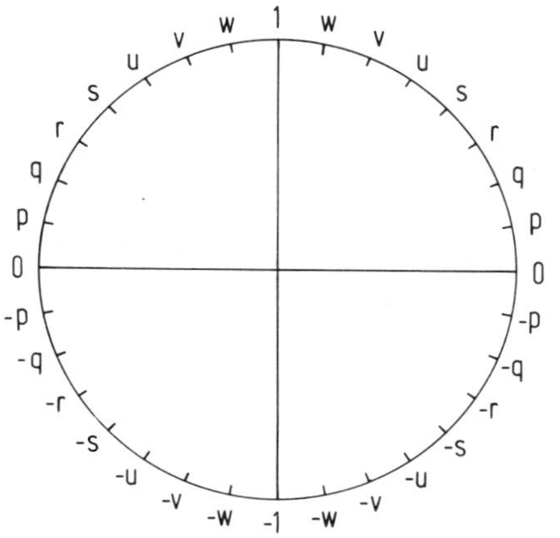


FIG. 3.2-1. Determination of the samples 0, $\pm p$, $\pm q$, ..., $\pm 1$ of the first 32 functions of the Fourier series. $p = \sin \pi/16$, $q = \sin \pi/8$, $r = \sin 3\pi/16$, $s = \sin \pi/4 = \sqrt{2}/2$, $u = \sin 5\pi/16$, $v = \sin 3\pi/8$, $w = \sin 7\pi/16$.

The purpose of this listing is to make Table 3.2-2 half as long as it would be if the listing were done as in Table 3.1-3.

According to Table 3.2-2, we obtain the sums and differences of four voltages each at the output terminals 32, 33, ..., 63, if we feed 32 input voltages A, ..., **F** to the input terminals 0, 1, ..., 31. The respective summing circuits are shown on the left in Fig. 3.2-3. Deviating from the presentation of the circuit in Fig. 3.1-3, we do not show the voltages A, B, ..., **F**, in order to simplify the drafting, but one can readily verify with the help of the top part of Table 3.2-2 that the voltages (A + Q) + (I + Y) to (H − X) − (J − Z) are produced at the terminals 32 to 63 if the voltages A to **F** are fed to the input terminals 0 to 31.

One could obtain the Fourier transform according to Table 3.2-2 by multiplying eight voltages at the terminals 32 to 63 with coefficients 0, $\pm p$, $\pm q$, ..., $\pm 1$ and summing the products. Very good operational amplifiers permit the summation of eight voltages, but we will reduce Table 3.2-2 so that only four voltages have to be summed by an operational amplifier. This is possible because there are still several symmetries in Table 3.2-2. For instance, the function s4 = sin $8\pi x$ shows the coefficient $\pm s$ at the terminals 33, 35, 37, and 39. Symmetries of this type occur for all functions 0, s2, c2, ..., c16 in Table 3.2-2. The functions s1, c1, ..., c15 do not have this type of symmetry, but they have another one. All functions s1, s3, ..., s15 have in column 52 the coefficient $\pm s$; in columns 50 and 54 they have only the coefficients $\pm q$ and $\pm v$. A similar statement holds for the coefficients of the functions c1, c3, ..., c15 in columns 60, or 58 and 62. Using these symmetries we produce the sums and differences of four voltages

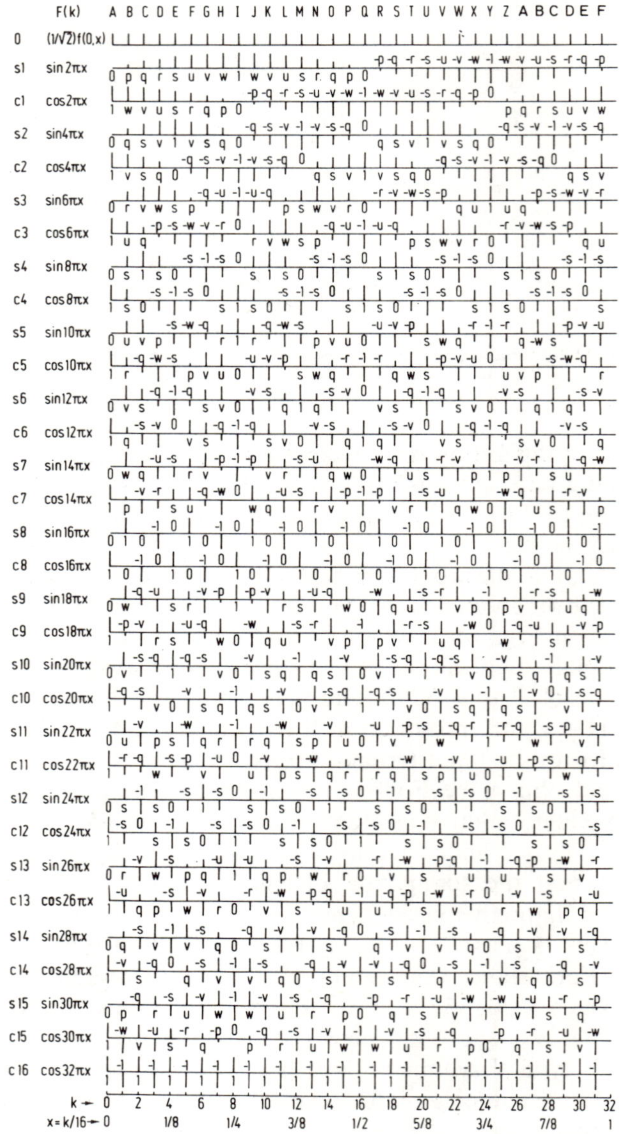

FIG. 3.2-2. The first 32 functions of the Fourier series represented by the samples derived from Fig. 3.2-1.

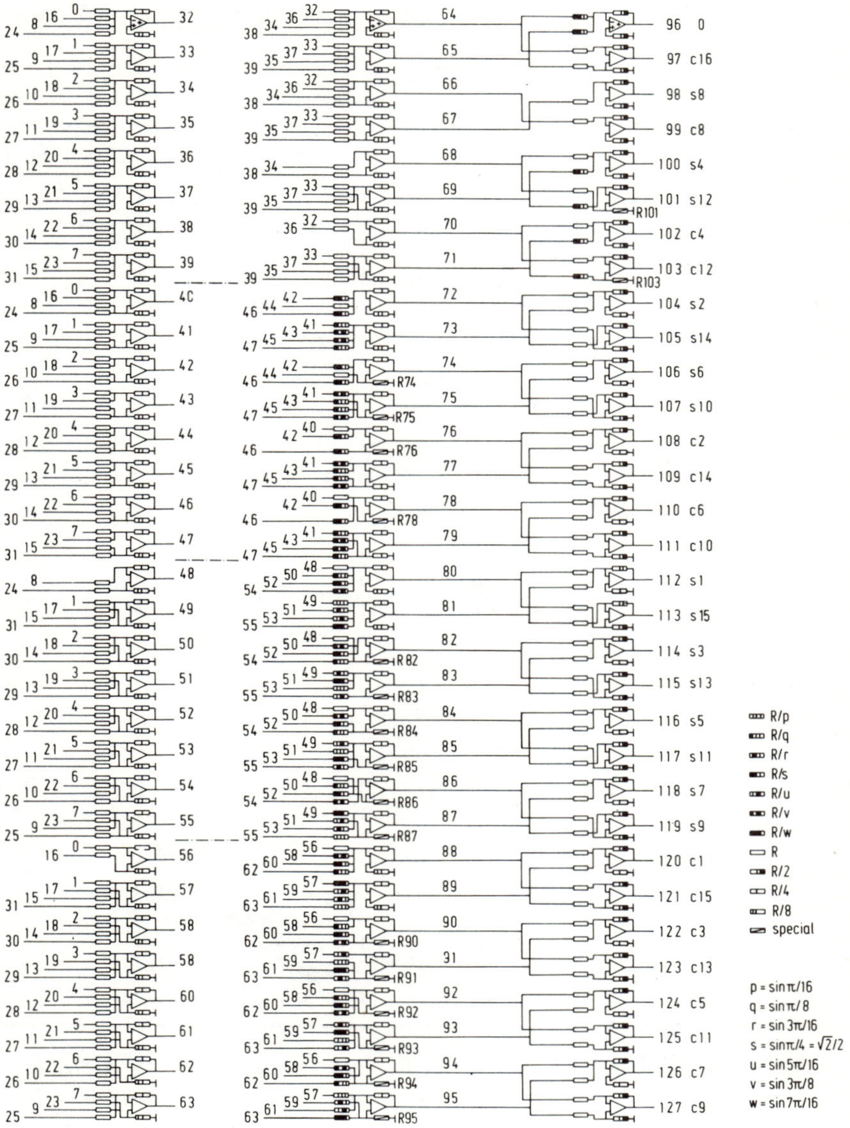


FIG. 3.2-3. Circuit performing a Fourier transform of 32 input voltages at the terminals 0, 1, ..., 31. A combination of fast Fourier transforms producing the sum of four or two voltages at a time is used.

TABLE

THE VALUES OF THE SAMPLES OF THE 32 FUNCTIONS

| $k$ | 0 | 1 | 2 | 3 | 4 | 5 | 6 | 7 | 8 | 9 | 10 | 11 | 12 | 13 | 14 | 15 |
|---|---|---|---|---|---|---|---|---|---|---|---|---|---|---|---|---|
| $F(k)$ | A | B | C | D | E | F | G | H | I | J | K | L | M | N | O | P |
| 0 | $s$ | $s$ | $s$ | $s$ | $s$ | $s$ | $s$ | $s$ | $s$ | $s$ | $s$ | $s$ | $s$ | $s$ | $s$ | $s$ |
| s1 | 0 | $p$ | $q$ | $r$ | $s$ | $u$ | $v$ | $w$ | 1 | $w$ | $v$ | $u$ | $s$ | $r$ | $q$ | $p$ |
| c1 | 1 | $w$ | $v$ | $u$ | $s$ | $r$ | $q$ | $p$ | 0 | $-p$ | $-q$ | $-r$ | $-s$ | $-u$ | $-v$ | $-w$ |
| s2 | 0 | $q$ | $s$ | $v$ | 1 | $v$ | $s$ | $q$ | 0 | $-q$ | $-s$ | $-v$ | $-1$ | $-v$ | $-s$ | $-q$ |
| c2 | 1 | $v$ | $s$ | $q$ | 0 | $-q$ | $-s$ | $-v$ | $-1$ | $-v$ | $-s$ | $-q$ | 0 | $q$ | $s$ | $v$ |
| s3 | 0 | $r$ | $v$ | $w$ | $s$ | $p$ | $-q$ | $-u$ | $-1$ | $-u$ | $-q$ | $p$ | $s$ | $w$ | $v$ | $r$ |
| c3 | 1 | $u$ | $q$ | $-p$ | $-s$ | $-w$ | $-v$ | $-r$ | 0 | $r$ | $v$ | $w$ | $s$ | $p$ | $-q$ | $-u$ |
| s4 | 0 | $s$ | 1 | $s$ | 0 | $-s$ | $-1$ | $-s$ | 0 | $s$ | 1 | $s$ | 0 | $-s$ | $-1$ | $-s$ |
| c4 | 1 | $s$ | 0 | $-s$ | $-1$ | $-s$ | 0 | $s$ | 1 | $s$ | 0 | $-s$ | $-1$ | $-s$ | 0 | $s$ |
| s5 | 0 | $u$ | $v$ | $p$ | $-s$ | $-w$ | $-q$ | $r$ | 1 | $r$ | $-q$ | $-w$ | $-s$ | $p$ | $v$ | $u$ |
| c5 | 1 | $r$ | $-q$ | $-w$ | $-s$ | $p$ | $v$ | $u$ | 0 | $-u$ | $-v$ | $-p$ | $s$ | $w$ | $q$ | $-r$ |
| s6 | 0 | $v$ | $s$ | $-q$ | $-1$ | $-q$ | $s$ | $v$ | 0 | $-v$ | $-s$ | $q$ | 1 | $q$ | $-s$ | $-v$ |
| c6 | 1 | $q$ | $-s$ | $-v$ | 0 | $v$ | $s$ | $-q$ | $-1$ | $-q$ | $s$ | $v$ | 0 | $-v$ | $-s$ | $q$ |
| s7 | 0 | $w$ | $q$ | $-u$ | $-s$ | $r$ | $v$ | $-p$ | $-1$ | $-p$ | $v$ | $r$ | $-s$ | $-u$ | $q$ | $w$ |
| c7 | 1 | $p$ | $-v$ | $-r$ | $s$ | $u$ | $-q$ | $-w$ | 0 | $w$ | $q$ | $-u$ | $-s$ | $r$ | $v$ | $-p$ |
| s8 | 0 | 1 | 0 | $-1$ | 0 | 1 | 0 | $-1$ | 0 | 1 | 0 | $-1$ | 0 | 1 | 0 | $-1$ |
| c8 | 1 | 0 | $-1$ | 0 | 1 | 0 | $-1$ | 0 | 1 | 0 | $-1$ | 0 | 1 | 0 | $-1$ | 0 |
| s9 | 0 | $w$ | $-q$ | $-u$ | $s$ | $r$ | $-v$ | $-p$ | 1 | $-p$ | $-v$ | $r$ | $s$ | $-u$ | $-q$ | $w$ |
| c9 | 1 | $-p$ | $-v$ | $r$ | $s$ | $-u$ | $-q$ | $w$ | 0 | $-w$ | $q$ | $u$ | $-s$ | $-r$ | $v$ | $p$ |
| s10 | 0 | $v$ | $-s$ | $-q$ | 1 | $-q$ | $-s$ | $v$ | 0 | $-v$ | $s$ | $q$ | $-1$ | $q$ | $s$ | $-v$ |
| c10 | 1 | $-q$ | $-s$ | $v$ | 0 | $-v$ | $s$ | $q$ | $-1$ | $q$ | $s$ | $-v$ | 0 | $v$ | $-s$ | $-q$ |
| s11 | 0 | $u$ | $-v$ | $p$ | $s$ | $-w$ | $q$ | $r$ | $-1$ | $r$ | $q$ | $-w$ | $s$ | $p$ | $-v$ | $u$ |
| c11 | 1 | $-r$ | $-q$ | $w$ | $-s$ | $-p$ | $v$ | $-u$ | 0 | $u$ | $-v$ | $p$ | $s$ | $-w$ | $q$ | $r$ |
| s12 | 0 | $s$ | $-1$ | $s$ | 0 | $-s$ | 1 | $-s$ | 0 | $s$ | $-1$ | $s$ | 0 | $-s$ | 1 | $-s$ |
| c12 | 1 | $-s$ | 0 | $s$ | $-1$ | $s$ | 0 | $-s$ | 1 | $-s$ | 0 | $s$ | $-1$ | $s$ | 0 | $-s$ |
| s13 | 0 | $r$ | $-v$ | $w$ | $-s$ | $p$ | $q$ | $-u$ | 1 | $-u$ | $q$ | $p$ | $-s$ | $w$ | $-v$ | $r$ |
| c13 | 1 | $-u$ | $q$ | $p$ | $-s$ | $w$ | $-v$ | $r$ | 0 | $-r$ | $v$ | $-w$ | $s$ | $-p$ | $-q$ | $u$ |
| s14 | 0 | $q$ | $-s$ | $v$ | $-1$ | $v$ | $-s$ | $q$ | 0 | $-q$ | $s$ | $-v$ | 1 | $-v$ | $s$ | $-q$ |
| c14 | 1 | $-v$ | $s$ | $-q$ | 0 | $q$ | $-s$ | $v$ | $-1$ | $v$ | $-s$ | $q$ | 0 | $-q$ | $s$ | $-v$ |
| s15 | 0 | $p$ | $-q$ | $r$ | $-s$ | $u$ | $-v$ | $w$ | $-1$ | $w$ | $-v$ | $u$ | $-s$ | $r$ | $-q$ | $p$ |
| c15 | 1 | $-w$ | $v$ | $-u$ | $s$ | $-r$ | $q$ | $-p$ | 0 | $p$ | $-q$ | $r$ | $-s$ | $u$ | $-v$ | $w$ |
| c16 | 1 | $-1$ | 1 | $-1$ | 1 | $-1$ | 1 | $-1$ | 1 | $-1$ | 1 | $-1$ | 1 | $-1$ | 1 | $-1$ |

[a] $p = \sin \pi/8$, $r = \sin 3\pi/16$, $s = \sin \pi/4 = \sqrt{2}/2$, $u = \sin 5\pi/16$, $v = \sin 3\pi/8$, $w = \sin 7\pi/16$.

3.2-1

OF THE FOURIER SERIES IN FIG. 3.2-2[a]

| 16 | 17 | 18 | 19 | 20 | 21 | 22 | 23 | 24 | 25 | 26 | 27 | 28 | 29 | 30 | 31 |
|---|---|---|---|---|---|---|---|---|---|---|---|---|---|---|---|
| **Q** | **R** | **S** | **T** | **U** | **V** | **W** | **X** | **Y** | **Z** | **A** | **B** | **C** | **D** | **E** | **F** |
| *s* | *s* | *s* | *s* | *s* | *s* | *s* | *s* | *s* | *s* | *s* | *s* | *s* | *s* | *s* | *s* |
| 0 | −*p* | −*q* | −*r* | −*s* | −*u* | −*v* | −*w* | −1 | −*w* | −*v* | −*u* | −*s* | −*r* | −*q* | −*p* |
| −1 | −*w* | −*v* | −*u* | −*s* | −*r* | −*q* | −*p* | 0 | *p* | *q* | *r* | *s* | *u* | *v* | *w* |
| 0 | *q* | *s* | *v* | 1 | *v* | *s* | *q* | 0 | −*q* | −*s* | −*v* | −1 | −*v* | −*s* | −*q* |
| 1 | *v* | *s* | *q* | 0 | −*q* | −*s* | −*v* | −1 | −*v* | −*s* | −*q* | 0 | *q* | *s* | *v* |
| 0 | −*r* | −*v* | −*w* | −*s* | −*p* | *q* | *u* | 1 | *u* | *q* | −*p* | −*s* | −*w* | −*v* | −*r* |
| −1 | −*u* | −*q* | *p* | *s* | *w* | *v* | *r* | 0 | −*r* | −*v* | −*w* | −*s* | −*p* | *q* | *u* |
| 0 | *s* | 1 | *s* | 0 | −*s* | −1 | −*s* | 0 | *s* | 1 | *s* | 0 | −*s* | −1 | −*s* |
| 1 | *s* | 0 | −*s* | −1 | −*s* | 0 | *s* | 1 | *s* | 0 | −*s* | −1 | −*s* | 0 | *s* |
| 0 | −*u* | −*v* | −*p* | *s* | *w* | *q* | −*r* | −1 | −*r* | *q* | +*w* | *s* | −*p* | −*v* | −*u* |
| −1 | −*r* | *q* | *w* | *s* | −*p* | −*v* | −*u* | 0 | *u* | *v* | *p* | −*s* | −*w* | −*q* | *r* |
| 0 | *v* | *s* | −*q* | −1 | −*q* | *s* | *v* | 0 | −*v* | −*s* | *q* | 1 | *q* | −*s* | −*v* |
| 1 | *q* | −*s* | −*v* | 0 | *v* | *s* | −*q* | −1 | −*q* | *s* | *v* | 0 | −*v* | −*s* | *q* |
| 0 | −*w* | −*q* | *u* | *s* | −*r* | −*v* | *p* | 1 | *p* | −*v* | −*r* | *s* | *u* | −*q* | −*w* |
| −1 | −*p* | *v* | *r* | −*s* | −*u* | *q* | *w* | 0 | −*w* | −*q* | *u* | *s* | −*r* | −*v* | *p* |
| 0 | 1 | 0 | −1 | 0 | 1 | 0 | −1 | 0 | 1 | 0 | −1 | 0 | 1 | 0 | −1 |
| 1 | 0 | −1 | 0 | 1 | 0 | −1 | 0 | 1 | 0 | −1 | 0 | 1 | 0 | −1 | 0 |
| 0 | −*w* | *q* | *u* | −*s* | −*r* | *v* | *p* | −1 | *p* | *v* | −*r* | −*s* | *u* | *q* | −*w* |
| −1 | *p* | *v* | −*r* | −*s* | *u* | *q* | −*w* | 0 | *w* | −*q* | −*u* | *s* | *r* | −*v* | −*p* |
| 0 | *v* | −*s* | −*q* | 1 | −*q* | −*s* | *v* | 0 | −*v* | *s* | *q* | −1 | *q* | *s* | −*v* |
| 1 | −*q* | −*s* | *v* | 0 | −*v* | *s* | *q* | −1 | *q* | *s* | −*v* | 0 | *v* | −*s* | −*q* |
| 0 | −*u* | *v* | −*p* | −*s* | *w* | −*q* | −*r* | 1 | −*r* | −*q* | *w* | −*s* | −*p* | *v* | −*u* |
| −1 | *r* | *q* | −*w* | *s* | *p* | −*v* | *u* | 0 | −*u* | *v* | −*p* | −*s* | *w* | −*q* | −*r* |
| 0 | *s* | −1 | *s* | 0 | −*s* | 1 | −*s* | 0 | *s* | −1 | *s* | 0 | −*s* | 1 | −*s* |
| 1 | −*s* | 0 | *s* | −1 | *s* | 0 | −*s* | 1 | −*s* | 0 | *s* | −1 | *s* | 0 | −*s* |
| 0 | −*r* | *v* | −*w* | *s* | −*p* | −*q* | *u* | −1 | *u* | −*q* | −*p* | *s* | −*w* | *v* | −*r* |
| −1 | *u* | −*q* | −*p* | *s* | −*w* | *v* | −*r* | 0 | *r* | −*v* | *w* | −*s* | *p* | *q* | −*u* |
| 0 | *q* | −*s* | *v* | −1 | *v* | −*s* | *q* | 0 | −*q* | *s* | −*v* | 1 | −*v* | *s* | −*q* |
| 1 | −*v* | *s* | −*q* | 0 | *q* | −*s* | *v* | −1 | *v* | −*s* | *q* | 0 | −*q* | *s* | −*v* |
| 0 | −*p* | *q* | −*r* | *s* | −*u* | *v* | −*w* | 1 | −*w* | *v* | −*u* | *s* | −*r* | *q* | −*p* |
| −1 | *w* | −*v* | *u* | −*s* | *r* | −*q* | *p* | 0 | −*p* | *q* | −*r* | *s* | −*u* | *v* | −*w* |
| 1 | −1 | 1 | −1 | 1 | −1 | 1 | −1 | 1 | −1 | 1 | −1 | 1 | −1 | 1 | −1 |

TABLE

SUMS AND DIFFERENCES OF FOUR SAMPLES A, ..., F OF THE FUNCTION $F(k)$, AND THE VALUES 0,

SERIES BY WHICH

| | 32 | 33 | 34 | 35 | 36 | 37 | 38 | 39 | 40 | 41 | 42 | 43 | 44 | 45 | 46 | 47 |
|---|---|---|---|---|---|---|---|---|---|---|---|---|---|---|---|---|
| | 0 16 8 24<br>(A + Q) + (I + Y) | 1 17 9 25<br>(B + R) + (J + Z) | 2 18 10 26<br>(C + S) + (K + A) | 3 19 11 27<br>(D + T) + (L + B) | 4 20 12 28<br>(E + U) + (M + C) | 5 21 13 29<br>(F + V) + (N + D) | 6 22 14 30<br>(G + W) + (O + E) | 7 23 15 31<br>(H + X) + (P + F) | 0 16 8 24<br>(A + Q) − (I + Y) | 1 17 9 25<br>(B + R) − (J + Z) | 2 18 10 26<br>(C + S) − (K + A) | 3 19 11 27<br>(D + T) − (L + B) | 4 20 12 28<br>(E + U) − (M + C) | 5 21 13 29<br>(F + V) − (N + D) | 6 22 14 30<br>(G + W) − (O + E) | 7 23 15 31<br>(H + X) − (P + F) |
| 0 | *s* | *s* | *s* | *s* | *s* | *s* | *s* | *s* | | | | | | | | |
| s2 | | | | | | | | | 0 | *q* | *s* | *v* | 1 | *v* | *s* | *q* |
| c2 | | | | | | | | | 1 | *v* | *s* | *q* | 0 | −*q* | −*s* | −*v* |
| s4 | 0 | *s* | 1 | *s* | 0 | −*s* | −1 | −*s* | | | | | | | | |
| c4 | 1 | *s* | 0 | −*s* | −1 | −*s* | 0 | *s* | | | | | | | | |
| s6 | | | | | | | | | 0 | *v* | *s* | −*q* | −1 | −*q* | *s* | *v* |
| c6 | | | | | | | | | 1 | *q* | −*s* | −*v* | 0 | *v* | *s* | −*q* |
| s8 | 0 | 1 | 0 | −1 | 0 | 1 | 0 | −1 | | | | | | | | |
| c8 | 1 | 0 | −1 | 0 | 1 | 0 | −1 | 0 | | | | | | | | |
| s10 | | | | | | | | | 0 | *v* | −*s* | −*q* | 1 | −*q* | −*s* | *v* |
| c10 | | | | | | | | | 1 | −*q* | −*s* | *v* | 0 | −*v* | *s* | *q* |
| s12 | 0 | *s* | −1 | *s* | 0 | −*s* | 1 | −*s* | | | | | | | | |
| c12 | 1 | −*s* | 0 | *s* | −1 | *s* | 0 | −*s* | | | | | | | | |
| s14 | | | | | | | | | 0 | *q* | −*s* | *v* | −1 | *v* | −*s* | *q* |
| c14 | | | | | | | | | 1 | −*v* | *s* | −*q* | 0 | *q* | −*s* | *v* |
| c16 | 1 | −1 | 1 | −1 | 1 | −1 | 1 | −1 | | | | | | | | |

3.2-2

$\pm p$, $\pm q$, $\pm r$, $\pm s$, $\pm u$, $\pm v$, $\pm w$, $\pm 1$ OF THE SAMPLES OF THE FIRST 32 FUNCTIONS OF THE FOURIER
THEY ARE MULTIPLIED

| | 48 | 49 | 50 | 51 | 52 | 53 | 54 | 55 | 56 | 57 | 58 | 59 | 60 | 61 | 62 | 63 |
|---|---|---|---|---|---|---|---|---|---|---|---|---|---|---|---|---|
| | 0 16 8 24 / I − Y | 1 17 15 31 / (B − R) + (P − F) | 2 18 14 30 / (C − S) + (O − E) | 3 19 13 29 / (D − T) + (N − D) | 4 20 12 28 / (E − U) + (M − C) | 5 21 11 27 / (F − V) + (L − B) | 6 22 10 26 / (G − W) + (K − A) | 7 23 9 25 / (H − X) + (J − Z) | 0 16 8 24 / A − Q | 1 17 15 31 / (B − R) − (P − F) | 2 18 14 30 / (C − S) − (O − E) | 3 19 13 29 / (D − T) − (N − D) | 4 20 12 28 / (E − U) − (M − C) | 5 21 11 27 / (F − V) − (L − B) | 6 22 10 26 / (G − W) − (K − A) | 7 23 9 25 / (H − X) − (J − Z) |
| s1 | 1 | $p$ | $q$ | $r$ | $s$ | $u$ | $v$ | $w$ | | | | | | | | |
| c1 | | | | | | | | | 1 | $w$ | $v$ | $u$ | $s$ | $r$ | $q$ | $p$ |
| s3 | −1 | $r$ | $v$ | $w$ | $s$ | $p$ | $-q$ | $-u$ | | | | | | | | |
| c3 | | | | | | | | | 1 | $u$ | $q$ | $-p$ | $-s$ | $-w$ | $-v$ | $-r$ |
| s5 | 1 | $u$ | $v$ | $p$ | $-s$ | $-w$ | $-q$ | $r$ | | | | | | | | |
| c5 | | | | | | | | | 1 | $r$ | $-q$ | $-w$ | $-s$ | $p$ | $v$ | $w$ |
| s7 | −1 | $w$ | $q$ | $-u$ | $-s$ | $r$ | $v$ | $-p$ | | | | | | | | |
| c7 | | | | | | | | | 1 | $p$ | $-v$ | $-r$ | $s$ | $u$ | $-q$ | $-w$ |
| s9 | 1 | $w$ | $-q$ | $-u$ | $s$ | $r$ | $-v$ | $-p$ | | | | | | | | |
| c9 | | | | | | | | | 1 | $-p$ | $-v$ | $r$ | $s$ | $-u$ | $-q$ | $w$ |
| s11 | −1 | $u$ | $-v$ | $p$ | $s$ | $-w$ | $q$ | $r$ | | | | | | | | |
| c11 | | | | | | | | | 1 | $-r$ | $-q$ | $w$ | $-s$ | $-p$ | $v$ | $-u$ |
| s13 | 1 | $r$ | $-v$ | $w$ | $-s$ | $p$ | $q$ | $-u$ | | | | | | | | |
| c13 | | | | | | | | | 1 | $-u$ | $q$ | $p$ | $-s$ | $w$ | $-v$ | $r$ |
| s15 | −1 | $p$ | $-q$ | $r$ | $-s$ | $u$ | $-v$ | $w$ | | | | | | | | |
| c15 | | | | | | | | | 1 | $-w$ | $v$ | $-u$ | $s$ | $-r$ | $q$ | $-p$ |

each multiplied with the coefficients 0, $\pm p$, $\pm q$, ..., $\pm 1$ as shown on top of Table 3.2-3. The input voltages for this process come from the terminals 32 to 63, while the output voltages are supplied to the terminals 64 to 95. The circuits implementing the 32 sums are shown in the middle of Fig. 3.2-3.

Let us turn to the lower part of Table 3.2-3. The Fourier transform for the function s4 ($= \sum_{k=0}^{15} F(k) \sin 8\pi k/32$) is obtained by multiplying the voltage at terminal 68 by 1, the voltage at terminal 69 by $s$, and by summing the two products; the sum is delivered to terminal 100. One may see on the right side of Fig. 3.2-3 that terminal 100 receives its input voltages from the terminals 68 and 69. There should be no difficulty in associating the rest of the lower part of Table 3.2-3 with the circuits on the right side of Fig. 3.2-3,

Let us turn to the Fourier transform circuit for 64 samples. The first step is to determine the $64^2 = 4096$ samples of the first 64 functions of the Fourier series consisting of the constant $f(0, k/64)$, $si = \sin 2\pi lk/64$ and $ci = \cos 2\pi lk/64$ with $k = 0, 1, \ldots, 63$ and $l = 1, 2, \ldots, 31$, as well as c32. We obtain them from the circle in Fig. 3.2-4, divided into 64 equal sections, by using every $l$th term, starting at 0 for s and at $+1$ for c, and going counterclockwise. The obtained samples are listed in Table 3.2-4. Only the samples for $k = 0, \ldots, 32$ are shown, since the table is symmetric for the functions $\cos 2\pi lk/64$ and skew-symmetric for the functions $\sin 2\pi lk/64$ for $k = 33, \ldots, 63$.

The next step is to produce the sums and differences of four voltages each out of the 64 voltages A, ..., $L$ in analogy to Table 3.2-2. We note that Table 3.2-2 can be divided into four distinct parts. On the left are entries for the functions s$l$, c$l$ with $l$ divisible by four, and entries with $l$ divisible by 2; on the right are entries for s$l$ with odd values of $l$, and entries for c$l$ with odd values of $l$. Table 3.2-5 is divided in this way into four parts, which makes it more manageable to write.

Since the Fourier transform for each function 0, s$l$, c$l$ in Table 3.2-5 calls for multiplication and summation of 16 voltages, we use again the symmetries to replace these summations by summations of four voltages at a time. The two types of symmetries discussed for the Fourier transform of 32 samples may again be recognized in Table 3.2-5. Hence, there is no particular difficulty in deriving Table 3.2-6, although it takes time to do so without error.

Let us turn to the implementation of the circuit. We need, in essence, three columns of 64 summing amplifiers each if we designed the circuit according to Figs. 3.1-3 and 3.2-3; these circuits have been arranged so that a layout for a printed circuit card can be derived directly. The circuit of Fig. 3.2-3 yields already a card that is rather long and narrow. A better

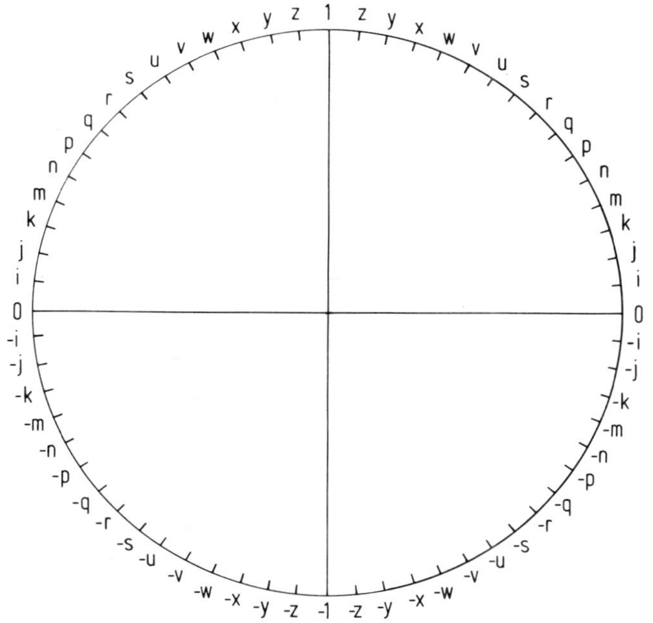


FIG. 3.2-4. Determination of the samples 0, $\pm i$, $\pm j$, ..., $\pm 1$ of the first 64 functions of the Fourier series. $i = \sin \pi/32$, $j = \sin \pi/16$, $k = \sin 3\pi/32$, $m = \sin \pi/8$, $n = \sin 5\pi/32$, $p = \sin 3\pi/16$, $q = \sin 7\pi/32$, $r = \sin \pi/4 = \sqrt{2}/2$, $s = \sin 9\pi/32$, $u = \sin 5\pi/16$, $v = \sin 11\pi/32$, $w = \sin 3\pi/8$, $x = \sin 13\pi/32$, $y = \sin 7\pi/16$, $z = \sin 15\pi/32$.

shape is achieved by using a layout of 6 columns of essentially 32 summing amplifiers each for the Fourier transform of 64 samples. Figure 3.2-5 shows this arrangement in six columns. The three columns to the right implement Tables 3.2-5a,b and 3.2-6a,b while the ones to the left implement Tables 3.2-5c,d and 3.2-6c,d. The sum and differences of four voltages according to Tables 3.2-5a–c are produced at the terminals 128, ..., 159a and 160, ..., 191a; the sums of the lower part of Tables 3.2-6a–c are produced at the terminals 192, ..., 223 and 224, ..., 255.

A printed circuit card producing the Fourier transform of 32 input voltages according to Fig. 3.2-3 is shown in Fig. 3.2-6. The layout follows closely the circuit diagram of Fig. 3.2-3; one only has to observe that each dual-in-line package contains two operational amplifiers. One decoupling capacitor CAP is required per amplifier package, and two more are used at the input terminals of the card.

The technological difficulties to be overcome in acoustic imaging with electronic circuits may be inferred from the printed circuit card of Fig. 3.2-6.

## TABLE 3.2-3

Sums and Differences of $4^2$ Samples A, ..., F of the Function $F(k)$ Multiplied with the Values $\pm p$, $\pm q$, $\pm r$, $\pm s$, $\pm u$, $\pm v$, $\pm w$, $\pm 1$ of the Samples of the First 32 Functions of the Fourier Series, and Values 0, $\pm s$, $\pm 1$, by Which These Weighted Sums Are Multiplied

| | | | | |
|---|---|---|---|---|
| | 32 | 36 | 34 | 38 |
| 64 | (A + Q + I + Y) | + (E + U + M + C) | + (C + S + K + A) | + (G + W + O + E) |
| 66 | (A + Q + I + Y) | + (E + U + M + C) | − (C + S + K + A) | − (G + W + O + E) |
| 68 | | | (C + S + K + A) | − (G + W + O + E) |
| 70 | (A + Q + I + Y) | − (E + U + M + C) | | |
| | 33 | 37 | 35 | 39 |
| 65 | (B + R + J + Z) | + (F + V + N + D) | + (D + T + L + B) | + (H + X + P + F) |
| 67 | (B + R + J + Z) | + (F + V + N + D) | − (D + T + L + B) | − (H + X + P + F) |
| 69 | (B + R + J + Z) | − (F + V + N + D) | + (D + T + L + B) | − (H + X + P + F) |
| 71 | (B + R + J + Z) | − (F + V + N + D) | − (D + T + L + B) | + (H + X + P + F) |
| | 40 | 44 | 42 | 46 |
| 72 | 0(A + Q − I − Y) | + (E + U − M − C) | + s(C + S − K − A) | + s(G + W − O − E) |
| 74 | | − (E + U − M − C) | + s(C + S − K − A) | + s(G + W − O − E) |
| 76 | (A + Q − I − Y) | | + s(C + S − K − A) | − s(G + W − O − E) |
| 78 | (A + Q − I − Y) | | − s(C + S − K − A) | + s(G + W − O − E) |
| | 41 | 45 | 43 | 47 |
| 73 | q(B + R − J − Z) | + v(F + V − N − D) | + v(D + T − L − B) | + q(H + X − P − F) |
| 75 | v(B + R − J − Z) | − q(F + V − N − D) | − q(D + T − L − B) | + v(H + X − P − F) |
| 77 | v(B + R − J − Z) | − q(F + V − N − D) | + q(D + T − L − B) | − v(H + X − P − F) |
| 79 | q(B + R − J − Z) | + v(F + V − N − D) | − v(D + T − L − B) | − q(H + X − P − F) |

| | | | | | | | | | | | | | | | | | |
|---|---|---|---|---|---|---|---|---|---|---|---|---|---|---|---|---|---|
| 96 | 0 | *s* | | | | *s* | | | | | | | | | | | |
| 100 | s4 | | | 1 | | | | *s* | | | | | | | | | |
| 102 | c4 | | | | 1 | | | | *s* | | | | | | | | |
| 98 | s8 | | 0 | | | | 1 | | | | | | | | | | |
| 99 | c8 | | 1 | | | | 0 | | | | | | | | | | |
| 101 | s12 | | | −1 | | | | *s* | | | | | | | | | |
| 103 | c12 | | | | 1 | | | | −*s* | | | | | | | | |
| 97 | c16 | 1 | | | | −1 | | | | | | | | | | | |
| 104 | s2 | | | | | | | | | 1 | | | | 1 | | | |
| 106 | s6 | | | | | | | | | | 1 | | | | 1 | | |
| 107 | s10 | | | | | | | | | | −1 | | | | 1 | | |
| 105 | s14 | | | | | | | | | −1 | | | | 1 | | | |
| 108 | c2 | | | | | | | | | | | 1 | | | | 1 | |
| 110 | c6 | | | | | | | | | | | | 1 | | | | 1 |
| 111 | c10 | | | | | | | | | | | | 1 | | | | −1 |
| 109 | c14 | | | | | | | | | | | 1 | | | | −1 | |

(*continued*)

TABLE 3.2-3 (*continued*)

| | | | | |
|---|---|---|---|---|
| 48 | 52 | 50 | 54 | |
| (I − Y) | + *s*(E − U + **M** − **C**) + | *q*(C − S + O − **E**) + | *v*(G − W + K − **A**) | 80 |
| −(I − Y) | + *s*(E − U + **M** − **C**) + | *v*(C − S + O − **E**) − | *q*(G − W + K − **A**) | 82 |
| (I − Y) | − *s*(E − U + **M** − **C**) + | *v*(C − S + O − **E**) − | *q*(G − W + K − **A**) | 84 |
| −(I − Y) | − *s*(E − U + **M** − **C**) + | *q*(C − S + O − **E**) + | *v*(G − W + K − **A**) | 86 |
| 49 | 53 | 51 | 55 | |
| *p*(B − R + P − **F**) + | *u*(F − V + L − **B**) + | *r*(D − T + N − **D**) + | *w*(H − X + J − Z) | 81 |
| *r*(B − R + P − **F**) + | *p*(F − V + L − **B**) + | *w*(D − T + N − **D**) − | *u*(H − X + J − Z) | 83 |
| *u*(B − R + P − **F**) − | *w*(F − V + L − **B**) + | *p*(D − T + N − **D**) + | *r*(H − X + J − Z) | 85 |
| *w*(B − R + P − **F**) + | *r*(F − V + L − **B**) − | *u*(D − T + N − **D**) − | *p*(H − X + J − Z) | 87 |
| 56 | 60 | 58 | 62 | |
| (A − Q) | + *s*(E − U − **M** + **C**) + | *v*(C − S − O + **E**) + | *q*(G − W − K + **A**) | 88 |
| (A − Q) | − *s*(E − U − **M** + **C**) + | *q*(C − S − O + **E**) − | *v*(G − W − K + **A**) | 90 |
| (A − Q) | − *s*(E − U − **M** + **C**) − | *q*(C − S − O + **E**) + | *v*(G − W − K + **A**) | 92 |
| (A − Q) | + *s*(E − U − **M** + **C**) − | *v*(C − S − O + **E**) − | *q*(G − W − K + **A**) | 94 |
| 57 | 61 | 59 | 63 | |
| *w*(B − R − P + **F**) + | *r*(F − V − L + **B**) + | *u*(D − T − N + **D**) + | *p*(H − X − J + Z) | 89 |
| *u*(B − R − P + **F**) − | *w*(F − V − L + **B**) − | *p*(D − T − N + **D**) − | *r*(H − X − J + Z) | 91 |
| *r*(B − R − P + **F**) + | *p*(F − V − L + **B**) − | *w*(D − T − N + **D**) + | *u*(H − X − J + Z) | 93 |
| *p*(B − R − P + **F**) + | *u*(F − V − L + **B**) − | *r*(D − T − N + **D**) − | *w*(H − X − J + Z) | 95 |

| | | | | | | | | | | | | | | | | | |
|---|---|---|---|---|---|---|---|---|---|---|---|---|---|---|---|---|---|
| 112 | s1 | 1 | | | | 1 | | | | | | | | | | | |
| 114 | s3 | | 1 | | | | 1 | | | | | | | | | | |
| 116 | s5 | | | 1 | | | | 1 | | | | | | | | | |
| 118 | s7 | | | | 1 | | | | 1 | | | | | | | | |
| 119 | s9 | | | | −1 | | | | 1 | | | | | | | | |
| 117 | s11 | | | −1 | | | | 1 | | | | | | | | | |
| 115 | s13 | | −1 | | | | 1 | | | | | | | | | | |
| 113 | s15 | −1 | | | | 1 | | | | | | | | | | | |
| 120 | c1 | | | | | | | | | 1 | | | | 1 | | | |
| 122 | c3 | | | | | | | | | | 1 | | | | 1 | | |
| 124 | c5 | | | | | | | | | | | 1 | | | | 1 | |
| 126 | c7 | | | | | | | | | | | | 1 | | | | 1 |
| 127 | c9 | | | | | | | | | | | | 1 | | | | −1 |
| 125 | c11 | | | | | | | | | | | 1 | | | | −1 | |
| 123 | c13 | | | | | | | | | | 1 | | | | −1 | | |
| 121 | c15 | | | | | | | | | 1 | | | | −1 | | | |

TABLE

THE VALUES OF THE SAMPLES OF THE FIRST

| $k$ | 0 | 1 | 2 | 3 | 4 | 5 | 6 | 7 | 8 | 9 | 10 | 11 | 12 | 13 | 14 | 15 |
|---|---|---|---|---|---|---|---|---|---|---|---|---|---|---|---|---|
| $F(k)$ | A | B | C | D | E | F | G | H | I | J | K | L | M | N | O | P |
| 0 | r | r | r | r | r | r | r | r | r | r | r | r | r | r | r | r |
| s1 | 0 | i | j | k | m | n | p | q | r | s | u | v | w | x | y | z |
| c1 | 1 | z | y | x | w | v | u | s | r | q | p | n | m | k | j | i |
| s2 | 0 | j | m | p | r | u | w | y | 1 | y | w | u | r | p | m | j |
| c2 | 1 | y | w | u | r | p | m | j | 0 | −j | −m | −p | −r | −u | −w | −y |
| s3 | 0 | k | p | s | w | z | y | v | r | n | j | −i | −m | −q | −u | −x |
| c3 | 1 | x | u | q | m | i | −j | −n | −r | −v | −y | −z | −w | −s | −p | −k |
| s4 | 0 | m | r | w | 1 | w | r | m | 0 | −m | −r | −w | −1 | −w | −r | −m |
| c4 | 1 | w | r | m | 0 | −m | −r | −w | −1 | −w | −r | −m | 0 | m | r | w |
| s5 | 0 | n | u | z | w | q | j | −k | −r | −x | −y | −s | −m | i | p | v |
| c5 | 1 | v | p | i | −m | −s | −y | −x | −r | −k | j | q | w | z | u | n |
| s6 | 0 | p | w | y | r | j | −m | −u | −1 | −u | −m | j | r | y | w | p |
| c6 | 1 | u | m | −j | −r | −y | −w | −p | 0 | p | w | y | r | j | −m | −u |
| s7 | 0 | q | y | v | m | −k | −u | −z | −r | −i | p | x | w | n | −j | −s |
| c7 | 1 | s | j | −n | −w | −x | −p | i | r | z | u | k | −m | −v | −y | −q |
| s8 | 0 | r | 1 | r | 0 | −r | −1 | −r | 0 | r | 1 | r | 0 | −r | −1 | −r |
| c8 | 1 | r | 0 | −r | −1 | −r | 0 | r | 1 | r | 0 | −r | −1 | −r | 0 | r |
| s9 | 0 | s | y | n | −m | −x | −u | −i | r | z | p | −k | −w | −v | −j | q |
| c9 | 1 | q | −j | −v | −w | −k | p | z | r | −i | −u | −x | −m | n | y | s |
| s10 | 0 | u | w | j | −r | −y | −m | p | 1 | p | −m | −y | −r | j | w | u |
| c10 | 1 | p | −m | −y | −r | j | w | u | 0 | −u | −w | −j | r | y | m | −p |
| s11 | 0 | v | u | −i | −w | −s | j | x | r | −k | −y | −q | m | z | p | −n |
| c11 | 1 | n | −p | −z | −m | q | y | k | −r | −x | −j | s | w | i | −u | −v |
| s12 | 0 | w | r | −m | −1 | −m | r | w | 0 | −w | −r | m | 1 | m | −r | −w |
| c12 | 1 | m | −r | −w | 0 | w | r | −m | −1 | −m | r | w | 0 | −w | −r | m |
| s13 | 0 | x | p | −q | −w | i | y | n | −r | −v | j | z | m | −s | −u | k |
| c13 | 1 | k | −u | −s | m | z | j | −v | −r | n | y | i | −w | −q | p | x |
| s14 | 0 | y | m | −u | −r | p | w | −j | −1 | −j | w | p | −r | −u | m | y |
| c14 | 1 | j | −w | −p | r | u | −m | −y | 0 | y | m | −u | −r | p | w | −j |
| s15 | 0 | z | j | −x | −m | v | p | −s | −r | q | u | −n | −w | k | y | −i |
| c15 | 1 | i | −y | −k | w | n | −u | −q | r | s | p | −v | m | x | −j | −z |
| s16 | 0 | 1 | 0 | −1 | 0 | 1 | 0 | −1 | 0 | 1 | 0 | −1 | 0 | 1 | 0 | −1 |
| c16 | 1 | 0 | −1 | 0 | 1 | 0 | −1 | 0 | 1 | 0 | −1 | 0 | 1 | 0 | −1 | 0 |
| s17 | 0 | z | −j | −x | m | v | −p | −s | r | q | −u | −n | w | k | −y | −i |
| c17 | 1 | −i | −y | k | w | −n | −u | q | r | −s | −p | v | m | −x | −j | z |
| s18 | 0 | y | −m | −u | r | p | −w | −j | 1 | −j | −w | p | r | −u | −m | y |
| c18 | 1 | −j | −w | p | r | −u | −m | y | 0 | −y | m | u | −r | −p | w | j |
| s19 | 0 | x | −p | −q | w | i | −y | n | r | −v | −j | z | −m | −s | u | k |
| c19 | 1 | −k | −u | s | m | −z | j | v | −r | −n | y | −i | −w | q | p | −x |
| s20 | 0 | w | −r | −m | 1 | −m | −r | w | 0 | −w | r | m | −1 | m | r | −w |
| c20 | 1 | −m | −r | w | 0 | −w | r | m | −1 | m | r | −w | 0 | w | −r | −m |
| s21 | 0 | v | −u | −i | w | −s | −j | x | −r | −k | y | −q | −m | z | −p | −n |
| c21 | 1 | −n | −p | z | −m | −q | y | −k | −r | x | −j | −s | w | −i | −u | v |
| s22 | 0 | u | −w | j | r | −y | m | p | −1 | p | m | −y | r | j | −w | u |
| c22 | 1 | −p | −m | y | −r | −j | w | −u | 0 | u | −w | j | r | −y | m | p |
| s23 | 0 | s | −y | n | m | −x | u | −i | −r | z | −p | −k | w | −v | j | q |
| c23 | 1 | −q | −j | v | −w | k | p | −z | r | i | −u | x | −m | −n | y | −s |
| s24 | 0 | r | −1 | r | 0 | −r | 1 | −r | 0 | r | −1 | r | 0 | −r | 1 | −r |
| c24 | 1 | −r | 0 | r | −1 | r | 0 | −r | 1 | −r | 0 | r | −1 | r | 0 | −r |
| s25 | 0 | q | −y | v | −m | −k | u | −z | r | −i | −p | x | −w | n | j | −s |
| c25 | 1 | −s | j | n | −w | x | −p | −i | r | −z | u | −k | −m | v | −y | q |
| s26 | 0 | p | −w | y | −r | j | m | −u | 1 | −u | m | j | −r | y | −w | p |
| c26 | 1 | −u | m | j | −r | y | −w | p | 0 | −p | w | −y | r | −j | −m | u |
| s27 | 0 | n | −u | z | −w | q | −j | −k | r | −x | y | −s | m | i | −p | v |
| c27 | 1 | −v | p | −i | −m | s | −y | x | −r | k | j | −q | w | −z | u | −n |
| s28 | 0 | m | −r | w | −1 | w | −r | m | 0 | −m | r | −w | 1 | −w | r | −m |
| c28 | 1 | −w | r | −m | 0 | m | −r | w | −1 | w | −r | m | 0 | −m | r | −w |
| s29 | 0 | k | −p | s | −w | z | −y | v | −r | n | −j | −i | m | −q | u | −x |
| c29 | 1 | −x | u | −q | m | −i | −j | n | −r | v | −y | z | −w | s | −p | k |
| s30 | 0 | j | −m | p | −r | u | −w | y | −1 | y | −w | u | −r | p | −m | j |
| c30 | 1 | −y | w | −u | r | −p | m | −j | 0 | j | −m | p | −r | u | −w | y |
| s31 | 0 | i | −j | k | −m | n | −p | q | −r | s | −u | v | −w | x | −y | z |
| c31 | 1 | −z | y | −x | w | −v | u | −s | r | −q | p | −n | m | −k | j | −i |
| c32 | 1 | −1 | 1 | −1 | 1 | −1 | 1 | −1 | 1 | −1 | 1 | −1 | 1 | −1 | 1 | −1 |

[a] The samples are continued to the right of column $k = 32$ with equal signs for the symmetric functions of Fig. 3.2-4 for the values of the samples $i, j, \ldots, z$.

3.2-4

64 FUNCTIONS OF THE FOURIER SERIES[a]

| 16 | 17 | 18 | 19 | 20 | 21 | 22 | 23 | 24 | 25 | 26 | 27 | 28 | 29 | 30 | 31 | 32 |
|---|---|---|---|---|---|---|---|---|---|---|---|---|---|---|---|---|
| Q | R | S | T | U | V | W | X | Y | Z | A | B | C | D | E | F | G |
| r | r | r | r | r | r | r | r | r | r | r | r | r | r | r | r | r |
| 1 | z | y | x | w | v | u | s | r | q | p | n | m | k | j | i | 0 |
| 0 | −i | −j | −k | −m | −n | −p | −q | −r | −s | −u | −v | −w | −x | −y | −z | −1 |
| 0 | −j | −m | −p | −r | −u | −w | −y | −1 | −y | −w | −u | −r | −p | −m | −j | 0 |
| −1 | −y | −w | −u | −r | −p | −m | −j | 0 | j | m | p | r | u | w | y | 1 |
| −1 | −x | −u | −q | −m | −i | j | n | r | v | y | z | w | s | p | k | 0 |
| 0 | k | p | s | w | z | y | v | r | n | j | −i | −m | −q | −u | −x | −1 |
| 0 | m | r | w | 1 | w | r | m | 0 | −m | −r | −w | −1 | −w | −r | −m | 0 |
| 1 | w | r | m | 0 | −m | −r | −w | −1 | −w | −r | −m | 0 | m | r | w | 1 |
| 1 | v | p | i | −m | −s | −y | −x | −r | −k | j | q | w | z | u | n | 0 |
| 0 | −n | −u | −z | −w | −q | −j | k | r | x | y | s | m | −i | −p | −v | −1 |
| 0 | −p | −w | −y | −r | −j | m | u | 1 | u | m | −j | −r | −y | −w | −p | 0 |
| −1 | −u | −m | j | r | y | w | p | 0 | −p | −w | −y | −r | −j | m | u | 1 |
| −1 | −s | −j | n | w | x | p | −i | −r | −z | −u | −k | m | v | y | q | 0 |
| 0 | q | y | v | m | −k | −u | −z | −r | −i | p | x | w | n | −j | −s | −1 |
| 0 | r | 1 | r | 0 | −r | −1 | −r | 0 | r | 1 | r | 0 | −r | −1 | −r | 0 |
| 1 | r | 0 | −r | −1 | −r | 0 | r | 1 | r | 0 | −r | −1 | −r | 0 | r | 1 |
| 1 | q | −j | −v | −w | −k | p | z | r | −i | −u | −x | −m | n | y | s | 0 |
| 0 | −s | −y | −n | m | x | u | i | −r | −z | −p | k | w | v | j | −q | −1 |
| 0 | −u | −w | −j | r | y | m | −p | −1 | −p | m | y | r | −j | −w | −u | 0 |
| −1 | −p | m | y | r | −j | −w | −u | 0 | u | w | j | −r | −y | −m | p | 1 |
| −1 | −n | p | z | m | −q | −y | −k | r | x | j | −s | −w | −i | u | v | 0 |
| 0 | v | u | −i | −w | −s | j | x | r | −k | −y | −q | m | z | p | −n | −1 |
| 0 | w | r | −m | −1 | −m | r | w | 0 | −w | −r | m | 1 | m | −r | −w | 0 |
| 1 | m | −r | −w | 0 | w | r | −m | −1 | −m | r | w | 0 | −w | −r | m | 1 |
| 1 | k | −u | −s | m | z | j | −v | −r | n | y | i | −w | −q | p | x | 0 |
| 0 | −x | −p | q | w | −i | −y | −n | r | v | −j | −z | −m | s | u | −k | 1 |
| 0 | −y | −m | u | r | −p | −w | j | 1 | j | −w | −p | r | u | −m | −y | 0 |
| −1 | −j | w | p | −r | −u | m | y | 0 | −y | −m | u | r | −p | −w | j | 1 |
| −1 | −i | y | k | −w | −n | u | q | −r | −s | p | v | −m | −x | j | z | 0 |
| 0 | z | j | −x | −m | v | p | −s | −r | q | u | −n | −w | k | y | −i | −1 |
| 0 | 1 | 0 | −1 | 0 | 1 | 0 | −1 | 0 | 1 | 0 | −1 | 0 | 1 | 0 | −1 | 0 |
| 1 | 0 | −1 | 0 | 1 | 0 | −1 | 0 | 1 | 0 | −1 | 0 | 1 | 0 | −1 | 0 | 1 |
| 1 | −i | −y | k | w | −n | −u | q | r | −s | −p | v | m | −x | −j | z | 0 |
| 0 | −z | j | x | −m | −v | p | s | −r | −q | u | n | −w | −k | −y | i | −1 |
| 0 | −y | m | u | −r | −p | w | j | −1 | j | w | −p | −r | u | m | −y | 0 |
| −1 | j | w | −p | −r | u | m | −y | 0 | y | −m | −u | r | p | −w | −j | 1 |
| −1 | k | u | −s | −m | z | −j | −v | r | n | −y | i | w | −q | −p | x | 0 |
| 0 | x | −p | −q | w | i | −y | n | r | −v | −j | z | −m | −s | u | k | −1 |
| 0 | w | −r | −m | 1 | −m | −r | w | 0 | −w | r | m | −1 | m | r | −w | 0 |
| 1 | −m | −r | w | 0 | −w | r | m | −1 | m | r | −w | 0 | w | −r | −m | 1 |
| 1 | −n | −p | z | −m | −q | y | −k | −r | x | −j | −s | w | −i | −u | v | 0 |
| 0 | −v | u | i | −w | s | j | −x | r | k | −y | q | m | −z | p | n | −1 |
| 0 | −u | w | −j | −r | y | −m | −p | 1 | −p | −m | y | −r | −j | w | −u | 0 |
| −1 | p | m | −y | r | j | −w | u | 0 | −u | w | −j | −r | y | −m | −p | 1 |
| −1 | q | j | −v | w | −k | −p | z | −r | −i | u | −x | m | n | −y | s | 0 |
| 0 | s | −y | n | m | −x | u | −i | −r | z | −p | −k | w | −v | j | q | −1 |
| 0 | r | −1 | r | 0 | −r | 1 | −r | 0 | r | −1 | r | 0 | −r | 1 | −r | 0 |
| 1 | −r | 0 | r | −1 | r | 0 | −r | 1 | −r | 0 | r | −1 | r | 0 | −r | 1 |
| 1 | −s | j | n | −w | x | −p | −i | r | −z | u | −k | −m | v | −y | q | 0 |
| 0 | −q | y | −v | m | k | −u | z | −r | i | p | −x | w | −n | −j | s | −1 |
| 0 | −p | w | −y | r | −j | −m | u | −1 | u | −m | −j | r | −y | w | −p | 0 |
| −1 | u | −m | −j | r | −y | w | −p | 0 | p | −w | y | −r | j | m | −u | 1 |
| −1 | v | −p | i | m | −s | y | −x | r | −k | −j | q | −w | z | −u | n | 0 |
| 0 | n | −u | z | −w | q | −j | −k | r | −x | y | −s | m | i | −p | v | −1 |
| 0 | m | −r | w | −1 | w | −r | m | 0 | −m | r | −w | 1 | −w | r | −m | 0 |
| 1 | −w | r | −m | 0 | m | −r | w | −1 | w | −r | m | 0 | −m | r | −w | 1 |
| 1 | −x | u | −q | m | −i | −j | n | −r | v | −y | z | −w | s | −p | k | 0 |
| 0 | −k | p | −s | w | −z | y | −v | r | −n | j | i | −m | q | −u | x | −1 |
| 0 | −j | m | −p | r | −u | w | −y | 1 | −y | w | −u | r | −p | m | −j | 0 |
| −1 | y | −w | u | −r | p | −m | j | 0 | −j | m | −p | r | −u | w | −y | 1 |
| −1 | z | −y | x | −w | v | −u | s | −r | q | −p | n | −m | k | −j | i | 0 |
| 0 | i | −j | k | −m | n | −p | q | −r | s | −u | v | −w | x | −y | z | −1 |
| 1 | −1 | 1 | −1 | 1 | −1 | 1 | −1 | 1 | −1 | 1 | −1 | 1 | −1 | 1 | −1 | 1 |

(0, c1, c2, ..., c32) and with reversed signs for the skew-symmetric functions (s1, s2, ..., s31). See the caption

TABLE 3.2-5

SUMS AND DIFFERENCES OF FOUR SAMPLES OF THE FUNCTION $F(k)$, AND THE VALUES $0, \pm i, \pm j, \ldots, \pm 1$ OF THE SAMPLES OF THE FIRST 64 FUNCTIONS OF THE FOURIER SERIES BY WHICH THEY ARE MULTIPLIED

TABLE 3.2-5a[a]

| | 64 | 65 | 66 | 67 | 68 | 69 | 70 | 71 | 72 | 73 | 74 | 75 | 76 | 77 | 78 | 79 |
|---|---|---|---|---|---|---|---|---|---|---|---|---|---|---|---|---|
| | 0 32 16 48<br>**A**+**G**+**Q**+**W** | 1 33 17 49<br>**B**+**H**+**R**+**X** | 2 34 18 50<br>**C**+**I**+**S**+**Y** | 3 35 19 51<br>**D**+**J**+**T**+**Z** | 4 36 20 52<br>**E**+**K**+**U**+*A* | 5 37 21 53<br>**F**+**L**+**V**+*B* | 6 38 22 54<br>**G**+**M**+**W**+*C* | 7 39 23 55<br>**H**+**N**+**X**+*D* | 8 40 24 56<br>**I**+**O**+**Y**+*E* | 9 41 25 57<br>**J**+**P**+**Z**+*F* | 10 42 26 58<br>**K**+**Q**+**A**+*G* | 11 43 27 59<br>**L**+**R**+**B**+*H* | 12 44 28 60<br>**M**+**S**+**C**+*I* | 13 45 29 61<br>**N**+**T**+**D**+*J* | 14 46 30 62<br>**O**+**U**+**E**+*K* | 15 47 31 63<br>**P**+**V**+**F**+*L* |
| 0 | *r* | *r* | *r* | *r* | *r* | *r* | *r* | *r* | *r* | *r* | *r* | *r* | *r* | *r* | *r* | *r* |
| s4 | 0 | *m* | *r* | *w* | 1 | *w* | *r* | *m* | 0 | −*m* | −*r* | −*w* | −1 | −*w* | −*r* | −*m* |
| c4 | 1 | *w* | *r* | *m* | 0 | −*m* | −*r* | −*w* | −1 | −*w* | −*r* | −*m* | 0 | *m* | *r* | *w* |
| s8 | 0 | *r* | 1 | *r* | 0 | −*r* | −1 | −*r* | 0 | *r* | 1 | *r* | 0 | −*r* | −1 | −*r* |
| c8 | 1 | *r* | 0 | −*r* | −1 | −*r* | 0 | *r* | 1 | *r* | 0 | −*r* | −1 | −*r* | 0 | *r* |
| s12 | 0 | *w* | *r* | −*m* | −1 | −*m* | *r* | *w* | 0 | −*w* | −*r* | *m* | 1 | *m* | −*r* | −*w* |
| c12 | 1 | *m* | −*r* | −*w* | 0 | *w* | *r* | −*m* | −1 | −*m* | *r* | *w* | 0 | −*w* | −*r* | *m* |
| s16 | 0 | 1 | 0 | −1 | 0 | 1 | 0 | −1 | 0 | 1 | 0 | −1 | 0 | 1 | 0 | −1 |
| c16 | 1 | 0 | −1 | 0 | 1 | 0 | −1 | 0 | 1 | 0 | −1 | 0 | 1 | 0 | −1 | 0 |
| s20 | 0 | *w* | −*r* | −*m* | 1 | −*m* | −*r* | *w* | 0 | −*w* | *r* | *m* | −1 | *m* | *r* | −*w* |
| c20 | 1 | −*m* | −*r* | *w* | 0 | −*w* | *r* | *m* | −1 | *m* | *r* | −*w* | 0 | *w* | −*r* | −*m* |
| s24 | 0 | *r* | −1 | *r* | 0 | −*r* | 1 | −*r* | 0 | *r* | −1 | *r* | 0 | −*r* | 1 | −*r* |
| c24 | 1 | −*r* | 0 | *r* | −1 | *r* | 0 | −*r* | 1 | −*r* | 0 | *r* | −1 | *r* | 0 | −*r* |
| s28 | 0 | *m* | −*r* | *w* | −1 | *w* | −*r* | *m* | 0 | −*m* | *r* | −*w* | 1 | −*w* | *r* | −*m* |
| c28 | 1 | −*w* | *r* | −*m* | 0 | *m* | −*r* | *w* | −1 | *w* | −*r* | *m* | 0 | −*m* | *r* | −*w* |
| c32 | 1 | −1 | 1 | −1 | 1 | −1 | 1 | −1 | 1 | −1 | 1 | −1 | 1 | −1 | 1 | −1 |

[a] The table is arranged in four parts for $sl = \sin 2\pi lk/64$ and $cl = \cos 2\pi lk/64$. (a) $l$ divisible by 4; (b) $l$ divisible by 2 but not by 4; (c) $l$ odd, functions $sl$; (d) $l$ odd, functions $cl$.

TABLE 3.2-5b[a]

| | 80 | 81 | 82 | 83 | 84 | 85 | 86 | 87 | 88 | 89 | 90 | 91 | 92 | 93 | 94 | 95 |
|---|---|---|---|---|---|---|---|---|---|---|---|---|---|---|---|---|
| | 0 32 16 48 (**A** + **G**) − (**Q** + **W**) | 1 33 17 49 (**B** + **H**) − (**R** + **X**) | 2 34 18 50 (**C** + **I**) − (**S** + **Y**) | 3 35 19 51 (**D** + **J**) − (**T** + **Z**) | 4 36 20 52 (**E** + **K**) − (**U** + *A*) | 5 37 21 53 (**F** + **L**) − (**V** + *B*) | 6 38 22 54 (**G** + **M**) − (**W** + *C*) | 7 39 23 55 (**H** + **N**) − (**X** + *D*) | 8 40 24 56 (**I** + **O**) − (**Y** + *E*) | 9 41 25 57 (**J** + **P**) − (**Z** + *F*) | 10 42 26 58 (**K** + **Q**) − (**A** + *G*) | 11 43 27 59 (**L** + **R**) − (**B** + *H*) | 12 44 28 60 (**M** + **S**) − (**C** + *I*) | 13 45 29 61 (**N** + **T**) − (**D** + *J*) | 14 46 30 62 (**O** + **U**) − (**E** + *K*) | 15 47 31 63 (**P** + **V**) − (**F** + *L*) |
| s2 | 0 | $j$ | $m$ | $p$ | $r$ | $u$ | $w$ | $y$ | 1 | $y$ | $w$ | $u$ | $r$ | $p$ | $m$ | $j$ |
| c2 | 1 | $y$ | $w$ | $u$ | $r$ | $p$ | $m$ | $j$ | 0 | $-j$ | $-m$ | $-p$ | $-r$ | $-u$ | $-w$ | $-y$ |
| s6 | 0 | $p$ | $w$ | $y$ | $r$ | $j$ | $-m$ | $-u$ | $-1$ | $-u$ | $-m$ | $j$ | $r$ | $y$ | $w$ | $p$ |
| c6 | 1 | $u$ | $m$ | $-j$ | $-r$ | $-y$ | $-w$ | $-p$ | 0 | $p$ | $w$ | $y$ | $r$ | $j$ | $-m$ | $-u$ |
| s10 | 0 | $u$ | $w$ | $j$ | $-r$ | $-y$ | $-m$ | $p$ | 1 | $p$ | $-m$ | $-y$ | $-r$ | $j$ | $w$ | $u$ |
| c10 | 1 | $p$ | $-m$ | $-y$ | $-r$ | $j$ | $w$ | $u$ | 0 | $-u$ | $-w$ | $-j$ | $r$ | $y$ | $m$ | $-p$ |
| s14 | 0 | $y$ | $m$ | $-u$ | $-r$ | $p$ | $w$ | $-j$ | $-1$ | $-j$ | $w$ | $p$ | $-r$ | $-u$ | $m$ | $y$ |
| c14 | 1 | $j$ | $-w$ | $-p$ | $r$ | $u$ | $-m$ | $-y$ | 0 | $y$ | $m$ | $-u$ | $-r$ | $p$ | $w$ | $-j$ |
| s18 | 0 | $y$ | $-m$ | $-u$ | $r$ | $p$ | $-w$ | $-j$ | 1 | $-j$ | $-w$ | $p$ | $r$ | $-u$ | $-m$ | $y$ |
| c18 | 1 | $-j$ | $-w$ | $p$ | $r$ | $-u$ | $-m$ | $y$ | 0 | $-y$ | $m$ | $u$ | $-r$ | $-p$ | $w$ | $j$ |
| s22 | 0 | $u$ | $-w$ | $j$ | $r$ | $-y$ | $m$ | $p$ | $-1$ | $p$ | $m$ | $-y$ | $r$ | $j$ | $-w$ | $u$ |
| c22 | 1 | $-p$ | $-m$ | $y$ | $-r$ | $-j$ | $w$ | $-u$ | 0 | $u$ | $-w$ | $j$ | $r$ | $-y$ | $m$ | $p$ |
| s26 | 0 | $p$ | $-w$ | $y$ | $-r$ | $j$ | $m$ | $-u$ | 1 | $-u$ | $m$ | $j$ | $-r$ | $y$ | $-w$ | $p$ |
| c26 | 1 | $-u$ | $m$ | $j$ | $-r$ | $y$ | $-w$ | $p$ | 0 | $-p$ | $w$ | $-y$ | $r$ | $-j$ | $-m$ | $u$ |
| s30 | 0 | $j$ | $-m$ | $p$ | $-r$ | $u$ | $-w$ | $y$ | $-1$ | $y$ | $-w$ | $u$ | $-r$ | $p$ | $-m$ | $j$ |
| c30 | 1 | $-y$ | $w$ | $-u$ | $r$ | $-p$ | $m$ | $-j$ | 0 | $j$ | $-m$ | $p$ | $-r$ | $u$ | $-w$ | $y$ |

[a] The table is arranged in four parts for s$l$ = sin $2\pi lk/64$ and c$l$ = cos $2\pi lk/64$. (a) $l$ divisible by 4; (b) $l$ divisible by 2 but not by 4; (c) $l$ odd, functions s$l$; (d) $l$ odd, functions c$l$.

TABLE 3.2-5c[a]

| | 96 | 97 | 98 | 99 | 100 | 101 | 102 | 103 | 104 | 105 | 106 | 107 | 108 | 109 | 110 | 111 |
|---|---|---|---|---|---|---|---|---|---|---|---|---|---|---|---|---|
| | 0 32 16 48<br>Q − W | 1 33 31 63<br>(B − H) + (F − L) | 2 34 30 62<br>(C − I) + (E − K) | 3 35 29 61<br>(D − J) + (D − J) | 4 36 28 60<br>(E − K) + (C − I) | 5 37 27 59<br>(F − L) + (B − H) | 6 38 26 58<br>(G − M) + (A − G) | 7 39 25 57<br>(H − N) + (Z − F) | 8 40 24 56<br>(I − O) + (Y − E) | 9 41 23 55<br>(J − P) + (X − D) | 10 42 22 54<br>(K − Q) + (W − C) | 11 43 21 53<br>(L − R) + (V − B) | 12 44 20 52<br>(M − S) + (U − A) | 13 45 19 51<br>(N − T) + (T − Z) | 14 46 18 50<br>(O − U) + (S − Y) | 15 47 17 49<br>(P − V) + (R − X) |
| s1 | 1 | i | j | k | m | n | p | q | r | s | u | v | w | x | y | z |
| s3 | −1 | k | p | s | w | z | y | v | r | n | j | −i | −m | −q | −u | −x |
| s5 | 1 | n | u | z | w | q | j | −k | −r | −x | −y | −s | −m | i | p | v |
| s7 | −1 | q | y | v | m | −k | −u | −z | −r | −i | p | x | w | n | −j | −s |
| s9 | 1 | s | y | n | −m | −x | −u | −i | r | z | p | −k | −w | −v | −j | q |
| s11 | −1 | v | u | −i | −w | −s | j | x | r | −k | −y | −q | m | z | p | −n |
| s13 | 1 | x | p | −q | −w | i | y | n | −r | −v | j | z | m | −s | −u | k |
| s15 | −1 | z | j | −x | −m | v | p | −s | −r | q | u | −n | −w | k | y | −i |
| s17 | 1 | z | −j | −x | m | v | −p | −s | r | q | −u | −n | w | k | −y | −i |
| s19 | −1 | x | −p | −q | w | i | −y | n | r | −v | −j | z | −m | −s | u | k |
| s21 | 1 | v | −u | −i | w | −s | −j | x | −r | −k | y | −q | −m | z | −p | −n |
| s23 | −1 | s | −y | n | m | −x | u | −i | −r | z | −p | −k | w | −v | j | q |
| s25 | 1 | q | −y | v | −m | −k | u | −z | r | −i | −p | x | −w | n | j | −s |
| s27 | −1 | n | −u | z | −w | q | −j | −k | r | −x | y | −s | m | i | −p | v |
| s29 | 1 | k | −p | s | −w | z | −y | v | −r | n | −j | −i | m | −q | u | −x |
| s31 | −1 | i | −j | k | −m | n | −p | q | −r | s | −u | v | −w | x | −y | z |

[a] The table is arranged in four parts for s$l$ = sin $2\pi lk/64$ and c$l$ = cos $2\pi lk/64$. (a) $l$ divisible by 4; (b) $l$ divisible by 2 but not by 4; (c) $l$ odd, functions s$l$; (d) $l$ odd, functions c$l$.

TABLE 3.2-5d[a]

| | 112 | 113 | 114 | 115 | 116 | 117 | 118 | 119 | 120 | 121 | 122 | 123 | 124 | 125 | 126 | 127 |
|---|---|---|---|---|---|---|---|---|---|---|---|---|---|---|---|---|
| | 0 32 16 48 A − G | 1 33 31 63 (B − H) − (F − L) | 2 34 30 62 (C − I) − (E − K) | 3 35 29 61 (D − J) − (D − J) | 4 36 28 60 (E − K) − (C − I) | 5 37 27 59 (F − L) − (B − H) | 6 38 26 58 (G − M) − (A − G) | 7 39 25 57 (H − N) − (Z − F) | 8 40 24 56 (I − O) − (Y − E) | 9 41 23 55 (J − P) − (X − D) | 10 42 22 54 (K − Q) − (W − C) | 11 43 21 53 (L − R) − (V − B) | 12 44 20 52 (M − S) − (U − A) | 13 45 19 51 (N − T) − (T − Z) | 14 46 18 50 (O − U) − (S − Y) | 15 47 17 49 (P − V) − (R − X) |
| c1 | 1 | $z$ | $y$ | $x$ | $w$ | $v$ | $u$ | $s$ | $r$ | $q$ | $p$ | $n$ | $m$ | $k$ | $j$ | $i$ |
| c3 | 1 | $x$ | $u$ | $q$ | $m$ | $i$ | $-j$ | $-n$ | $-r$ | $-v$ | $-y$ | $-z$ | $-w$ | $-s$ | $-p$ | $-k$ |
| c5 | 1 | $v$ | $p$ | $i$ | $-m$ | $-s$ | $-y$ | $-x$ | $-r$ | $-k$ | $j$ | $q$ | $w$ | $z$ | $u$ | $n$ |
| c7 | 1 | $s$ | $j$ | $-n$ | $-w$ | $-x$ | $-p$ | $i$ | $r$ | $z$ | $u$ | $k$ | $-m$ | $-v$ | $-y$ | $-q$ |
| c9 | 1 | $q$ | $-j$ | $-v$ | $-w$ | $-k$ | $p$ | $z$ | $r$ | $-i$ | $-u$ | $-x$ | $-m$ | $n$ | $y$ | $s$ |
| c11 | 1 | $n$ | $-p$ | $-z$ | $-m$ | $q$ | $y$ | $k$ | $-r$ | $-x$ | $-j$ | $s$ | $w$ | $i$ | $-u$ | $-v$ |
| c13 | 1 | $k$ | $-u$ | $-s$ | $m$ | $z$ | $j$ | $-v$ | $-r$ | $n$ | $y$ | $i$ | $-w$ | $-q$ | $p$ | $x$ |
| c15 | 1 | $i$ | $-y$ | $-k$ | $w$ | $n$ | $-u$ | $-q$ | $r$ | $s$ | $-p$ | $-v$ | $m$ | $x$ | $-j$ | $-z$ |
| c17 | 1 | $-i$ | $-y$ | $k$ | $w$ | $-n$ | $-u$ | $q$ | $r$ | $-s$ | $-p$ | $v$ | $m$ | $-x$ | $-j$ | $z$ |
| c19 | 1 | $-k$ | $-u$ | $s$ | $m$ | $-z$ | $j$ | $v$ | $-r$ | $-n$ | $y$ | $-i$ | $-w$ | $q$ | $p$ | $-x$ |
| c21 | 1 | $-n$ | $-p$ | $z$ | $-m$ | $-q$ | $y$ | $-k$ | $-r$ | $x$ | $-j$ | $-s$ | $w$ | $-i$ | $-u$ | $v$ |
| c23 | 1 | $-q$ | $-j$ | $v$ | $-w$ | $k$ | $p$ | $-z$ | $r$ | $i$ | $-u$ | $x$ | $-m$ | $-n$ | $y$ | $-s$ |
| c25 | 1 | $-s$ | $j$ | $n$ | $-w$ | $x$ | $-p$ | $-i$ | $r$ | $-z$ | $u$ | $-k$ | $-m$ | $v$ | $-y$ | $q$ |
| c27 | 1 | $-v$ | $p$ | $-i$ | $-m$ | $s$ | $-y$ | $x$ | $-r$ | $k$ | $j$ | $-q$ | $w$ | $-z$ | $u$ | $-n$ |
| c29 | 1 | $-x$ | $u$ | $-q$ | $m$ | $-i$ | $-j$ | $n$ | $-r$ | $v$ | $-y$ | $z$ | $-w$ | $s$ | $-p$ | $k$ |
| c31 | 1 | $-z$ | $y$ | $-x$ | $w$ | $-v$ | $u$ | $-s$ | $r$ | $-q$ | $p$ | $-n$ | $m$ | $-k$ | $j$ | $-i$ |

[a] The table is arranged in four parts for $sl = \sin 2\pi lk/64$ and $cl = \cos 2\pi lk/64$. (a) $l$ divisible by 4; (b) $l$ divisible by 2 but not by 4; (c) $l$ odd, functions $sl$; (d) $l$ odd, functions $cl$.

TABLE 3.2-6

SUMS AND DIFFERENCES OF $4^2$ SAMPLES A, ..., $L$ OF THE FUNCTION $F(k)$ MULTIPLIED WITH THE VALUES $\pm i$, $\pm j$, ..., $\pm 1$ OF THE SAMPLES OF THE FIRST 64 FUNCTIONS OF THE FOURIER SERIES, AND THE VALUES $\pm m$, $\pm r$, $\pm w$, $\pm 1$ BY WHICH THE WEIGHTED SUMS ARE MULTIPLIED

TABLE 3.2-6a[a]

| | | | | | | | |
|---|---|---|---|---|---|---|---|
| | 64 | | 68 | | 72 | | 76 |
| 128 | (A + G + Q + W) | + | (E + K + U + *A*) | + | (I + O + Y + *E*) | + | (M + S + C + *I*) |
| 129 | (A + G + Q + W) | − | (E + K + U + *A*) | + | (I + O + Y + *E*) | − | (M + S + C + *I*) |
| 130 | | | (E + K + U + *A*) | | | − | (M + S + C + *I*) |
| 131 | (A + G + Q + W) | | | − | (I + O + Y + *E*) | | |
| | 65 | | 69 | | 73 | | 77 |
| 132 | (B + H + R + X) | + | (F + L + V + *B*) | + | (J + P + Z + *F*) | + | (N + T + D + *J*) |
| 133 | (B + H + R + X) | − | (F + L + V + *B*) | + | (J + P + Z + *F*) | − | (N + T + D + *J*) |
| 134 | (B + H + R + X) | + | (F + L + V + *B*) | − | (J + P + Z + *F*) | − | (N + T + D + *J*) |
| 135 | (B + H + R + X) | − | (F + L + V + *B*) | − | (J + P + Z + *F*) | + | (N + T + D + *J*) |
| | 66 | | 70 | | 74 | | 78 |
| 136 | (C + I + S + Y) | + | (G + M + W + *C*) | + | (K + Q + A + *G*) | + | (O + U + E + *K*) |
| 137 | (C + I + S + Y) | − | (G + M + W + *C*) | + | (K + Q + A + *G*) | − | (O + U + E + *K*) |
| 138 | (C + I + S + Y) | + | (G + M + W + *C*) | − | (K + Q + A + *G*) | − | (O + U + E + *K*) |
| 139 | (C + I + S + Y) | − | (G + M + W + *C*) | − | (K + Q + A + *G*) | + | (O + U + E + *K*) |
| | 67 | | 71 | | 75 | | 79 |
| 140 | (D + J + T + Z) | + | (H + N + X + *D*) | + | (L + R + B + *H*) | + | (P + V + F + *L*) |
| 141 | (D + J + T + Z) | − | (H + N + X + *D*) | + | (L + R + B + *H*) | − | (P + V + F + *L*) |
| 142 | (D + J + T + Z) | + | (H + N + X + *D*) | − | (L + R + B + *H*) | − | (P + V + F + *L*) |
| 143 | (D + J + T + Z) | − | (H + N + X + *D*) | − | (L + R + B + *H*) | + | (P + V + F + *L*) |

| | | | | | | | | | | | | | | | | | |
|---|---|---|---|---|---|---|---|---|---|---|---|---|---|---|---|---|---|
| 192 | 0 | $r$ | | | | $r$ | | | | $r$ | | | | $r$ | | | |
| 193 | s4 | | | 1 | | | | $m$ | | | | $r$ | | | | $w$ | |
| 194 | c4 | | | | 1 | | | | $w$ | | | | $r$ | | | | $m$ |
| 195 | s8 | | | | | | $r$ | | | | 1 | | | | $r$ | | |
| 196 | c8 | | 1 | | | | $r$ | | | | | | | | $-r$ | | |
| 197 | s12 | | | $-1$ | | | | $w$ | | | | $r$ | | | | $-m$ | |
| 198 | c12 | | | | 1 | | | | $m$ | | | | $-r$ | | | | $-w$ |
| 199 | s16 | | | | | 1 | | | | | | | | $-1$ | | | |
| 200 | c16 | 1 | | | | | | | | $-1$ | | | | | | | |
| 201 | s20 | | | 1 | | | | $w$ | | | | $-r$ | | | | $-m$ | |
| 202 | c20 | | | | 1 | | | | $-m$ | | | | $-r$ | | | | $w$ |
| 203 | s24 | | 1 | | | | $r$ | | | | $-1$ | | | | $r$ | | |
| 204 | c24 | | | | | | $-r$ | | | | | | | | $r$ | | |
| 205 | s28 | | | $-1$ | | | | $m$ | | | | $-r$ | | | | $w$ | |
| 206 | c28 | | | | 1 | | | | $-w$ | | | | $r$ | | | | $-m$ |
| 207 | c32 | 1 | | | | $-1$ | | | | 1 | | | | $-1$ | | | |

[a] The table is arranged in four parts that correspond to the four parts of Table 3.2-5.

TABLE 3.2-6b[a]

| | | | | |
|---|---|---|---|---|
| 80 | 84 | 88 | 92 | |
| 0(A + G − Q − W) + | r(E + K − U − A) + | (I + O − Y − E) + | r(M + S − C − I) | 144 |
| | r(E + K − U − A) − | (I + O − Y − E) + | r(M + S − C − I) | 145 |
| (A + G − Q − W) + | r(E + K − U − A) | | − r(M + S − C − I) | 152 |
| (A + G − Q − W) − | r(E + K − U − A) | | + r(M + S − C − I) | 153 |
| 82 | 86 | 90 | 94 | |
| m(C + I − S − Y) + | w(G + M − W − C) + | w(K + Q − A − G) + | m(O + U − E − K) | 146 |
| w(C + I − S − Y) − | m(G + M − W − C) − | m(K + Q − A − G) + | w(O + U − E − K) | 147 |
| w(C + I − S − Y) + | m(G + M − W − C) − | m(K + Q − A − G) − | w(O + U − E − K) | 154 |
| m(C + I − S − Y) − | w(G + M − W − C) + | w(K + Q − A − G) − | m(O + U − E − K) | 155 |
| 81 | 83 | 85 | 87 | |
| j(B + H − R − X) + | p(D + J − T − Z) + | u(F + L − V − B) + | y(H + N − X − D) | 148 |
| p(B + H − R − X) + | y(D + J − T − Z) + | j(F + L − V − B) − | u(H + N − X − D) | 149 |
| u(B + H − R − X) + | j(D + J − T − Z) − | y(F + L − V − B) + | p(H + N − X − D) | 150 |
| y(B + H − R − X) − | u(D + J − T − Z) + | p(F + L − V − B) − | j(H + N − X − D) | 151 |
| y(B + H − R − X) + | u(D + J − T − Z) + | p(F + L − V − B) + | j(H + N − X − D) | 156 |
| u(B + H − R − X) − | j(D + J − T − Z) − | y(F + L − V − B) − | p(H + N − X − D) | 157 |
| p(B + H − R − X) − | y(D + J − T − Z) + | j(F + L − V − B) + | u(H + N − X − D) | 158 |
| j(B + H − R − X) − | p(D + J − T − Z) + | u(F + L − V − B) − | y(H + N − X − D) | 159 |
| 89 | 91 | 93 | 95 | |
| y(J + P − Z − F) + | u(L + R − B − H) + | p(N + T − D − J) + | j(P + V − F − L) | 148a |
| −u(J + P − Z − F) + | j(L + R − B − H) + | y(N + T − D − J) + | p(P + V − F − L) | 149a |
| p(J + P − Z − F) − | y(L + R − B − H) + | j(N + T − D − J) + | u(P + V − F − L) | 150a |
| −j(J + P − Z − F) + | p(L + R − B − H) − | u(N + T − D − J) + | y(P + V − F − L) | 151a |
| −j(J + P − Z − F) − | p(L + R − B − H) − | u(N + T − D − J) − | y(P + V − F − L) | 156a |
| p(J + P − Z − F) + | y(L + R − B − H) + | j(N + T − D − J) − | u(P + V − F − L) | 157a |
| −u(J + P − Z − F) − | j(L + R − B − H) + | y(N + T − D − J) − | p(P + V − F − L) | 158a |
| y(J + P − Z − F) − | u(L + R − B − H) + | p(N + T − D − J) − | j(P + V − F − L) | 159a |

| | | | | | | | | | | | | | | | | | | | | | | | | | |
|---|---|---|---|---|---|---|---|---|---|---|---|---|---|---|---|---|---|---|---|---|---|---|---|---|---|
| 208 | s2 | 1 | | | | 1 | | | | 1 | | | | | | | | 1 | | | | | | | |
| 209 | s6 | | 1 | | | | 1 | | | | 1 | | | | | | | | 1 | | | | | | |
| 210 | s10 | | −1 | | | | 1 | | | | | 1 | | | | | | | | 1 | | | | | |
| 211 | s14 | −1 | | | | 1 | | | | | | | 1 | | | | | | | | 1 | | | | |
| 212 | s18 | 1 | | | | −1 | | | | | | | 1 | | | | | | | | 1 | | | | |
| 213 | s22 | | 1 | | | | −1 | | | | | 1 | | | | | | | | 1 | | | | | |
| 214 | s26 | | −1 | | | | −1 | | | | 1 | | | | | | | | 1 | | | | | | |
| 215 | s30 | −1 | | | | −1 | | | | 1 | | | | | | | | 1 | | | | | | | |
| 216 | c2 | | | 1 | | | | 1 | | | | | | 1 | | | | | | | | 1 | | | |
| 217 | c6 | | | | 1 | | | | 1 | | | | | | 1 | | | | | | | | 1 | | |
| 218 | c10 | | | | 1 | | | | −1 | | | | | | | 1 | | | | | | | | 1 | |
| 219 | c14 | | | 1 | | | | −1 | | | | | | | | | 1 | | | | | | | | 1 |
| 220 | c18 | | | 1 | | | | −1 | | | | | | | | | −1 | | | | | | | | −1 |
| 221 | c22 | | | | 1 | | | | −1 | | | | | | | −1 | | | | | | | | −1 | |
| 222 | c26 | | | | 1 | | | | 1 | | | | | | −1 | | | | | | | | −1 | | |
| 223 | c30 | | | 1 | | | | 1 | | | | | | −1 | | | | | | | | −1 | | | |

[a] The table is arranged in four parts that correspond to the four parts of Table 3.2-5.

TABLE 3.2-6c[a]

| | | | | |
|---|---|---|---|---|
| 96 | 100 | 104 | 108 | |
| (Q − **W**) | + *m*(E − **K** + **C** − *I*) + | *r*(I − **O** + Y − *E*) | + *w*(**M** − **S** + U − *A*) | 160 |
| −(Q − **W**) | + *w*(E − **K** + **C** − *I*) + | *r*(I − **O** + Y − *E*) | − *m*(**M** − **S** + U − *A*) | 161 |
| (Q − **W**) | + *w*(E − **K** + **C** − *I*) − | *r*(I − **O** + Y − *E*) | − *m*(**M** − **S** + U − *A*) | 162 |
| −(Q − **W**) | + *m*(E − **K** + **C** − *I*) − | *r*(I − **O** + Y − *E*) | + *w*(**M** − **S** + U − *A*) | 163 |
| 98 | 102 | 106 | 110 | |
| *j*(C − **I** + **E** − *K*) | + *p*(G − **M** + **A** − *G*) + | *u*(K − **Q** + W − *C*) + | *y*(O − **U** + S − **Y**) | 164 |
| *p*(C − **I** + **E** − *K*) | + *y*(G − **M** + **A** − *G*) + | *j*(*K* − **Q** + W − *C*) − | *u*(O − **U** + S − **Y**) | 165 |
| *u*(C − **I** + **E** − *K*) | + *j*(G − **M** + **A** − *G*) − | *y*(K − **Q** + W − *C*) + | *p*(O − **U** + S − **Y**) | 166 |
| *y*(C − **I** + **E** − *K*) | − *u*(G − **M** + **A** − *G*) + | *p*(K − **Q** + W − *C*) − | *j*(O − **U** + S − **Y**) | 167 |
| 97 | 99 | 101 | 103 | |
| *i*(B − **H** + **F** − *L*) | + *k*(D − **J** + **D** + *J*) + | *n*(F − **L** + **B** − *H*) | + *q*(H − **N** + Z − *F*) | 168 |
| *k*(B − **H** + **F** − *L*) | + *s*(D − **J** + **D** + *J*) + | *z*(F − **L** + **B** − *H*) | + *v*(H − **N** + Z − *F*) | 169 |
| *n*(B − **H** + **F** − *L*) | + *z*(D − **J** + **D** + *J*) + | *q*(F − **L** + **B** − *H*) | − *k*(H − **N** + Z − *F*) | 170 |
| *q*(B − **H** + **F** − *L*) | + *v*(D − **J** + **D** + *J*) − | *k*(F − **L** + **B** − *H*) | − *z*(H − **N** + Z − *F*) | 171 |
| *s*(B − **H** + **F** − *L*) | + *n*(D − **J** + **D** + *J*) − | *x*(F − **L** + **B** − *H*) | − *i*(H − **N** + Z − *F*) | 172 |
| *v*(B − **H** + **F** − *L*) | − *i*(D − **J** + **D** + *J*) − | *s*(F − **L** + **B** − *H*) | + *x*(H − **N** + Z − *F*) | 173 |
| *x*(B − **H** + **F** − *L*) | − *q*(D − **J** + **D** + *J*) + | *i*(F − **L** + **B** − *H*) | + *n*(H − **N** + Z − *F*) | 174 |
| *z*(B − **H** + **F** − *L*) | − *x*(D − **J** + **D** + *J*) + | *v*(F − **L** + **B** − *H*) | − *s*(H − **N** + Z − *F*) | 175 |
| 105 | 107 | 109 | 111 | |
| *s*(J − **P** + X − *D*) + | *v*(L − **R** + V − *B*) + | *x*(N − **T** + T − **Z**) + | *z*(**P** − **V** + R − **X**) | 168a |
| *n*(J − **P** + X − *D*) − | *i*(L − **R** + V − *B*) − | *q*(N − **T** + T − **Z**) − | *x*(**P** − **V** + R − **X**) | 169a |
| −*x*(J − **P** + X − *D*) − | *s*(L − **R** + V − *B*) + | *i*(N − **T** + T − **Z**) + | *v*(**P** − **V** + R − **X**) | 170a |
| −*i*(J − **P** + X − *D*) + | *x*(L − **R** + V − *B*) + | *n*(N − **T** + T − **Z**) − | *s*(**P** − **V** + R − **X**) | 171a |
| *z*(J − **P** + X − *D*) − | *k*(L − **R** + V − *B*) − | *v*(N − **T** + T − **Z**) + | *q*(**P** − **V** + R − **X**) | 172a |
| −*k*(J − **P** + X − *D*) − | *q*(L − **R** + V − *B*) + | *z*(N − **T** + T − **Z**) − | *n*(**P** − **V** + R − **X**) | 173a |
| −*v*(J − **P** + X − *D*) + | *z*(L − **R** + V − *B*) − | *s*(N − **T** + T − **Z**) + | *k*(**P** − **V** + R − **X**) | 174a |
| *q*(J − **P** + X − *D*) − | *n*(L − **R** + V − *B*) + | *k*(N − **T** + T − **Z**) − | *i*(**P** − **V** + R − **X**) | 175a |

| | | | | | | | | | | | | | | | | | | | | | | | | | |
|---|---|---|---|---|---|---|---|---|---|---|---|---|---|---|---|---|---|---|---|---|---|---|---|---|---|
| 224 | s1 | 1 | | | | 1 | | | | 1 | | | | | | | | 1 | | | | | | | |
| 225 | s3 | | 1 | | | | 1 | | | | 1 | | | | | | | | 1 | | | | | | |
| 226 | s5 | | | 1 | | | | 1 | | | | 1 | | | | | | | | 1 | | | | | |
| 227 | s7 | | | | 1 | | | | 1 | | | | 1 | | | | | | | | 1 | | | | |
| 228 | s9 | | | | −1 | | | | 1 | | | | | 1 | | | | | | | | 1 | | | |
| 229 | s11 | | | −1 | | | | 1 | | | | | | | 1 | | | | | | | | 1 | | |
| 230 | s13 | | −1 | | | | 1 | | | | | | | | | 1 | | | | | | | | 1 | |
| 231 | s15 | −1 | | | | 1 | | | | | | | | | | | 1 | | | | | | | | 1 |
| 232 | s17 | 1 | | | | −1 | | | | | | | | | | | 1 | | | | | | | | 1 |
| 233 | s19 | | 1 | | | | −1 | | | | | | | | | 1 | | | | | | | | 1 | |
| 234 | s21 | | | 1 | | | | −1 | | | | | | | 1 | | | | | | | | 1 | | |
| 235 | s23 | | | | 1 | | | | −1 | | | | | 1 | | | | | | | | 1 | | | |
| 236 | s25 | | | | −1 | | | | −1 | | | | 1 | | | | | | | | 1 | | | | |
| 237 | s27 | | | −1 | | | | −1 | | | | 1 | | | | | | | | 1 | | | | | |
| 238 | s29 | | −1 | | | | −1 | | | | 1 | | | | | | | | 1 | | | | | | |
| 239 | s31 | −1 | | | | −1 | | | | 1 | | | | | | | | 1 | | | | | | | |

[a] The table is arranged in four parts that correspond to the four parts of Table 3.2-5.

TABLE 3.2-6d[a]

| | | | | |
|---|---|---|---|---|
| 112 | 116 | 120 | 124 | |
| (A − G) | + w(E − K − C + I) | + r(I − O − Y + E) | + m(M − S − U + A) | 176 |
| (A − G) | + m(E − K − C + I) | − r(I − O − Y + E) | − w(M − S − U + A) | 177 |
| (A − G) | − m(E − K − C + I) | − r(I − O − Y + E) | + w(M − S − U + A) | 178 |
| (A − G) | − w(E − K − C + I) | + r(I − O − Y + E) | − m(M − S − U + A) | 179 |
| 114 | 118 | 122 | 126 | |
| y(C − I − E + K) | + u(G − M − A + G) | + p(K − Q − W + C) | + j(O − U − S + Y) | 180 |
| u(C − I − E + K) | − j(G − M − A + G) | − y(K − Q − W + C) | − p(O − U − S + Y) | 181 |
| p(C − I − E + K) | − y(G − M − A + G) | + j(K − Q − W + C) | + u(O − U − S + Y) | 182 |
| j(C − I − E + K) | − p(G − M − A + G) | + u(K − Q − W + C) | − y(O − U − S + Y) | 183 |
| 113 | 115 | 117 | 119 | |
| z(B − H − F + L) | + x(D − J − D + J) | + v(F − L − B + H) | + s(H − N − Z + F) | 184 |
| x(B − H − F + L) | + q(D − J − D + J) | + i(F − L − B + H) | − n(H − N − Z + F) | 185 |
| v(B − H − F + L) | + i(D − J − D + J) | − s(F − L − B + H) | − x(H − N − Z + F) | 186 |
| s(B − H − F + L) | − n(D − J − D + J) | − x(F − L − B + H) | + i(H − N − Z + F) | 187 |
| q(B − H − F + L) | − v(D − J − D + J) | − k(F − L − B + H) | + z(H − N − Z + F) | 188 |
| n(B − H − F + L) | − z(D − J − D + J) | + q(F − L − B + H) | + k(H − N − Z + F) | 189 |
| k(B − H − F + L) | − s(D − J − D + J) | + z(F − L − B + H) | − v(H − N − Z + F) | 190 |
| i(B − H − F + L) | − k(D − J − D + J) | + n(F − L − B + H) | − q(H − N − Z + F) | 191 |
| 121 | 123 | 125 | 127 | |
| q(J − P − X + D) | + n(L − R − V + B) | + k(N − T − T + Z) | + i(P − V − R + X) | 184a |
| −v(J − P − X + D) | − z(L − R − V + B) | − s(N − T − T + Z) | − k(P − V − R + X) | 185a |
| −k(J − P − X + D) | + q(L − R − V + B) | + z(N − T − T + Z) | + n(P − V − R + X) | 186a |
| +z(J − P − X + D) | + k(L − R − V + B) | − v(N − T − T + Z) | − q(P − V − R + X) | 187a |
| −i(J − P − X + D) | − x(L − R − V + B) | + n(N − T − T + Z) | + s(P − V − R + X) | 188a |
| −x(J − P − X + D) | + s(L − R − V + B) | + i(N − T − T + Z) | − v(P − V − R + X) | 189a |
| n(J − P − X + D) | + i(L − R − V + B) | − q(N − T − T + Z) | + x(P − V − R + X) | 190a |
| s(J − P − X + D) | − v(L − R − V + B) | + x(N − T − T + Z) | − z(P − V − R + X) | 191a |

| | | | | | | | | | | | | | | | | | | | | | | | | | |
|---|---|---|---|---|---|---|---|---|---|---|---|---|---|---|---|---|---|---|---|---|---|---|---|---|---|
| 240 | c1 | 1 | | | | 1 | | | | 1 | | | | | | | | 1 | | | | | | | |
| 241 | c3 | | 1 | | | | 1 | | | | 1 | | | | | | | | 1 | | | | | | |
| 242 | c5 | | | 1 | | | | 1 | | | | 1 | | | | | | | | 1 | | | | | |
| 243 | c7 | | | | 1 | | | | 1 | | | | 1 | | | | | | | | 1 | | | | |
| 244 | c9 | | | | 1 | | | | −1 | | | | | 1 | | | | | | | | 1 | | | |
| 245 | c11 | | | 1 | | | | −1 | | | | | | | 1 | | | | | | | | 1 | | |
| 246 | c13 | | 1 | | | | −1 | | | | | | | | | 1 | | | | | | | | 1 | |
| 247 | c15 | 1 | | | | −1 | | | | | | | | | | | 1 | | | | | | | | 1 |
| 248 | c17 | 1 | | | | −1 | | | | | | | | | | | −1 | | | | | | | | −1 |
| 249 | c19 | | 1 | | | | −1 | | | | | | | | | −1 | | | | | | | | −1 | |
| 250 | c21 | | | 1 | | | | −1 | | | | | | | −1 | | | | | | | | −1 | | |
| 251 | c23 | | | | 1 | | | | −1 | | | | | −1 | | | | | | | | −1 | | | |
| 252 | c25 | | | | 1 | | | | 1 | | | | −1 | | | | | | | | −1 | | | | |
| 253 | c27 | | | 1 | | | | 1 | | | | −1 | | | | | | | | −1 | | | | | |
| 254 | c29 | | 1 | | | | 1 | | | | −1 | | | | | | | | −1 | | | | | | |
| 255 | c31 | 1 | | | | 1 | | | | −1 | | | | | | | | −1 | | | | | | | |

[a] The table is arranged in four parts that correspond to the four parts of Table 3.2-5.

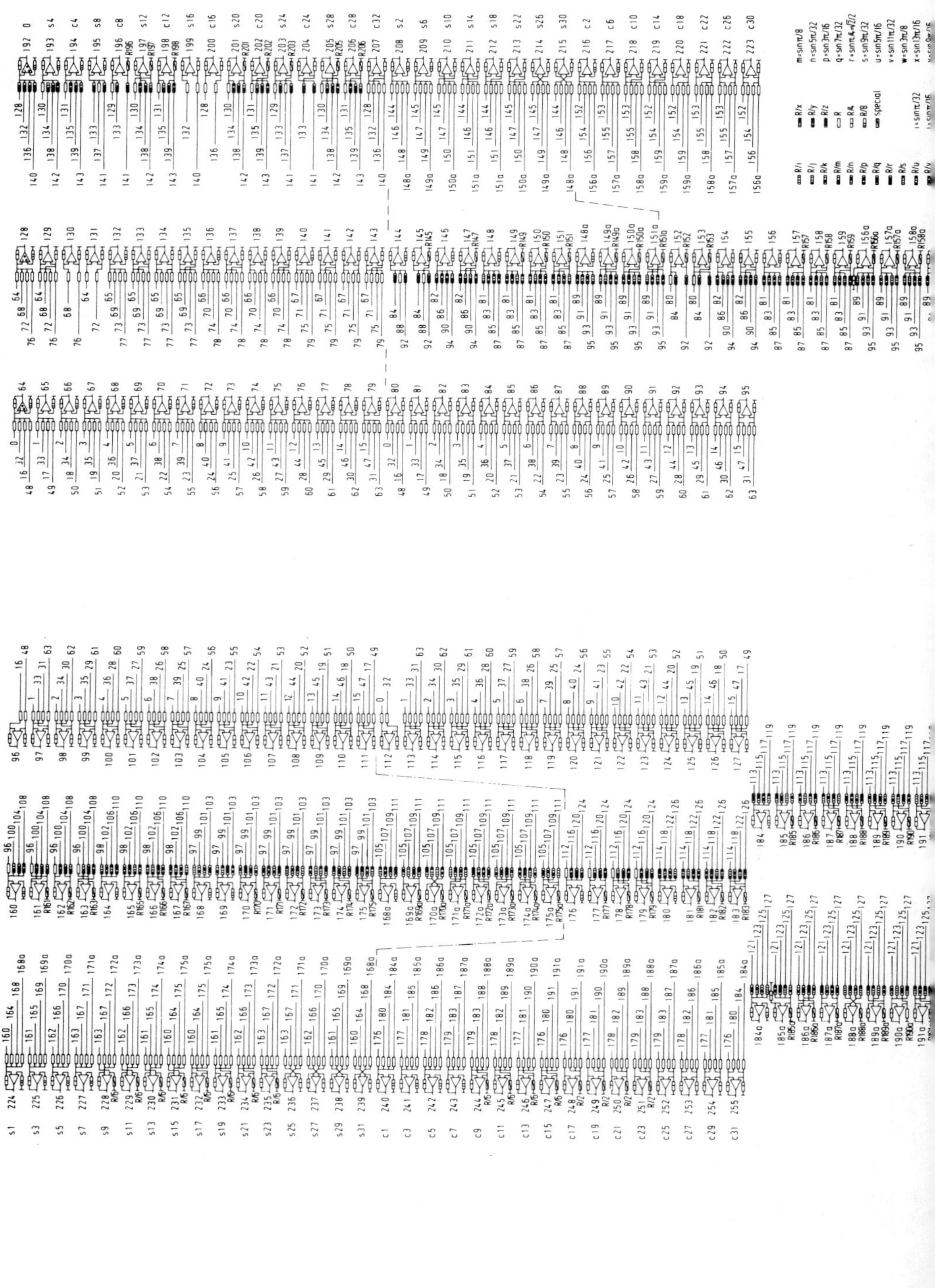

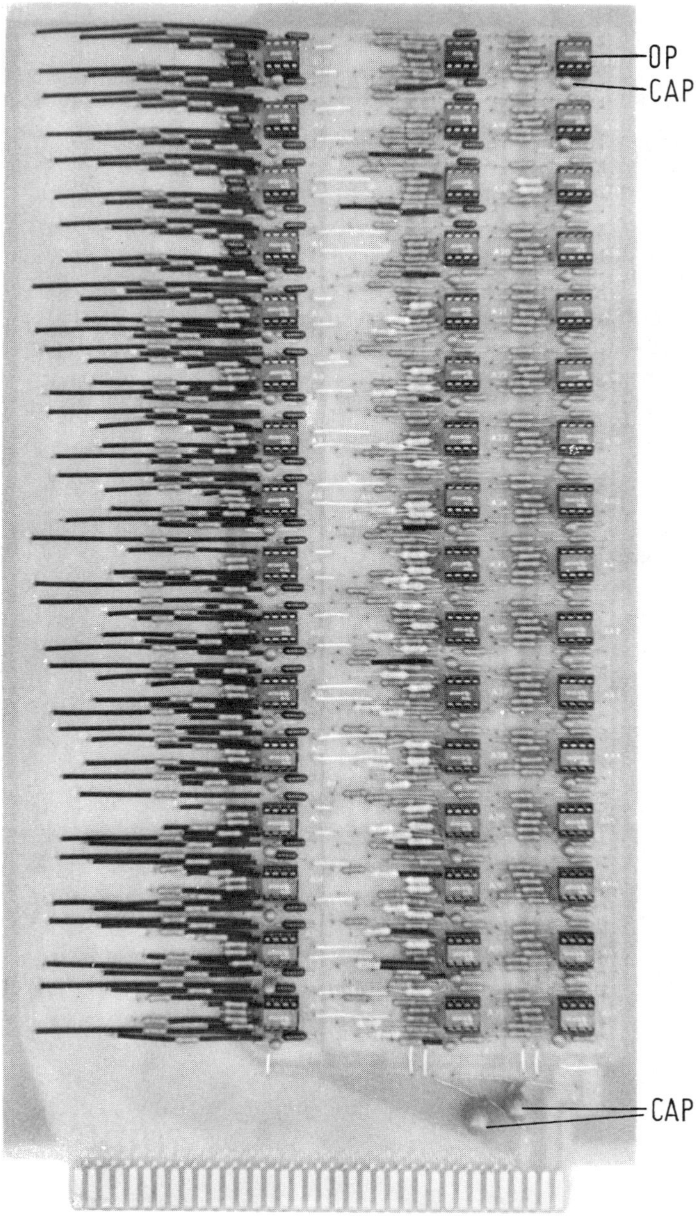


FIG. 3.2-6. Printed circuit card producing a Fourier transform of 32 input voltages according to Fig. 3.2-3. The input terminals are on the left and the output terminals on the right. Each dual-in-line package OP contains two operational amplifiers (RC-4558P); hence, there are $2 \times 3 \times 16 = 96$ operational amplifiers on this card. The decoupling capacitors CAP for the power supply are not shown in Fig. 3.2-3.

---

FIG. 3.2-5. Circuit performing a Fourier transform of 64 input voltages at the input terminals 0, ..., 63. The transformed voltages are obtained at the terminals 192, ..., 255.

The card contains 96 operational amplifiers. To produce $32^2 = 1024$ beams or resolved points one needs $2 \times 32 = 64$ such cards with a total of 6144 amplifiers; the complete equipment requires well in excess of 10,000 amplifiers. These numbers are within our current technological capabilities, particularly since dual-in-line packages with four amplifiers are now available. In order to achieve the resolution of TV images one needs $256^2$ to $512^2$ beams, and this implies equipment with about one million operational amplifiers. An advancement in linear integrated circuits, comparable to the advancement from four gates per dual-in-line package to the microprocessor, will be needed before such equipment can be built.

## 3.3 Fast Fourier Transform Circuits for $2^k 3^m 5^n$ . . . Samples

The fast Fourier transform is almost exclusively performed for a number of samples that equals a power of 2. The main reason for this is that digital circuits usually sum two numbers at a time. There is no reason why the analog circuits discussed here should be used in such a restricted fashion. If we can sum four voltages at a time, we certainly can sum three voltages with even less error. This permits one to extend the fast Fourier transform to a number of samples equal to $2^k 3^m$, $2^k 5^n$, or $2^k 3^m 5^n$, where $k, m, n = 0, 1, 2, \ldots$. Generally speaking, fast Fourier transform circuits can be worked out for any number $M$ of samples that can be decomposed into a product of small prime numbers. To demonstrate this, consider $12 = 2^2 3^1$ samples. Figure 3.3-1 shows the circle divided into 12 equal sections, and Table 3.3-1 lists the samples of the first 12 functions of the Fourier series derived from Fig. 3.3-1. Sums and differences of four samples are shown in Table 3.3-2 in complete analogy to Table 3.1-3. The only difference is that three rather than four of these sums and differences are now summed, after multiplication with 0, $\pm p$, $\pm q$, $r$, or $\pm 1$. Figure 3.3-2 shows a circuit that implements

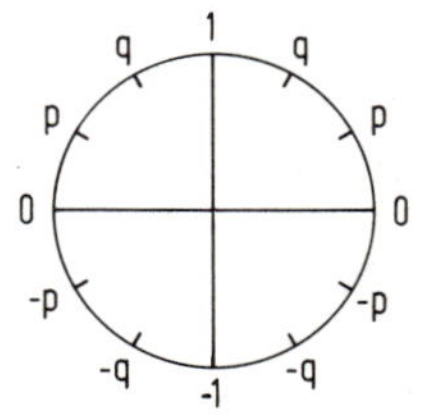


FIG. 3.3-1. Determination of the samples 0, $\pm p$, $\pm q$, $\pm 1$ of the first 12 functions of the Fourier series. $p = \sin \pi/6$, $q = \sin \pi/3$.

TABLE 3.3-1

THE VALUES OF THE SAMPLES OF THE FIRST 12 FUNCTIONS OF THE FOURIER SERIES[a]

| $k$ | 0 | 1 | 2 | 3 | 4 | 5 | 6 | 7 | 8 | 9 | 10 | 11 |
|---|---|---|---|---|---|---|---|---|---|---|---|---|
| $F(k)$ | A | B | C | D | E | F | G | H | I | J | K | L |
| 0 | $r$ | $r$ | $r$ | $r$ | $r$ | $r$ | $r$ | $r$ | $r$ | $r$ | $r$ | $r$ |
| s1 | 0 | $p$ | $q$ | 1 | $q$ | $p$ | 0 | $-p$ | $-q$ | $-1$ | $-q$ | $-p$ |
| c1 | 1 | $q$ | $p$ | 0 | $-p$ | $-q$ | $-1$ | $-q$ | $-p$ | 0 | $p$ | $q$ |
| s2 | 0 | $q$ | $q$ | 0 | $-q$ | $-q$ | 0 | $q$ | $q$ | 0 | $-q$ | $-q$ |
| c2 | 1 | $p$ | $-p$ | $-1$ | $-p$ | $p$ | 1 | $p$ | $-p$ | $-1$ | $-p$ | $p$ |
| s3 | 0 | 1 | 0 | $-1$ | 0 | 1 | 0 | $-1$ | 0 | 1 | 0 | $-1$ |
| c3 | 1 | 0 | $-1$ | 0 | 1 | 0 | $-1$ | 0 | 1 | 0 | $-1$ | 0 |
| s4 | 0 | $q$ | $-q$ | 0 | $q$ | $-q$ | 0 | $q$ | $-q$ | 0 | $q$ | $-q$ |
| c4 | 1 | $-p$ | $-p$ | 1 | $-p$ | $-p$ | 1 | $-p$ | $-p$ | 1 | $-p$ | $-p$ |
| s5 | 0 | $p$ | $-q$ | 1 | $-q$ | $p$ | 0 | $-p$ | $q$ | $-1$ | $q$ | $-p$ |
| c5 | 1 | $-q$ | $p$ | 0 | $-p$ | $q$ | $-1$ | $q$ | $-p$ | 0 | $p$ | $-q$ |
| c6 | 1 | $-1$ | 1 | $-1$ | 1 | $-1$ | 1 | $-1$ | 1 | $-1$ | 1 | $-1$ |

[a] $p = \sin \pi/6$, $q = \sin \pi/3$, $r = \sqrt{2}/2$.

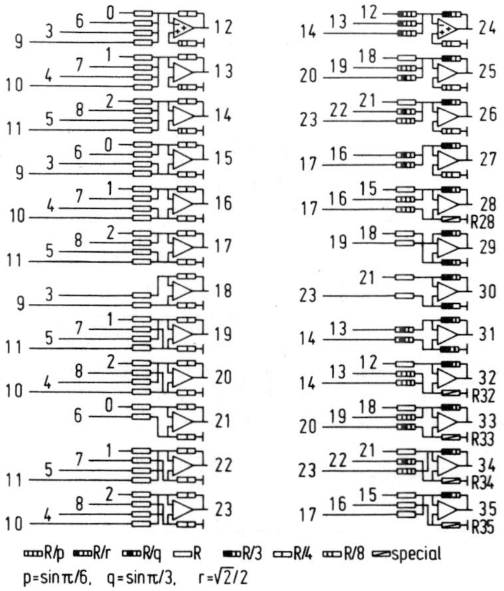


FIG. 3.3-2. Circuit performing a Fourier transform of 12 input voltages at the terminals 0, ..., 11. The transformed voltages are obtained at the terminals 24, ..., 35.

Table 3.3-2. There should be no difficulty—other than the tedious writing of tables—in designing circuits for the more general case of $2^k 3^m 5^n$ samples.

Let us see how much one saves by using the fast Fourier transform compared with the ordinary transform. The saving shall be characterized by the number of operational amplifiers used.

TABLE 3.3-2

SUMS AND DIFFERENCES OF FOUR SAMPLES A, ..., L OF THE FUNCTION $F(k)$, AND THE VALUES 0, $\pm p$, $\pm q$, $r$, $\pm 1$ OF THE 12 FUNCTIONS OF THE FOURIER SERIES BY WHICH THEY ARE MULTIPLIED

| | | 12 | 13 | 14 | 15 | 16 | 17 | 18 | 19 | 20 | 21 | 22 | 23 |
|---|---|---|---|---|---|---|---|---|---|---|---|---|---|
| | | (A + G) + (D + J) | (B + H) + (E + K) | (C + I) + (F + L) | (A + G) − (D + J) | (B + H) − (E + K) | (C + I) − (F + L) | D − J | (B − H) + (F − L) | (C − I) + (E − K) | A − G | (B − H) − (F − L) | (C − I) − (E − K) |
| 24 | 0 | $r$ | $r$ | $r$ | | | | | | | | | |
| 25 | s1 | | | | | | | 1 | $p$ | $q$ | | | |
| 26 | c1 | | | | | | | | | | 1 | $q$ | $p$ |
| 27 | s2 | | | | 0 | $q$ | $p$ | | | | | | |
| 28 | c2 | | | | 1 | $p$ | $-p$ | | | | | | |
| 29 | s3 | | | | | | | −1 | 1 | 0 | | | |
| 30 | c3 | | | | | | | | | | 1 | 0 | −1 |
| 31 | s4 | 0 | $q$ | $-q$ | | | | | | | | | |
| 32 | c4 | 1 | $-p$ | $-p$ | | | | | | | | | |
| 33 | s5 | | | | | | | 1 | $p$ | $-q$ | | | |
| 34 | c5 | | | | | | | | | | 1 | $-q$ | $p$ |
| 35 | c6 | | | | 1 | −1 | 1 | | | | | | |

If the circuit of Fig. 3.1-2 is implemented by summing nominally four voltages with one operational amplifier, one needs $4 + 1 = 5$ amplifiers. The Fourier transform for all 16 functions of Fig. 3.1-1 requires thus $16 \times 5 = 80$ amplifiers, while the fast transform of Fig. 3.1-3 requires $16 \times 2 = 32$. The numbers 80, 32, and the ratio $80/32 = 2.5$ are shown in Table 3.3-3. For 32 samples one requires $(32/4) + (8/4) + (2/2) = 8 + 2 + 1 = 11$ amplifiers for each function of the Fourier series, or a total of $32 \times 11 = 352$ amplifiers, if the ordinary transform is used. The fast transform of Fig. 3.2-3 requires $32 \times 3 = 96$ amplifiers. The numbers 352 96, and the ratio $352/96 = 3.7$ are shown again in Table 3.3-3. For 6 samples one obtains $(64/4) + (16/4) + (4/4) = 16 + 4 + 1 = 21$ amplifier per function for the ordinary transform, or a total of $64 \times 21 = 134$ amplifiers. The fast transform of Fig. 3.2-5 requires 216 amplifiers. Th

TABLE 3.3-3

THE NUMBER $N$ OF OPERATIONAL AMPLIFIERS REQUIRED BY THE ORDINARY FOURIER TRANSFORM, THE NUMBER $N_F$ REQUIRED BY THE FAST TRANSFORM, AND THE RATIO $N/N_F$ FOR 12, 16, 32, AND 64 SAMPLES

| Samples | 12 | 16 | 32 | 64 |
|---|---|---|---|---|
| $N$ | 48 | 80 | 352 | 1344 |
| $N_F$ | 24 | 32 | 96 | 216 |
| $N/N_F$ | 2 | 2.5 | 3.7 | 6.2 |

ratio 1344/216 equals 6.2. For 12 samples one needs $(12/4) + (3/3) = 3 + 1 = 4$ amplifiers per function of the Fourier series, or a total of $12 \times 4 = 48$ amplifiers; the fast transform circuit of Fig. 3.3-2 requires 24 amplifiers; the ratio 48/24 equals 2. These numbers show that the savings are always significant and that they increase rapidly with the number of samples.[1]

[1] One may measure the complexity of the circuits just as well by the number of resistors. Their number increases essentially like $M^2$ for the ordinary Fourier transform according to Fig. 3.1-2 and like $M \log M$ for the fast Fourier transform, but the exact number is harder to determine than that of the operational amplifiers.

### 3.4 Passive Beam Forming with Fourier Transform Circuits

Let us connect the output voltages of the circuit of Fig. 3.1-3 with the voltages $v(0, \beta, t)$, $v(\gamma, \beta, s, t)$ and $v(\gamma, \beta, c, t)$ of Eqs. (2.2-7)–(2.2-9). With the help of Fig. 3.1-1 one may readily infer the following:

(a) $v(0, \beta, t)$ obtained at output terminal 0
(b) $v(\gamma, \beta, s, t)$ obtained at output terminal s$l$
$l = (A/\lambda) \sin |\gamma| = 1, 2, \ldots, 2^n/2 - 1 = 7$
(c) $v(\gamma, \beta, c, t)$ obtained at output terminal c$l$
$l = (A/\lambda) \sin |\gamma| = 1, 2, \ldots, 2^n/2 = 8$

For $l = 0$ and $l = 2^n/2 = 8$ we have special cases. Let us consider the general case $0 < l < 2^n/2$ first. According to Eq. (2.2-10) we have to integrate $v(\gamma, \beta, s, t)$ and multiply by $-(2\pi v/\lambda)$. This operation is performed by the operational amplifier A1 in Fig. 3.4-1c. The potentiometer P1, the resistor

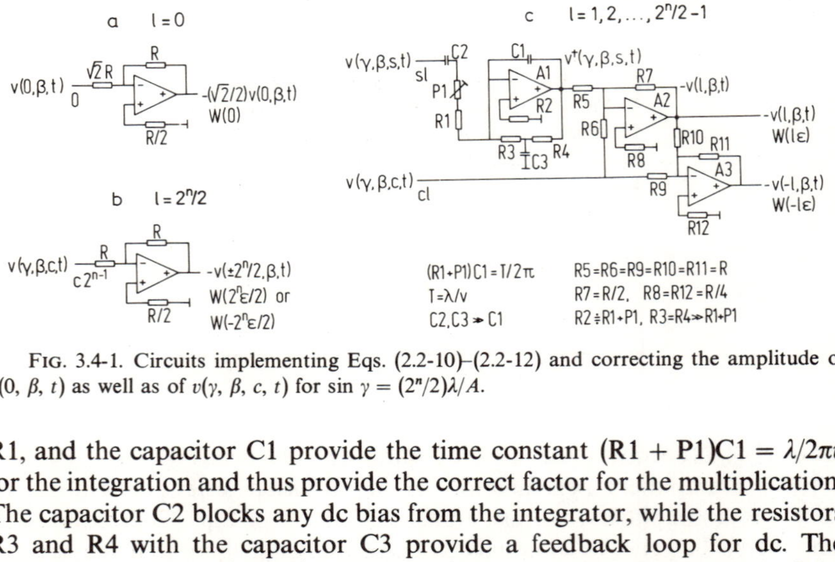


FIG. 3.4-1. Circuits implementing Eqs. (2.2-10)–(2.2-12) and correcting the amplitude of $v(0, \beta, t)$ as well as of $v(\gamma, \beta, c, t)$ for $\sin \gamma = (2^n/2)\lambda/A$.

R1, and the capacitor C1 provide the time constant $(R1 + P1)C1 = \lambda/2\pi v$ for the integration and thus provide the correct factor for the multiplication. The capacitor C2 blocks any dc bias from the integrator, while the resistors R3 and R4 with the capacitor C3 provide a feedback loop for dc. The negative sign in Eq. (2.2-10) is introduced by the amplitude reversal of the input signal by the operational amplifier A1.

The sum of $v(\gamma, \beta, c, t)$ and $v^{\dagger}(\gamma, \beta, s, t)$ multiplied by $\frac{1}{2}$, according to Eq. (2.2-11), is produced by the operational amplifier A2 in Fig. 3.4-1c, but the amplitude reversal of the amplifier yields $-v(l, \beta, t)$. This change of sign is of no consequence, because we are using sinusoidal waves and because the eye cannot see signs but only magnitudes.

The voltage $v(-l, \beta, t)$ of Eq. (2.2-12) is produced by the amplifier A3 in Fig. 3.4-1c; it produces the sum $-v(l, \beta, t) + v(\gamma, \beta, c, t)$ with reversed sign.

Consider next the special case $v(0, \beta, t)$. The amplitude of $v(0, \beta, t)$ in Eq. (2.2-7) is larger by the factor $\sqrt{2}$ than the amplitudes of $v(l, \beta, t)$ and $v(-l, \beta, t)$ in Eqs. (2.2-11) and (2.2-12). Hence, one must multiply it by $1/\sqrt{2}$. This is done by the circuit of Fig. 3.4-1a. This multiplication can, of course, be incorporated into the Fourier transform circuits of Figs. 3.1-3, 3.2-3, and 3.2-5.

The special case $v(\gamma, \beta, c, t)$ for $l = 2^n/2$ requires inherently no further processing than the Fourier transform of Eq. (2.2-9), since no voltage $v(\gamma, \beta, s, t)$ is available for $l = 2^n/2$. Figure 3.4-1b shows a circuit that reverses the amplitude of this voltage, but this is only necessary if one wants to have the same polarity for all voltages. Note that one cannot distinguish between a wavefront $W(2^n\varepsilon/2)$ and $W(-2^n\varepsilon/2)$ since the voltage $v(\gamma, \beta, s, t)$ is missing for $l = 2^n/2$.

The combination of a Fourier transform circuit with the *half-plane ambiguity resolving circuit*[1] is shown in Fig. 3.4-2. If the output voltages of

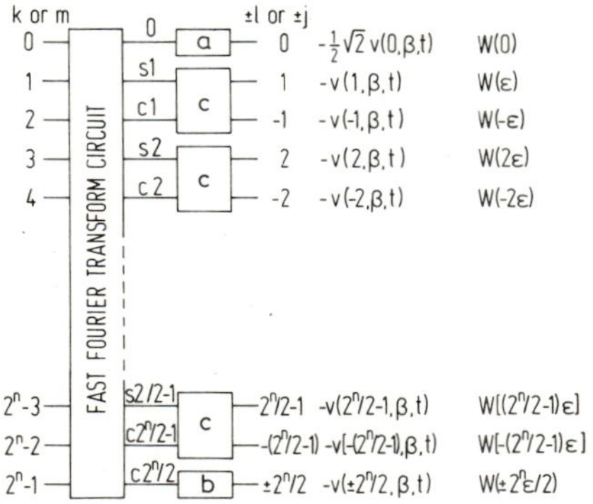


FIG. 3.4-2. Block diagram of a beam forming circuit using a fast Fourier transform circuit according to Figs. 3.1-3, 3.2-3, or 3.2-5 and the circuits of Fig. 3.4-1. The letter $k$ refers to the variable $x$ at the input, and the letter $l$ at the output. The letters $m$ and $j$ refer to the variable $y$ at the input and the output. Figure 2.1-5 shows this use of $k$, $m$ and $l$, $j$ in detail.

the receptor array of Fig. 2.1-1 are fed to the input terminals $k = 0, \ldots, 2^n - 1$, one obtains the voltages $-\frac{1}{2}\sqrt{2}\,v(0, \beta, t)$ to $-v(\pm 2^n/2, \beta, t)$ at the output terminals. These voltages represent the received wavefronts $W(\beta)$ of Fig. 2.1-1 for $\beta = 0, \pm\varepsilon, \pm 2\varepsilon, \ldots, \pm 2^n\varepsilon/2$.

[1] If the output voltages of the Fourier transform circuit are used without this circuit, one cannot distinguish between wavefronts $W(l\varepsilon)$ and $W(-l\varepsilon)$. Hence, there is an ambiguity between wavefronts from the right (or upper) and the left (or lower) half-plane.

The two-dimensional beam former of Fig. 2.1-5 requires printed circuit cards according to Fig. 3.4-2 both for the inverse transform of $W(\beta_x)$ and of $W(\beta_y)$. The letters $k$ and $l$ in Fig. 3.4-2 refer in this case to the transformation for $W(\beta_x)$ while the letters $m$ and $j$ refer to the transformation for $W(\beta_y)$. We will return to the two-dimensional transformation in Chapter 5.

## 3.5 The Inverse Fourier Transform

The functions in Fig. 3.1-1 and 3.2-2 are defined in the finite interval $0 \leq x \leq 1$. Hence, even though it is common to talk about the Fourier transform and the inverse transform one means the Fourier series and its inverse:

$$a(0) = \int_0^1 F(x) f(0, x)\, dx \tag{1}$$

$$a_s(l) = \sqrt{2} \int_0^1 F(x) \sin 2\pi l x \, dx \tag{2}$$

$$a_c(l) = \sqrt{2} \int_0^1 F(x) \cos 2\pi l x \, dx \tag{3}$$

$$F(x) = a(0) f(0, x) + \sqrt{2} \sum_{l=1}^{\infty} [a_s(l) \sin 2\pi l x + a_c(l) \cos 2\pi l x] \tag{4}$$

The function $f(0, x)$ has the constant value $+1$ in the interval $0 \leq x \leq 1$; it is used here strictly for tutorial purposes. To simplify writing, we frequently substitute $a(j)$ for the three kinds of coefficients $a(0)$, $a_s(l)$, and $a_c(l)$.

The sampled form of the functions in Figs. 3.1-1 and 3.2-2 replaces the integral in Eqs. (1)–(3) by sums and thus makes the transform and the inverse transform more equal:

$$a(0) = \sum_{k=0}^{2^n - 1} F(k) f(0, k) \tag{5}$$

$$a_s(l) = \sqrt{2} \sum_{k=0}^{2^n - 1} F(k) \sin(2\pi l k / 2^n) \tag{6}$$

$$a_c(l) = \sqrt{2} \sum_{k=0}^{2^n - 1} F(k) \cos(2\pi l k / 2^n) \tag{7}$$

$$a_c(2^n/2) = \sqrt{2} \sum_{k=0}^{2^n - 1} F(k) \cos \pi k \tag{8}$$

$$l = 1, 2, \ldots, 2^n/2 - 1$$

Note that the functions $\sqrt{2}\sin(2\pi lk/2^n)$, $\sqrt{2}\cos(2\pi lk/2^n)$, and $f(0, k)$ are used here, while Figs. 3.1-1 and 3.2-2 show these functions without the factor $\sqrt{2}$. The factor $\sqrt{2}$ normalizes the functions,

$$2^{-n}\sum_{k=0}^{2^n-1}(\sqrt{2}\sin 2\pi lk)^2 = 2^{-n}\sum_{k=0}^{2^n-1}(\sqrt{2}\cos 2\pi lk)^2 = 2^{-n}\sum_{k=0}^{2^n-1} f^2(0, k) = 1$$

but this point was not stressed in Sections 3.1–3.4; the factor $2^{-n} = 2^{-4}$, and its decomposition into two factors $2^{-2} \times 2^{-2}$ is clearly shown on top of Fig. 3.1-3.

Since both the original function $F(k)$ and the transformed function $a(j) = a(0), a_s(l), a_c(l)$ are represented by samples, one may write the Fourier transform in matrix notation. The samples of the functions of the Fourier series in Fig. 3.1-1 are represented by the matrix $\mathbf{M}$:

$$\mathbf{M} = \begin{pmatrix} q & q & q & \cdots & q & q \\ 0 & p & q & \cdots & -q & -p \\ 1 & r & q & \cdots & q & r \\ \vdots & \vdots & \vdots & & \vdots & \vdots \\ 1 & -r & q & \cdots & q & -r \\ 1 & -1 & 1 & \cdots & 1 & -1 \end{pmatrix} \tag{9}$$

The complete set of 16 × 16 coefficients of this matrix is shown in Table 3.1-1.

The samples of the function $F(k)$ in Fig. 3.1-1 can be written as a line vector $\mathbf{F}$:

$$\mathbf{F} = (\mathrm{A} \quad \mathrm{B} \quad \mathrm{C} \quad \cdots \quad \mathrm{O} \quad \mathrm{P}) \tag{10}$$

The transposed vector $\mathbf{F^*}$ is a column vector with the same coefficients but written in a column rather than a line. The product $\mathbf{MF^*}$ yields the Fourier transform $a(j)$ as a column vector $\mathbf{a^*}$:

$$\mathbf{MF^*} = \begin{pmatrix} q & q & q & \cdots & q & q & \mathrm{A} \\ 0 & p & q & \cdots & -q & -p & \mathrm{B} \\ 1 & r & q & \cdots & q & r & \mathrm{C} \\ \vdots & \vdots & \vdots & & \vdots & \vdots & \vdots \\ 1 & -r & q & \cdots & q & -r & \mathrm{O} \\ 1 & -1 & 1 & \cdots & 1 & -1 & \mathrm{P} \end{pmatrix} = \mathbf{a^*} = \begin{pmatrix} a(0) \\ a(1) \\ a(2) \\ \vdots \\ a(14) \\ a(15) \end{pmatrix} \tag{11}$$

The term $a(1) = a_s(1)$ of the vector $\mathbf{a^*}$ has the value

$$a(1) = 0\mathrm{A} + p\mathrm{B} + q\mathrm{C} + r\mathrm{D} + 1\mathrm{E} + r\mathrm{F} + q\mathrm{G} + p\mathrm{H}$$
$$+ 0\mathrm{I} - p\mathrm{J} - q\mathrm{K} - r\mathrm{L} - 1\mathrm{M} - r\mathrm{N} - q\mathrm{O} - p\mathrm{P}$$

which is the same as the sum shown in Fig. 3.1-2.

TABLE 3.5-1

THE 16 × 16 COEFFICIENTS OF THE TRANSPOSED MATRIX **M*** OF THE SAMPLED FUNCTIONS OF FIG. 3.1-1 AND TABLE 3.1-1

| $j$ | 0 | 1 | 2 | 3 | 4 | 5 | 6 | 7 | 8 | 9 | 10 | 11 | 12 | 13 | 14 | 15 |
|---|---|---|---|---|---|---|---|---|---|---|---|---|---|---|---|---|
| $a(j)$ | A* | B* | C* | D* | E* | F* | G* | H* | I* | J* | K* | L* | M* | N* | O* | P* |
| 0 | $q$ | 0 | 1 | 0 | 1 | 0 | 1 | 0 | 1 | 0 | 1 | 0 | 1 | 0 | 1 | 1 |
| s1* | $q$ | $p$ | $r$ | $q$ | $q$ | $r$ | $p$ | 1 | 0 | $r$ | $-p$ | $q$ | $-q$ | $p$ | $-r$ | $-1$ |
| c1* | $q$ | $q$ | $q$ | 1 | 0 | $q$ | $-q$ | 0 | $-1$ | $-q$ | $-q$ | $-1$ | 0 | $-q$ | $q$ | 1 |
| s2* | $q$ | $r$ | $p$ | $q$ | $-q$ | $-p$ | $-r$ | $-1$ | 0 | $-p$ | $r$ | $q$ | $q$ | $r$ | $-p$ | $-1$ |
| c2* | $q$ | 1 | 0 | 0 | $-1$ | $-1$ | 0 | 0 | 1 | 1 | 0 | 0 | $-1$ | $-1$ | 0 | 1 |
| s3* | $q$ | $r$ | $-p$ | $-q$ | $-q$ | $-p$ | $r$ | 1 | 0 | $-p$ | $-r$ | $-q$ | $q$ | $r$ | $p$ | $-1$ |
| c3* | $q$ | $q$ | $-q$ | $-1$ | 0 | $q$ | $q$ | 0 | $-1$ | $-q$ | $q$ | 1 | 0 | $-q$ | $-q$ | 1 |
| s4* | $q$ | $p$ | $-r$ | $-q$ | $q$ | $r$ | $-p$ | $-1$ | 0 | $r$ | $p$ | $-q$ | $-q$ | $p$ | $r$ | $-1$ |
| c4* | $q$ | 0 | $-1$ | 0 | 1 | 0 | $-1$ | 0 | 1 | 0 | $-1$ | 0 | 1 | 0 | $-1$ | 1 |
| s5* | $q$ | $-p$ | $-r$ | $q$ | $q$ | $-r$ | $-p$ | 1 | 0 | $-r$ | $p$ | $q$ | $-q$ | $-p$ | $r$ | $-1$ |
| c5* | $q$ | $-q$ | $-q$ | 1 | 0 | $-q$ | $q$ | 0 | $-1$ | $q$ | $q$ | $-1$ | 0 | $q$ | $-q$ | 1 |
| s6* | $q$ | $-r$ | $-p$ | $q$ | $-q$ | $p$ | $r$ | $-1$ | 0 | $p$ | $-r$ | $q$ | $q$ | $-r$ | $p$ | $-1$ |
| c6* | $q$ | $-1$ | 0 | 0 | $-1$ | 1 | 0 | 0 | 1 | $-1$ | 0 | 0 | $-1$ | 1 | 0 | 1 |
| s7* | $q$ | $-r$ | $p$ | $-q$ | $-q$ | $p$ | $-r$ | 1 | 0 | $p$ | $r$ | $-q$ | $q$ | $-r$ | $-p$ | $-1$ |
| c7* | $q$ | $-q$ | $q$ | $-1$ | 0 | $-q$ | $-q$ | 0 | $-1$ | $q$ | $-q$ | 1 | 0 | $q$ | $q$ | 1 |
| c8* | $q$ | $-p$ | $r$ | $-q$ | $q$ | $-r$ | $p$ | $-1$ | 0 | $-r$ | $-p$ | $-q$ | $-q$ | $-p$ | $-r$ | $-1$ |

Instead of writing the Fourier transform as $\mathbf{MF^*} = \mathbf{a^*}$ according to Eq. (11), one could also write it as $\mathbf{FM} = \mathbf{a}$, where $\mathbf{F}$ and $\mathbf{a}$ are $\mathbf{F^*}$ and $\mathbf{a^*}$ written as line vectors. This notation requires much more space if equations are written in the explicit form of Eq. (11).

The inverse transform from $\mathbf{a^*}$ to $\mathbf{F^*}$ is produced by a new matrix denoted $\mathbf{M^\dagger}$:

$$\mathbf{M^\dagger a^*} = \mathbf{F^*} \tag{12}$$

Equation (12) is multiplied on the left with $\mathbf{M}$:

$$\mathbf{MM^\dagger a^*} = \mathbf{MF^*} \tag{13}$$

If $\mathbf{MM^\dagger}$ yields the unit matrix $\mathbf{1}$ multiplied by a constant $C$,

$$\mathbf{MM^\dagger} = C\mathbf{1} \tag{14}$$

one obtains from Eq. (13) again Eq. (11) except for the factor $C$:

$$\mathbf{MF^*} = C\mathbf{a^*} \tag{15}$$

An important consequence of Eq. (14) is that **M** and **M**† can be commuted:

$$\mathbf{M}\mathbf{M}^{\dagger} = \mathbf{M}^{\dagger}\mathbf{M} = C\mathbf{1} \tag{16}$$

The determination of the inverse matrix $\mathbf{M}^{\dagger}$ from **M** is usually a lengthy process. However, if **M** represents a system of orthogonal sampled functions, the inverse matrix is either equal to the transposed matrix **M*** or at least very nearly so. The transposed matrix **M** follows from Eq. (9) by interchanging rows and columns:

$$\mathbf{M}^{*} = \begin{pmatrix} q & 0 & 1 & \cdots & 1 & 1 \\ q & p & r & \cdots & -r & -1 \\ q & q & q & \cdots & q & 1 \\ \vdots & \vdots & \vdots & & \vdots & \vdots \\ q & -q & q & \cdots & q & 1 \\ q & -p & r & \cdots & -r & -1 \end{pmatrix} \tag{17}$$

The complete set of 16 × 16 coefficients of **M*** is listed in Table 3.5-1.

The product **MM*** of Eqs. (9) and (17) yields almost the unit matrix multiplied by 8, but the last coefficient equals 2 rather than 1:

$$\mathbf{M}\mathbf{M}^{*} = 8\begin{pmatrix} 1 & 0 & 0 & \cdots & 0 & 0 \\ 0 & 1 & 0 & \cdots & 0 & 0 \\ 0 & 0 & 1 & \cdots & 0 & 0 \\ \vdots & \vdots & \vdots & & \vdots & \vdots \\ 0 & 0 & 0 & \cdots & 1 & 0 \\ 0 & 0 & 0 & \cdots & 0 & 2 \end{pmatrix} \tag{18}$$

Hence, the inverse matrix $\mathbf{M}^{\dagger}$ is obtained from the transposed matrix **M*** by multiplying the last column in Eq. (17) and Table 3.5-1 by $\frac{1}{2}$. A look at Fig. 3.1-1 shows where the problem comes from. The sum of the squares of the samples of all functions except $\cos 16\pi x$ yields 8, while $\cos 16\pi x$ yields 16:

$$\begin{aligned} \tfrac{1}{2}\sqrt{2}f(0,x) &\quad \text{yields} \quad 16(\tfrac{1}{2}\sqrt{2})^2 = 8 \\ \sin 2\pi x &\quad \text{yields} \quad 2(1^2) + 4(p^2 + q^2 + r^2) = 8 \\ &\vdots \\ \cos 14\pi x &\quad \text{yields} \quad 2(1^2) + 4(p^2 + q^2 + r^2) = 8 \\ \cos 16\pi x &\quad \text{yields} \quad 16(1^2) = 16 \end{aligned}$$

TABLE 3.5-2

SUMS AND DIFFERENCES OF THE SAMPLES A*, ..., P* OF THE FUNCTION $a(j)$, AND THE VALUES $\pm p$, $\pm q$, $\pm r$, $\pm\frac{1}{2}$, $\pm 1$ OF THE COEFFICIENTS OF THE INVERSE MATRIX M* OF THE FOURIER TRANSFORM BY WHICH THEY ARE MULTIPLIED[a]

| | | | 16 | 17 | 18 | 19 | 20 | | 21 | 22 | 23 | 24 | | 25 | 26 | 27 | 28 | | 29 | 30 | 31 | 32 | 33 | 34 | 35 |
|---|---|---|---|---|---|---|---|---|---|---|---|---|---|---|---|---|---|---|---|---|---|---|---|---|---|
| | | 0 4 8 12 | qA* + 1E* + 1I* + 1M* | qA* + qE* − qM* | qA* − 1I* | qA* − qE* + qM* | qA* − 1E* + 1I* − 1M* | 1 5 9 13 | pB* + rF* + rJ* + pN* | qB* + qF* − qJ* − qN* | rB* − pF* − pJ* + rN* | 1B* − 1F* + 1J* − 1N* | 2 6 10 14 | 1C* + 1G* + 1K* + 1O* | rC* + pG* − pK* − rO* | qC* − qG* − qK* + qO* | pC* − rG* + rK* − pO* | 3 7 11 15 | ½P* | qD* + 1H* + qL* − ½P* | 1D* − 1L* + ½P* | qD* − 1H* + qL* − ½P* | −qD* + 1H* − qL* − ½P* | −1D* + 1L* + ½P* | −qD* − 1H* − qL* − ½P* |
| 36 | 0* | | 1 | | | | | | | | | | | 1 | | | | | 1 | | | | | | |
| 37 | s1* | | | 1 | | | | | 1 | | | | | | 1 | | | | | 1 | | | | | |
| 38 | c1* | | | | 1 | | | | | 1 | | | | | | 1 | | | | | 1 | | | | |
| 39 | s2* | | | | | 1 | | | | | 1 | | | | | | 1 | | | | | 1 | | | |

| | | | | | | | | | | | | | | | | | | | | | |
|---|---|---|---|---|---|---|---|---|---|---|---|---|---|---|---|---|---|---|---|---|---|
| 40 | c2* | | | | | 1 | | | | 1 | | | | | 1 | | | | | | |
| 41 | s3* | | | | 1 | | | | 1 | | | | | −1 | | | | | 1 | | |
| 42 | c3* | | | 1 | | | | 1 | | | | | −1 | | | | | | | 1 | |
| 43 | s4* | | 1 | | | | 1 | | | | | −1 | | | | | | | | | 1 |
| 44 | c4* | 1 | | | | | | | | | −1 | | | | 1 | | | | | | |
| 45 | s5* | | 1 | | | | −1 | | | | | −1 | | | | 1 | | | | | |
| 46 | c5* | | | 1 | | | | −1 | | | | | −1 | | | | 1 | | | | |
| 47 | s6* | | | | 1 | | | | −1 | | | | | −1 | | | | 1 | | | |
| 48 | c6* | | | | | 1 | | | | −1 | | | | | 1 | | | | | | |
| 49 | s7* | | | | 1 | | | | −1 | | | | | 1 | | | | | 1 | | |
| 50 | c7* | | | 1 | | | | −1 | | | | | 1 | | | | | | | 1 | |
| 51 | c8* | | 1 | | | | −1 | | | | | 1 | | | | | | | | | 1 |

[a] Note that **P*** is multiplied by $+\frac{1}{2}$ rather than $+1$.

Table 3.5-1 can be rewritten into the form of Table 3.5-2 by replacing **A***, ..., **P*** by certain weighted sums shown on top of Table 3.5-2. Note that **P*** is multiplied by $\pm\frac{1}{2}$ rather than $\pm 1$ which takes the difference between $\mathbf{M}^{\dagger}$ and $\mathbf{M}^{*}$ into account.

A circuit implementing Table 3.5-2 is shown in Fig. 3.5-1. It is somewhat

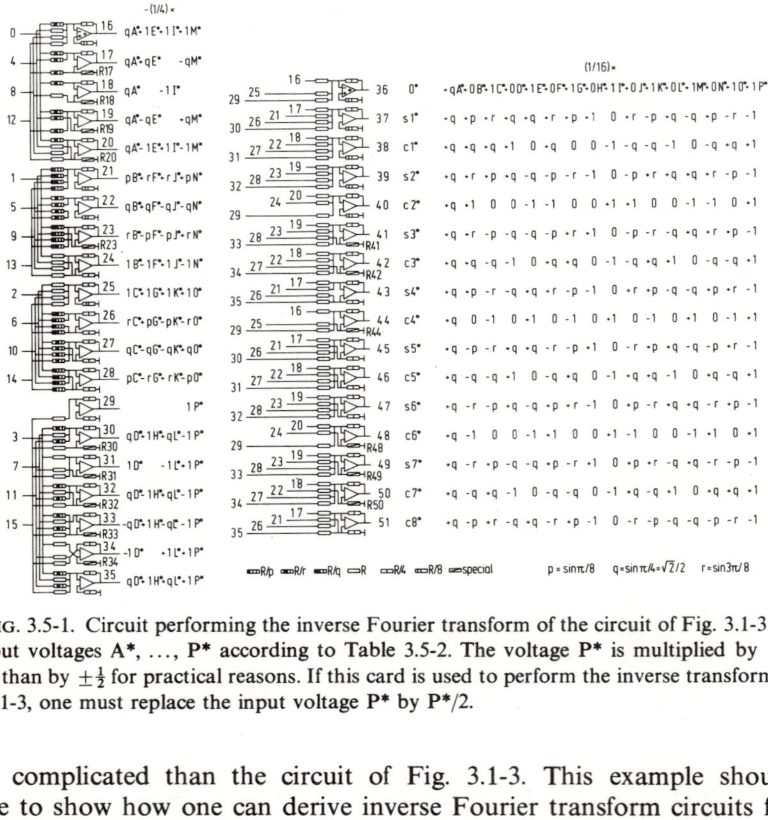


FIG. 3.5-1. Circuit performing the inverse Fourier transform of the circuit of Fig. 3.1-3 for 16 input voltages **A***, ..., **P*** according to Table 3.5-2. The voltage **P*** is multiplied by $\pm 1$ rather than by $\pm\frac{1}{2}$ for practical reasons. If this card is used to perform the inverse transform to Fig. 3.1-3, one must replace the input voltage **P*** by **P***/2.

more complicated than the circuit of Fig. 3.1-3. This example should suffice to show how one can derive inverse Fourier transform circuits for more general cases.

The inverse Fourier transform $F(x)$ of Eq. (4) can be written in the sampled notation by observing that $a_c(2^n/2)$ of Eq. (8) has to be multiplied by $\frac{1}{2}$, since it is the general form of **P*** in Table 3.5-1:

$$F(k) = a(0)f(0, k) + \sqrt{2} \sum_{l=1}^{2^n/2-1} [a_s(l)\sin(2\pi lk/2^n) + a_c(l)\cos(2\pi lk/2^n)] + \tfrac{1}{2}\sqrt{2}\, a_c(2^n/2)\cos \pi k \tag{18}$$

We have found it convenient in Section 2.2 to use the variable $i = k - 2^n/2 + 1$ instead of $k$, and we rewrite Eq. (18) as well as Eqs. (5)–(8) in terms of $i$ for use in the following section:

$$F(i) = a(0)f(0, i) + \sqrt{2} \sum_{l=1}^{2^n/2-1} [a_s(l) \sin(2\pi li/2^n) + a_c(l) \cos(2\pi li/2^n)] + \tfrac{1}{2}\sqrt{2}\, a_c(2^n/2) \cos \pi i \tag{19}$$

$$a(0) = \sum_{i=-2^n/2+1}^{2^n/2} F(i)f(0, i) \tag{20}$$

$$a_s(l) = \sqrt{2} \sum_{i=-2^n/2+1}^{2^n/2} F(i) \sin(2\pi li/2^n) \tag{21}$$

$$a_c(l) = \sqrt{2} \sum_{i=-2^n/2+1}^{2^n/2} F(i) \cos(2\pi li/2^n) \tag{22}$$

$$a_c(2^n/2) = \sqrt{2} \sum_{i=-2^n/2+1}^{2^n/2} F(i) \cos \pi i \tag{23}$$

## 3.6 Active Beam Forming with Inverse Fourier Transform Circuits

If the coefficients $a(0)$, $a_s(l)$, $a_c(l)$, and $a_c(2^n/2)$ of Eq. (3.5-19) are fed to the input terminals 0, s1 = 1, c1 = 2, s2 = 3, c2 = 4, ..., $c2^n/2 = 2^n - 1$ of an inverse Fourier transform circuit according to Fig. 3.5-1, one obtains at the output terminals 0* = 0, s1* = 1, c1* = 2, s2* = 3, c2* = 4, ..., $c2^n/2^* = 2^{n-1}$, the function $F(i)$. By choosing for these coefficients time variable voltages $V \sin(2\pi vt/\lambda)$ and $V \cos(2\pi vt/\lambda)$, one can produce output voltages $F(i) = F(i, t)$ that have the same time variation as the output voltages of the hydrophones of a line array according to Fig. 2.1-1. Hence, if the hydrophones in Fig. 2.1-1 are replaced by sound projectors that are driven by the voltages $F(i, t)$, one can radiate sound with the same beam pattern as shown by Eqs. (2.2-7), (2.2-11), and (2.2-12) for a received plane wave. We call this beam forming for radiation *active beam forming.*

Let us choose the following values for the coefficients in Eq. (3.5-19):

$$\begin{aligned} a(0) &= V \sin(2\pi vt/\lambda) \\ a_s(l) &= a_c(l) = a_c(2^n/2) = 0 \end{aligned} \tag{1}$$

One obtains:

$$F(i, t) = V \sin(2\pi vt/\lambda) f(0, i) \tag{2}$$

All voltages at the terminals $k = i + 2^n/2 - 1 = 0, 1, \ldots, 2^n - 1$—or 0, s1, c1, ..., $c2^n/2$—have the same phase. This implies that a beam in the broadside direction $\gamma = 0$ will be formed.

If a beam in the general direction $\gamma$ is to be formed, we infer from Eq. (2.2-1) that the voltages

$$F(i, t) = V \sin[2\pi(vt + id \sin \gamma)/\lambda] \tag{3}$$

are required at the terminals $k = i + 2^n/2 - 1 = 0, \ldots, 2^n - 1$. The comparison of Eq. (3) with Eq. (3.5-19) suggests that we choose $a_s(l)$ and $a_c(l)$ so that the following condition is satisfied:

$$\begin{aligned}\sqrt{2}[a_s(l) \sin(2\pi li/2^n) + a_c(l) \cos(2\pi li/2^n)] &= V \sin[2\pi(vt + id \sin \gamma)/\lambda] \\ &= V\{\sin(2\pi vt/\lambda) \cos[(2\pi id \sin \gamma)/\lambda] \\ &\quad + \cos(2\pi vt/\lambda) \sin[(2\pi id \sin \gamma)/\lambda)\}\end{aligned} \tag{4}$$

$$i = k - 2^n/2 + 1$$

The following values for $a_s(l)$ and $a_c(l)$ satisfy this equation:

$$\begin{array}{llll} \gamma > 0, & (d \sin \gamma)/\lambda = l/2^n & \text{or} & 2^n(d/\lambda) \sin \gamma = l \\ & a_s(l) = +\tfrac{1}{2}\sqrt{2}\, V \cos(2\pi vt/\lambda), & & a_c(l) = +\tfrac{1}{2}\sqrt{2}\, V \sin(2\pi vt/\lambda) \\ \gamma < 0, & [d \sin(-\gamma)]/\lambda = l/2^n & \text{or} & 2^n(d/\lambda) \sin(-\gamma) = l \\ & a_s(l) = -\tfrac{1}{2}\sqrt{2}\, V \cos(2\pi vt/\lambda), & & a_c(l) = +\tfrac{1}{2}\sqrt{2}\, V \sin(2\pi vt/\lambda) \end{array} \tag{5}$$

The final case, $a_c(2^n/2) \neq 0$ and all other coefficients zero, is a specialization of Eq. (4):

$$\tfrac{1}{2}\sqrt{2}\, a_c(2^n/2) \cos \pi i = V \sin[2\pi(vt + id \sin \gamma)/\lambda] \tag{6}$$

One obtains two solutions by comparison with Eq. (4):

$$\begin{array}{c} a_c(2^n/2) = \sqrt{2}\, V \sin(2\pi vt/\lambda) \\ (d \sin \gamma)/\lambda = \tfrac{1}{2} \quad \text{or} \quad [d \sin(-\gamma)]/\lambda = \tfrac{1}{2}, \qquad \sin \gamma = \pm\lambda/2d \end{array} \tag{7}$$

Two beams are formed in this last case, one in the direction $\sin \gamma = +\lambda/2d$ the other in the direction $\sin \gamma = -\lambda/2d$. In the case of the passive beam former, there was the corresponding ambiguity about the sign of the angle with which the plane wave arrived.

In order to form one beam at a time we can feed voltages according to Eqs. (2) or (5) to the input terminals 0, ..., 14 of the circuit in Fig. 3.5-1. The output terminals 36, ..., 51 are connected to 16 equally spaced sound projectors of a line array. This is shown in more detail in Fig. 3.6-1 for the three beams in the directions $\gamma = 0$, $+2\varepsilon$, and $-5\varepsilon$. There is no difficulty recognizing how this circuit can be used for scanning by forming a succession of beams for $\gamma = -7\varepsilon, -6\varepsilon, \ldots, +7\varepsilon$.

If the circuit of Fig. 3.6-1 is used to form all 15 beams $\gamma = -7\varepsilon, \ldots, +7\varepsilon$

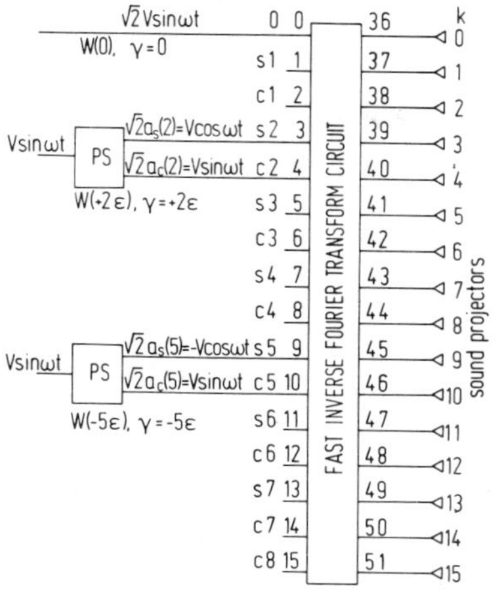


FIG. 3.6-1. Circuit for beam forming by means of the fast inverse Fourier transform circuit of Fig. 3.5-1. A typical circuit for the phase shifters PS is shown in Fig. 3.6-2c.

or all 17 beams $\gamma = -8\varepsilon, \ldots, +8\varepsilon$ at the same time, one obtains a wide beam with the fast drop-off of the narrow beams at the edges. Hence, the circuit may be used to produce an even insonification in a sharply defined sector. Only the voltages $a(0)$ and $a_c(l)$, $l = 1, \ldots, 8$, have to be fed to the fast inverse Fourier transform circuits in this case, since the beams in the direction $+\gamma$ and $-\gamma$ are formed simultaneously.

Of great practical importance is the possibility to form 15 beams simultaneously, each one at a slightly different frequency. The fast inverse Fourier transform circuit is nominally frequency independent, but the beam formed by a certain array depends on the frequency; hence the requirement for *slightly* different frequencies. In order to form a beam in the direction $l\varepsilon$ we need the voltages $a_s(l)$ and $a_c(l)$:

$$\begin{aligned} a_s(l) &= \tfrac{1}{2}\sqrt{2}\,V\cos(2\pi vt/\lambda_l) = \tfrac{1}{2}\sqrt{2}\,V\cos\omega_l t \\ a_c(l) &= \tfrac{1}{2}\sqrt{2}\,V\sin\omega_l t \end{aligned} \tag{8}$$

A beam in the direction $-l\varepsilon$ requires the voltages $a_s^*(l)$ and $a_c^*(l)$:

$$\begin{aligned} a_s^*(l) &= -\tfrac{1}{2}\sqrt{2}\,V\cos\omega_{-l}t \\ a_c^*(l) &= +\tfrac{1}{2}\sqrt{2}\,V\sin\omega_{-l}t \end{aligned} \tag{9}$$

If both beams are to be formed simultaneously, one needs the sum of Eqs. (8) and (9):

$$\begin{aligned} a_s(l) + a_s^*(l) &= \tfrac{1}{2}\sqrt{2}\,V(\cos\omega_l t - \cos\omega_{-l}t) \\ a_c(l) + a_c^*(l) &= \tfrac{1}{2}\sqrt{2}\,V(\sin\omega_l t + \sin\omega_{-l}t) \end{aligned} \tag{10}$$

A circuit producing these two voltages is shown in Fig. 3.6-2a. The complete

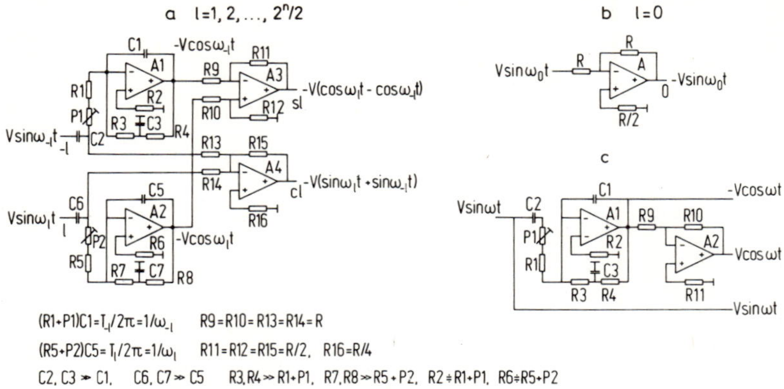


FIG. 3.6-2. (a) Phase shifter PS generating the voltages $a_s(l) + a_s^*(l)$ and $a_c(l) + a_c^*(l)$ of Eq. (10) required in Fig. 3.6-3. (b) Amplitude reverser AR of Fig. 3.6-3. (c) Phase shifter PS of Fig. 3.6-1.

beam former is shown in Fig. 3.6-3. The phase shifters PS contain the circuit of Fig. 3.6-2a, the amplitude reverser AR contains the circuit of Fig. 3.6-2b; this amplitude reverser is usually not needed.

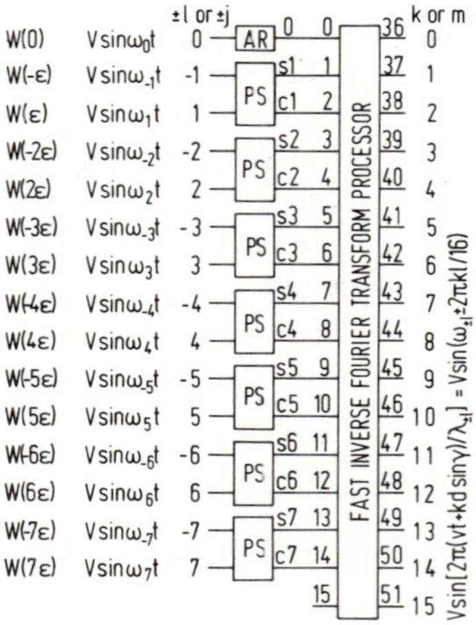


FIG. 3.6-3. Beam former producing the voltages for 15 simultaneous beams in the directions $-7\varepsilon, \ldots, +7\varepsilon$ with frequencies $\omega_{-7}, \ldots, \omega_{+7}$. The circuit of the phase shifters PS is shown in Fig. 3.6-2a, that of the amplitude reverser AR in Fig. 3.6-2b, and that of the fast inverse Fourier transform processor in Fig. 3.5-1.

# 4 Focusing for Spherical Wavefronts

## 4.1 Theoretical Foundations

We have so far considered planar wavefronts, which must come from points that are practically at an infinite distance. In reality one receives spherical wavefronts coming from points at a finite distance $R$. The problem of focusing is the correction of the deviations caused by the reception of spherical waves. Two methods are generally used in photography to solve the focusing problem. The film should be in the focal plane of the lens for $R = \infty$; an increase of the distance between lens and film permits focusing for finite distances. A second method is the reduction of the aperture of the lens, which increases the depth of field and thus provides indirectly a means for focusing. This second method is clearly not applicable to acoustic imaging. The hydrophone array is the costliest part of an acoustic imaging system, costing easily a thousand times as much as a photographic lens, and one must use all the hydrophones all the time rather than the ones in the center of the array only. Furthermore, the huge difference in the wavelengths used in optics and acoustics makes the geometric relations much more unfavorable in acoustics. Hence, one must try to find a focusing method that corresponds to the change of distance between lens and film in photography.

Consider Fig. 4.1-1 for the derivation of the basic formulas for focusing. The points on the plane of focus shall produce the same diffraction or beam pattern that so far has been produced by the points at infinity. The distance between this plane and the hydrophone array equals $R$. The point $W\beta$ on this plane has the distance

$$(R^2 + R^2 \tan^2 \beta)^{1/2} = R(1 + \tan^2 \beta)^{1/2} \tag{1}$$

from the hydrophone $i = 0$ in the center of the array. The distance between the point $W\beta$ and an arbitrary hydrophone $i$ is defined by the relation

$$[R^2 + (R \tan \beta - id)^2]^{1/2} = R[1 + (\tan \beta - id/R)^2]^{1/2} \tag{2}$$

The difference of the distances of Eqs. (2) and (1) divided by the wavelength $\lambda$ is called the relative phase or the relative propagation time $\psi(i, R, \beta)$:

$$\psi(i, R, \beta) = (R/\lambda)\{[1 + (\tan \beta - id/R)^2]^{1/2} - (1 + \tan^2 \beta)^{1/2}\} \tag{3}$$

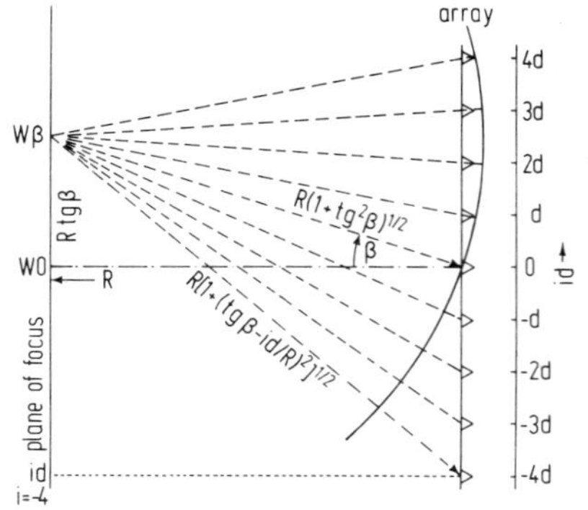


FIG. 4.1-1. Derivation of the exact values of the output voltages of a hydrophone array due to a spherical wavefront originating in the point $W\beta$ at the distance $R(1 + \tan^2 \beta)^{1/2}$ on a ray forming the angle $\beta$ with the axis of the array.

For small values of $\beta$ one may replace $(1 + \tan^2 \beta)^{1/2}$ by $1 + \frac{1}{2} \tan^2 \beta$. If, in addition, $id/R$ is small for the largest absolute value of $i$, which we denote by $2^n/2$, one may use the approximation

$$[1 + (\tan \beta - id/R)^2]^{1/2} \doteq 1 + \tfrac{1}{2}(\tan \beta - id/R)^2, \qquad |i| \leq 2^n/2$$

For small values of $\beta$ one may replace $\tan \beta$ by $\sin \beta$, and one thus obtains a first-order approximation of $\psi(i, R, \beta)$:

$$\psi(i, R, \beta) \doteq (-id \sin \beta + i^2 d^2/2R)/\lambda$$
$$|i| \leq 2^n/2, \qquad 2^n d/2 \ll R, \qquad \tan \beta \doteq \sin \beta \ll 1 \tag{4}$$

The output voltage of hydrophone $i$ at location $id$ in Fig. 4.1-1 is denoted $v(i, R, \beta, t)$:

$$v(i, R, \beta, t) = V \sin 2\pi[vt/\lambda - \psi(i, R, \beta)] \tag{5}$$

Using the approximation of Eq. (4), one obtains the following output voltages:

$$v(i, R, \beta, t) \doteq V \sin[2\pi(vt + id \sin \beta - i^2 d^2/2R)/\lambda] \tag{6}$$

To obtain a physical understanding of this voltage consider Fig. 4.1-2. It shows the output voltages of the hydrophones produced by a sound source located at $W0$ at the distance $R$ from the array and forming the angle $\beta = 0$ with the array axis. The approximation $(id/R)^2 \ll 1$ for all $i$ is made to derive the voltages, which is evidently equal to the condition $2^n d/2 \ll R$ in Eq. (4). The voltages in Fig. 4.1-2 equal the voltage of Eq. (6) for $\beta = 0$.

Consider now the sound source at $W\beta$ in Fig. 4.1-3. It has the same distance $R$ from the center of the array as $W0$ but it is seen from the hydrophone $i = 0$ with the angle $\beta$ rather than $\beta = 0$. It is left to the

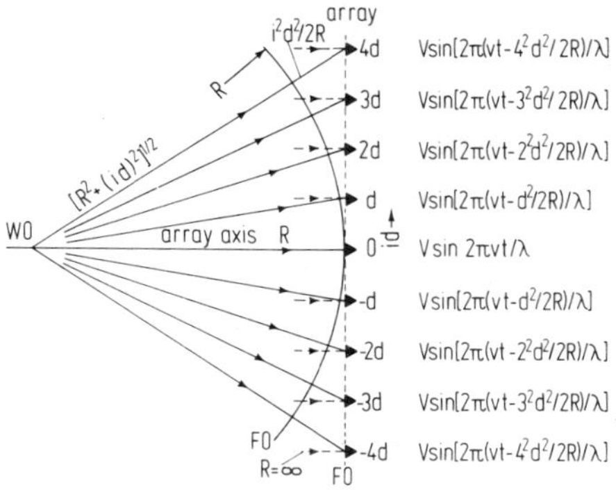


FIG. 4.1-2. Output voltages of a hydrophone array due to a spherical wave originating in the point $W0$ at the distance $R$ on the axis of the array.

reader to verify that the output voltages shown in Fig. 4.1-3 are obtained for small angles $\beta$ and a distance $R$ large compared with the aperture of the array; these are the conditions of Eq. (4). The voltages in Fig. 4.1-3 are evidently equal to the voltage $v(i, R, \beta, t)$ of Eq. (6). Hence, the use of Eq. (6) implies that the points in focus are not located on a straight line in Fig. 4.1-1 but on a circle.

The generalization of Fig. 4.1-3 from one to two dimensions is done in

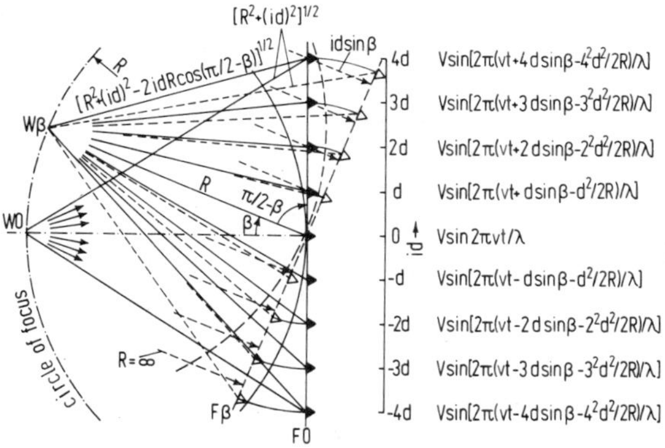


FIG. 4.1-3. Output voltages of a hydrophone array due to a spherical wave originating in the point $W\beta$ at the distance $R$ on a ray forming the angle $\beta$ with the axis of the array.

optics universally by rotating the illustration around the line from $W0$ to the center of the array at $id = 0$. This fits the polar coordinates for which lenses are made. It will be shown later on that the same generalization can be made for a beam former operating in cartesian coordinates, provided focusing is done in the first-order approximation of Fig. 4.1-3. In other words, one can use polar coordinates for focusing—and produce thus planar wavefronts from the spherical ones—but use cartesian coordinates for beam forming with the planar wavefronts.

The generalization of the circle of focus in Fig. 4.1-3 for cartesian coordinates is a cylinder. This cylinder is denoted *cylinder of focus for x* in Fig. 4.1-4. For the vertical coordinate one obtains the *cylinder of focus*

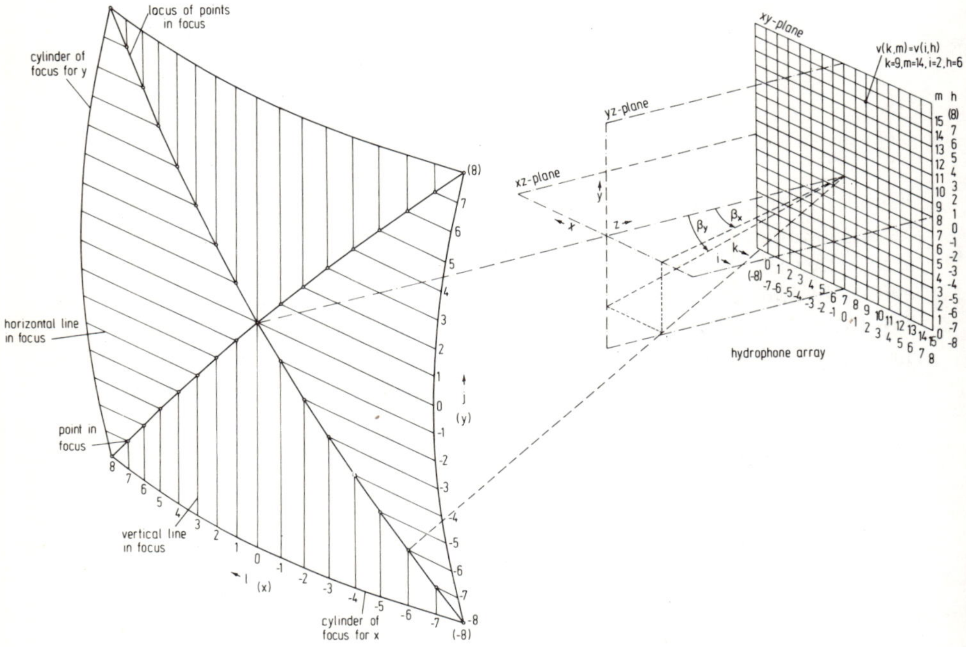


FIG. 4.1-4. Focusing in cartesian coordinates.

*for y*. The image of any straight line on either cylinder will be in focus. Two such lines are shown in Fig. 4.1-4 as *horizontal line in focus* and *vertical line in focus*. The image of the point where these two lines intersect will be in focus. The two curves denoted *locus of focused points* represent all the points that can be in focus. The surfaces of the horizontal and the vertical cylinder represent all the horizontal and the vertical lines that can be in focus.

These principles of optics in cartesian coordinates are somewhat confusing when first encountered, since optical imaging is universally based on polar coordinates. The reason is, of course, that a lens with rotational symmetry is the easiest one to grind, and this method of production suggests the use of a circular rather than a square piece of glass. The optical equivalent of our image generation in cartesian coordinates would be to use a cylindrical lens with horizontal axis followed by a cylindrical lens with vertical axis instead of one spherical lens.

Let us return to Eq. (6). We will first derive equations for electronic focusing from this approximation, but do the same later for the exact solution of Eq. (5). The integral of Eq. (6) multiplied by $-\lambda/2\pi v$ is denoted $v^*(i, R, \beta, t)$

$$\begin{aligned} v^*(i, R, \beta, t) &= -(2\pi v/\lambda) \int v(i, R, \beta, t)\, dt \\ &= -V \cos[2\pi(vt + id \sin\beta - i^2d^2/2R)/\lambda] \\ &= -V\{\cos(\pi i^2 d^2/R\lambda) \cos[2\pi(vt + id \sin\beta)/\lambda] \\ &\quad + \sin(\pi i^2 d^2/R\lambda) \sin[2\pi(vt + id \sin\beta)/\lambda]\} \end{aligned} \tag{7}$$

The voltage $v(i, R, \beta, t)$ is rewritten into a sum of two terms:

$$\begin{aligned} v(i, R, \beta, t) = V\{&\cos(\pi i^2 d^2/R\lambda) \sin[2\pi(vt + id \sin\beta)/\lambda] \\ &- \sin(\pi i^2 d^2/R\lambda) \cos[2\pi(vt + id \sin\beta)/\lambda]\} \end{aligned} \tag{8}$$

We multiply $v^*(i, R, \beta, t)$ by $-\sin(\pi i^2 d^2/R\lambda)$ and $v(i, R, \beta, t)$ by $\cos(\pi i^2 d^2/R\lambda)$ and sum the products:

$$\begin{aligned} v^*(i, R, \beta, t) \sin(\pi i^2 d^2/R\lambda) + v(i, R, \beta, t) \cos(\pi i^2 d^2/R\lambda) \\ = V \sin[2\pi(vt + id \sin\beta)/\lambda] \end{aligned} \tag{9}$$

This is already the voltage $v(i, t)$ representing a planar wavefront, as one can readily see from Fig. 4.1-3 or Eq. (2.2-1).

We must next generalize the one-dimensional arrays of Figs. 4.1-2 and 4.1-3 to a two-dimensional array in cartesian coordinates. The arriving wavefronts are still spherical, even though the array is not circular but square. Figure 4.1-5 shows such an array with 16 × 16 hydrophones. They are numbered $s = \pm 0, \pm 1, \ldots$ in the $x$-direction and $p = \pm 0, \pm 1, \ldots$ in the $y$-direction. The circles connect the hydrophones with the same distance from the center of the array. These distances replace the distance $id$ of the one-dimensional array of Fig. 4.1-2. The many circles, often with almost equal radius, look very complicated, and we must find a simpler representation. One can see that the four quadrants in Fig. 4.1-5 are symmetric, and we need thus to study only one of them. Furthermore, we are not interested in the distance of the hydrophones from the center of the array but in the distance from the four hydrophones that are closest to the center. These four hydrophones denoted by $s, p = -0, -0; -0, 0; 0, -0; 0, 0$ are shown as small circles rather than as dots in Fig. 4.1-5.

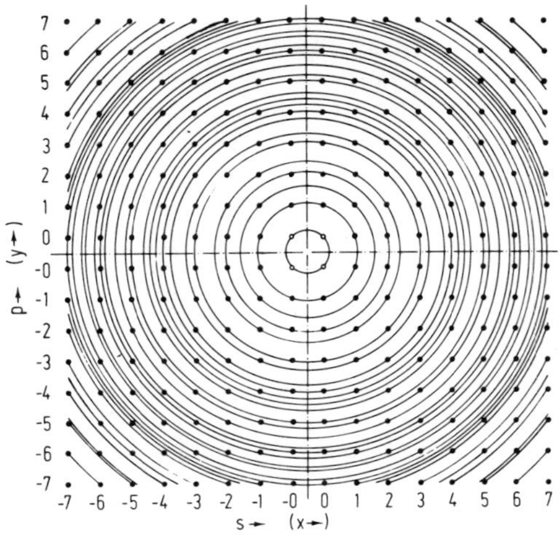


FIG. 4.1-5. Two-dimensional array of 16 × 16 hydrophones. The hydrophones with equal distance from the center of the array are connected by circles.

One quadrant of the array of Fig. 4.1-5 is shown in Fig. 4.1-6. The circles indicating the distance of the hydrophones are centered at the hydrophone $s = 0$, $p = 0$ rather than at the center of the array; only short sections of the circles are shown in order to avoid obscuring the picture. It is easy to see that a hydrophone located at a point with the coordinates $s$ and $p$ has the distance $(s^2 + p^2)^{1/2}d$ from the hydrophone located at

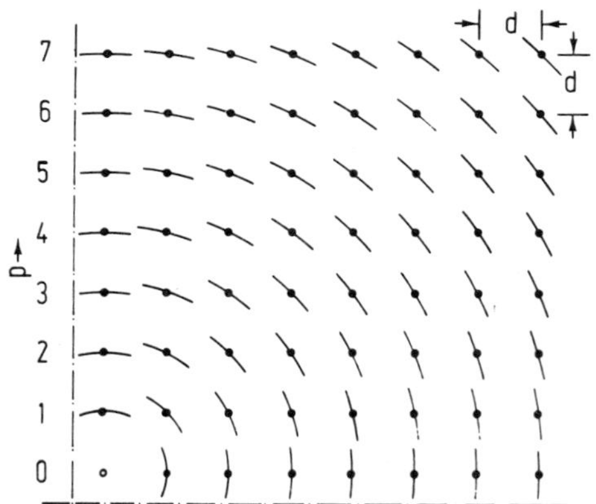


FIG. 4.1-6. The quadrant $s \geq 0$, $p \geq 0$ of the array of Fig. 4.1-5. The hydrophone $s$, $p$ has the distance $(s^2 + p^2)^{1/2}d$ from the hydrophone $s = 0$, $p = 0$.

$s = p = 0$. This distance replaces the distance $id$ in Eq. (6), while the delay time

$$t_i = i^2d^2/2Rv \tag{10}$$

is replaced by $t_{s,p}$:

$$t_{s,p} = (s^2 + p^2)d^2/2Rv \tag{11}$$

The largest delay time $t_{s,p}$ in Fig. 4.1-6 is obtained for $s = s' = 7$ and $p = p' = 7$:

$$t_{s',p'} = (s'^2 + p'^2)d^2/2Rv \tag{12}$$

The delay required to equalize the arrival time of a spherical wave at the hydrophone $s, p$ is given by the difference

$$\begin{gathered} t_{s',p'} - t_{s,p} = (s'^2 - s^2 + p'^2 - p^2)d^2/2Rv \\ s = 0, 1, \ldots, s'; \quad p = 0, 1, \ldots, p' \end{gathered} \tag{13}$$

If the focusing is done by delay devices, one must use Eq. (13); but one can use either Eq. (13) or Eq. (11) if one uses sinusoidal waves and replaces the delay devices by phase shifters. We can shift the phase of a sinusoidal voltage forward or backward, but more general waves can only be delayed, not advanced. Hence, the generalization from one to two dimensions for sinusoidal waves requires the replacement of $t_i$ of Eq. (10) by $t_{s,p}$ of Eq. (11). This means that $\sin(\pi i^2d^2/R\lambda)$ and $\cos(\pi i^2d^2/R\lambda)$ in Eq. (9) are replaced by

$$v_s(s, p) = \sin[\pi(s^2 + p^2)d^2/R\lambda] \tag{14}$$

and

$$v_c(s, p) = \cos[\pi(s^2 + p^2)d^2/R\lambda] \tag{15}$$

The numbers $v_s(s, p)$ and $v_c(s, p)$ can readily be produced by a microprocessor. Equations (7), (9), (14), and (15) represent the phase shifting method of focusing.

In addition to this phase shifting method of focusing, we have a second method using frequency conversion. We rewrite Eq. (6) with the help of the relation $v = f\lambda$:

$$v(i, R, \beta, t) \doteq V \sin 2\pi(ft + id\lambda^{-1} \sin \beta - i^2d^2/2R\lambda) \tag{16}$$

A voltage $e(i, t)$ is generated in the processor:

$$\begin{aligned} e(i, t) &= 2 \sin 2\pi[f_1(t - t_0) - i^2d^2/2R\lambda] \\ &= \cos(\pi i^2d^2/R\lambda) \sin 2\pi f_1(t - t_0) - \sin(\pi i^2d^2/R\lambda) \cos 2\pi f_1(t - t_0) \end{aligned} \tag{17}$$

Let $v(i, R, \beta, t)$ be multiplied by $e(i, t)$. One obtains a term with the sum frequency $f + f_1$, which shall be suppressed. The second term with the difference frequency $f' = f - f_1$ is denoted $v'(i, R, \beta, t)$:

$$\begin{aligned} v'(i, R, \beta, t) &= V \cos 2\pi(f't' + id\lambda^{-1} \sin \beta), \\ f' = f - f_1, &\qquad t' = t + t_0 f_1/f \end{aligned} \tag{18}$$

The term $i^2d^2/2R\lambda$ is eliminated, and the spherical wavefront has been transformed into a planar wavefront. The generalization from one to two dimensions requires again that $\sin(\pi i^2 d^2/R\lambda)$ and $\cos(\pi i^2 d^2/R\lambda)$ are replaced by $v_s(s, p)$ and $v_c(s, p)$ of Eqs. (14) and (15).

Let us turn to focusing without the approximations of Eq. (4). This means we use Eq. (5) rather than Eq. (6). We define the focusing phase or focusing time $\chi(i, R, \gamma)$ for the hydrophone $i$, a distance $R$ between the plane of focus and the array, and an observation angle $\gamma$:

$$\chi(i, R, \gamma) = (R/\lambda)\{[1 + (\tan \gamma - id/R)^2]^{1/2} - (1 + \tan^2 \gamma)^{1/2}\} \tag{19}$$

$\chi(i, R, \gamma)$ and $\psi(i, R, \beta)$ of Eq. (3) are equal except that the continuously variable angle of incidence $\beta$ is replaced by the observation angle $\gamma$, which assumes the values 0, $\pm\varepsilon$, $\pm 2\varepsilon$, ... only. The values of $\lambda$, $\gamma$, $i$, and $d$ are fixed, and a certain value of $R$ is chosen for focusing. Hence, the numerical value of $\chi(i, R, \gamma)$ is known. The value of $\psi(i, R, \beta)$, on the other hand, is not known since $\beta$ is not known but determined by the process of image generation.

For small values of $\gamma$ and large values of $R$, one obtains an approximation according to Eq. (4):

$$\chi(i, R, \gamma) \doteq (-id \sin \gamma + i^2d^2/2R)/\lambda \tag{20}$$

Since the numerical value of $\chi(i, R, \gamma)$ is known, one can produce the sine, $\sin[\chi(i, R, \gamma)]$, and the cosine, $\cos[\chi(i, R, \gamma)]$, of this numerical value. Furthermore, one can phase shift $v(i, R, \beta, t)$ of Eq. (5); the shifted voltage is denoted $v^*(i, R, \beta, t)$:

$$v^*(i, R, \beta, t) = V \cos 2\pi[vt/\lambda - \psi(i, R, \beta)] \tag{21}$$

The following expression is formed:

$$\begin{aligned} &\tfrac{1}{2}\{v(i, R, \beta, t) \cos[\chi(i, R, \gamma)] + v^*(i, R, \beta, t) \sin[\chi(i, R, \gamma)]\} \\ &\quad = \tfrac{1}{2}V \sin 2\pi[vt/\lambda - \psi(i, R, \beta) + \chi(i, R, \gamma)] \end{aligned} \tag{22}$$

Summation over all values of $i$ yields the voltage $v(R, \gamma, \beta, t)$:

$$v(R, \gamma, \beta, t) = \tfrac{1}{2}V \sum_i \sin 2\pi[vt/\lambda - \psi(i, R, \beta) + \chi(i, R, \gamma)] \tag{23}$$

This voltage represents the image of the point $W\beta$ in Fig. 4.1-1 in electric form. An electrooptic converter will transform it into a visible optical image. Let us show that this claim is true by substituting the approximations of Eqs. (4) and (20) for $\psi(i, R, \beta)$ and $\chi(i, R, \beta)$:

$$v(R, \gamma, \beta, t) \doteqdot \tfrac{1}{2}V \sum_i \sin\{2\pi[vt + id(\sin\beta - \sin\gamma)]/\lambda\}$$

$$\doteqdot \frac{V}{2 \cdot 2^n}\left\{\sin\frac{2\pi vt}{\lambda}\int_{-1/2}^{1/2}\cos[2\pi A\lambda^{-1}(\sin\beta - \sin\gamma)x]\,dx\right.$$

$$\left. + \cos\frac{2\pi vt}{\lambda}\int_{-1/2}^{1/2}\sin[2\pi A\lambda^{-1}(\sin\beta - \sin\gamma)x]\,dx\right\}$$

$$= \frac{VA}{2d}\frac{\sin[\pi A\lambda^{-1}(\sin\beta - \sin\gamma)]}{\pi A\lambda^{-1}(\sin\beta - \sin\gamma)}\sin\frac{2\pi vt}{\lambda}$$

$$2^n d = A, \qquad i/2^n = x \tag{24}$$

The last line of this equation is equal to Eqs. (2.2-11) and (2.2-12) if $\sin\gamma$ is inserted from Eq. (2.2-6).

One can rewrite Eq. (23) into a form that shows more clearly the diffraction pattern produced as the image of the point $W\beta$:

$$v(R, \gamma, \beta, t) = \tfrac{1}{2}V[A^2(R, \gamma, \beta) + B^2(R, \gamma, \beta)]^{1/2}\sin(2\pi vt/\lambda + \varphi)$$

$$A(R, \gamma, \beta) = \sum_i \cos[\chi(i, R, \gamma) - \psi(i, R, \beta)]$$

$$B(R, \gamma, \beta) = \sum_i \sin[\chi(i, R, \gamma) - \psi(i, R, \beta)] \tag{25}$$

$[A^2(R, \gamma, \beta) + B^2(R, \gamma, \beta)]^{1/2}$ is the diffraction pattern of the image of the point $W\beta$ in Fig. 4.1-1. For small values of $\beta$ and $\gamma$ and large values of $R$, one obtains readily the previously used approximation:

$$[A^2(R, \gamma, \beta) + B^2(R, \gamma, \beta)]^{1/2} \doteqdot A(R, \gamma, \beta) \doteqdot \frac{A}{d}\frac{\sin[\pi A\lambda^{-1}(\sin\beta - \sin\gamma)]}{\pi A\lambda^{-1}(\sin\beta - \sin\gamma)} \tag{26}$$

The generalization of Eq. (25) from one to two dimensions is not quite equal to the transition from $t_i$ in Eq. (10) to $t_{s,p}$ in Eq. (11). If we use separate circuits for focusing and for beam forming of the planarized waveforms, we can use the variables $s$ and $p$ of Figs. 4.1-5 and 4.1-6 for focusing, but the variables $k = i + 2^n/2 - 1$ and $m = h + 2^n/2 - 1$ of Fig. 2.1-5 for beam forming. The variables $s$ and $i$ as well as $p$ and $h$ are not equal; the center of the array for focusing is slightly different from the center for beam forming. We will derive the generalization of Eq. (25) to two dimensions with the help of Fig. 4.1-7.

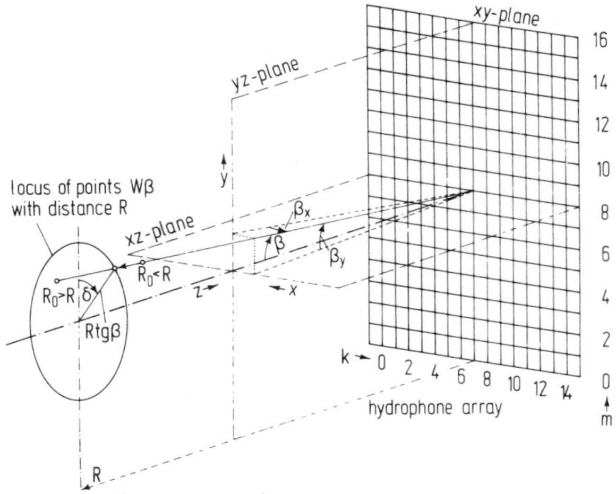


FIG. 4.1-7. Focusing with polar coordinates for the object plane but cartesian coordinates for the reception plane.

A point $W\beta$ on the circle "locus of points $W\beta$ with distance $R$" has the distance

$$(R^2 + R^2 \tan^2 \beta_x + R^2 \tan^2 \beta_y)^{1/2} = R(1 + \tan^2 \beta_x + \tan^2 \beta_y)^{1/2} \quad (27)$$

from the hydrophone $k = 7$, $m = 8$—or generally $i = 0$, $h = 0$—in the center of the array in Fig. 4.1-7. This is the generalization of Eq. (1). The distance between the point $W\beta$ and an arbitrary hydrophone $i$, $h$ is defined by the relation

$$[R^2 + (R \tan \beta_x - id)^2 + (R \tan \beta_y - hd)^2]^{1/2}$$
$$= R[1 + (\tan \beta_x - id/R)^2 + (\tan \beta_y - hd/R)^2]^{1/2} \quad (28)$$

The relative phase or propagation time $\psi(i, h, R, \beta_x, \beta_y)$ is the difference of the distances of Eqs. (28) and (27) divided by the wavelength $\lambda$:

$$\psi(i, h, R, \beta_x, \beta_y) = (R/\lambda)\{[1 + (\tan \beta_x - id/R)^2 + (\tan \beta_y - hd/R)^2]^{1/2} - (1 + \tan^2 \beta_x + \tan^2 \beta_y)^{1/2}\} \quad (29)$$

The generalization of the focusing phase or focusing time $\chi(i, R, \gamma)$ of Eq. (19) follows from analogy:

$$\chi(i, h, R, \gamma_x, \gamma_y) = (R/\lambda)\{[1 + (\tan \gamma_x - id/R)^2 + (\tan \gamma_y - hd/R)^2]^{1/2} - (1 + \tan^2 \gamma_x + \tan^2 \gamma_y)^{1/2}\} \quad (30)$$

The voltages $v(i, R, \beta, t)$ of Eq. (5) and $v^*(i, R, \beta, t)$ of Eq. (21) become:

$$v(i, h, R, \beta_x, \beta_y, t) = V \sin 2\pi[vt/\lambda - \psi(i, h, R, \beta_x, \beta_y)] \tag{31}$$

$$v^*(i, h, R, \beta_x, \beta_y, t) = V \cos 2\pi[vt/\lambda - \psi(i, h, R, \beta_x, \beta_y)] \tag{32}$$

Equations (22) and (23) are generalized as follows:

$$\begin{aligned}\tfrac{1}{2}v(i, h, R, \beta_x, \beta_y, t) &\cos[\chi(i, h, R, \gamma_x, \gamma_y)] \\ &+ v^*(i, h, R, \beta_x, \beta_y, t) \sin[\chi(i, h, R, \gamma_x, \gamma_y)] \\ &= \tfrac{1}{2}V \sin 2\pi[vt - \psi(i, h, R, \beta_x, \beta_y) + \chi(i, h, R, \gamma_x, \gamma_y)]\end{aligned} \tag{33}$$

$$\begin{aligned}v(R, &\gamma_x, \gamma_y, \beta_x, \beta_y, t) \\ &= \tfrac{1}{2}V \sum_i \sum_h \sin 2\pi[vt - \psi(i, h, R, \beta_x, \beta_y) + \chi(i, h, R, \gamma_x, \gamma_y)]\end{aligned} \tag{34}$$

Although Eq. (34) is a mathematically correct result, it is very difficult to implement practically, since the transformation cannot be broken up into two transformations, first one for $i$ then one for $h$. In order to do so, one must be able to write $\psi(i, h, R, \beta_x, \beta_y)$ and $\chi(i, h, R, \gamma_x, \gamma_y)$ as sums:

$$\psi(i, h, R, \beta_x, \beta_y) = \psi_x(i, R, \beta_x) + \psi_y(h, R, \beta_y) \tag{35}$$

$$\chi(i, h, R, \gamma_x, \gamma_y) = \chi_x(i, R, \gamma_x) + \chi_y(h, R, \gamma_y) \tag{36}$$

This is possible if we can expand the square roots in Eqs. (29) and (30) into series:

$$\begin{aligned}\psi(i, h, R, \beta_x, \beta_y) &= (R/2\lambda)[(\tan \beta_x - id/R)^2 - \tan^2 \beta_x \\ &\qquad + (\tan \beta_y - hd/R)^2 - \tan \beta_y^2] \\ &= (-id \tan \beta_x + i^2d^2/2R - hd \tan \beta_y + h^2d^2/2R)/\lambda\end{aligned} \tag{37}$$

A comparison with Eq. (4) shows that we are back to the approximation holding for large values of $R$ and small values of $\beta$. This is not an unexpected result. We have seen in Section 1.2, particularly Figs. 1.2-3 and 1.2-4, that the electronic equivalent of a lens is a device that performs a linear transformation first for the one variable, $x$ or $r$, then for the other variable, $y$ or $\varphi$. Since the lens essentially cannot focus for a short distance $R$ and a large angle of incidence $\beta$, one cannot expect this electronic circuit to perform the task either. [For a discussion of the difficulties encountered in electronic processing of voltages with two space variables if one cannot process one variable after the other, see Section 2.3.1 of Harmuth (1977).]

In order to show the diffraction pattern of Eq. (34) more clearly, we may rewrite this equation as follows:

$$v(R, \gamma_x, \gamma_y, \beta_x, \beta_y, t) = \tfrac{1}{2}V[A^2(R, \gamma_x, \gamma_y, \beta_x, \beta_y) + B^2(R, \gamma_x, \gamma_y, \beta_x, \beta_y)]^{1/2} \sin(2\pi vt/\lambda + \varphi)$$

$$A(R, \gamma_x, \gamma_y, \beta_x, \beta_y) = \sum_i \sum_h \cos[\chi(i, h, R, \gamma_x, \gamma_y) - \psi(i, h, R, \beta_x, \beta_y)] \quad (38)$$

$$B(R, \gamma_x, \gamma_y, \beta_x, \beta_y) = \sum_i \sum_h \sin[\chi(i, h, R, \gamma_x, \gamma_y) - \psi(i, h, R, \beta_x, \beta_y)]$$

The diffraction pattern $v(R, y_x, \gamma_y, \beta_x, \beta_y)$ is defined as follows:

$$v(R, \gamma_x, \gamma_y, \beta_x, \beta_y) = [A^2(R, \gamma_x, \gamma_y, \beta_x, \beta_y) + B^2(R, \gamma_x, \gamma_y, \beta_x, \beta_y)]^{1/2} \quad (39)$$

## 4.2 Computer Plots of Diffraction Patterns

In order to obtain a better insight into the diffraction pattern $v(R, \gamma_x, \gamma_y, \beta_x, \beta_y)$ of Eq. (4.1-39), one may use computer plots. Figure 4.2-1

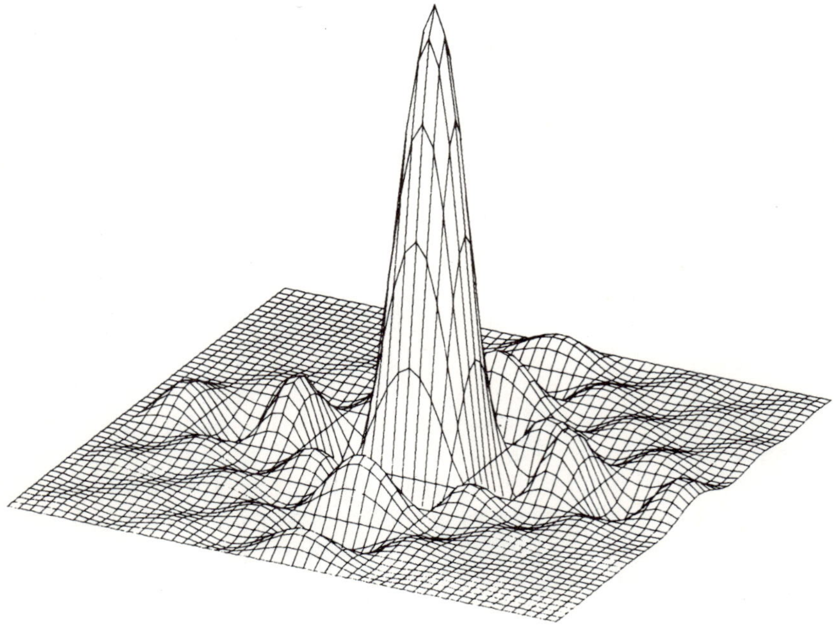

FIG. 4.2-1. Diffraction pattern $v(R, \gamma_x, \gamma_y, \beta_x, \beta_y)$ for $R = 10A$, $\gamma_x = \gamma_y = 0$ of a hydrophone array with $16 \times 16$ hydrophones and an aperture $A = 16d$ according to Fig. 4.1-7. (Courtesy N. Mohamed, Government of Kuwait.)

shows such a plot for a cartesian array with 16 × 16 hydrophones and an aperture $A = 16d$ according to Fig. 4.1-7. The distance $R$ to the object equals $10A$, and the observation angles $\gamma_x$ and $\gamma_y$ equal 0. The diffraction pattern is not circularly symmetric since the hydrophone array is a cartesian array.[1]

The excellent diffraction pattern of Fig. 4.2-1 degenerates rapidly if the object is not at the focusing distance $R$ but at another distance $R_0$. The focusing phase $\chi(i, h, R, 0, 0)$ remains unchanged but the relative phase of the received wave becomes $\psi(i, h, R_0, \beta_x, \beta_y)$. The resulting diffraction patterns are shown for $R_0 = 14A$, $13A$, $8A$, and $7A$ in Fig. 4.2-2. The pattern is not only widened but its center may become smaller than other

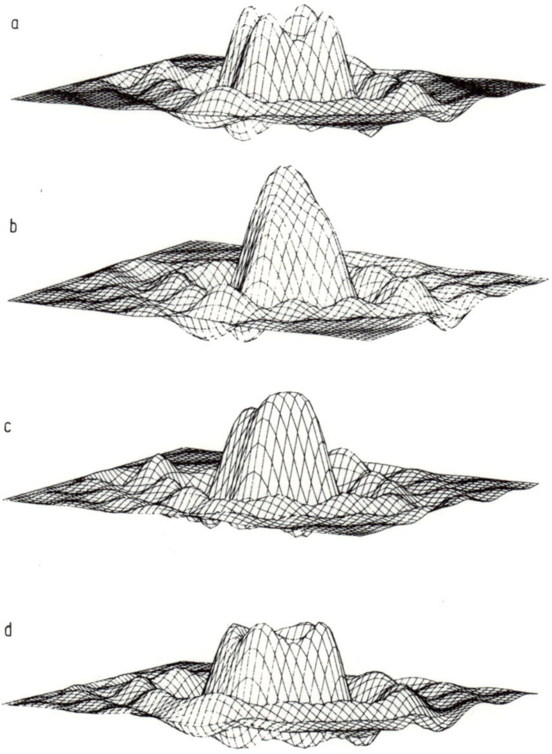


FIG. 4.2-2. Diffraction patterns according to Fig. 4.2-1 but with the imaged point at the actual distance $R_0$ instead of the focusing distance $R = 10A = 160d$. (a) $R_0 = 14A$, (b) $R_0 = 13A$, (c) $R_0 = 8A$, (d) $R_0 = 7A$. (Courtesy N. Mohamed, Government of Kuwait.)

[1] The computer plots of this section were obtained by Mohamed (1976) through simulation of the sound waves.

sections of the pattern if the focusing distance $R$ differs from the actual distance $R_0$. For instance, the image of the poorly focused point at $\beta_x = \beta_y = 0$ in Fig. 4.2-2a looks like a properly focused image of four adjacent points.

Some of the unsymmetries in Fig. 4.2-2 are due to the small number of $16 \times 16$ hydrophones, since an array with an even number of hydrophones is inherently unsymmetric if one wants to form the broadside beam $\gamma_x = \gamma_y = 0$. This type of unsymmetry decreases with an increasing number of hydrophones. Hence, diffraction patterns for an array with $64 \times 64$ hydrophones are shown in Fig. 4.2-3. The focusing distance equals $R = 5A$

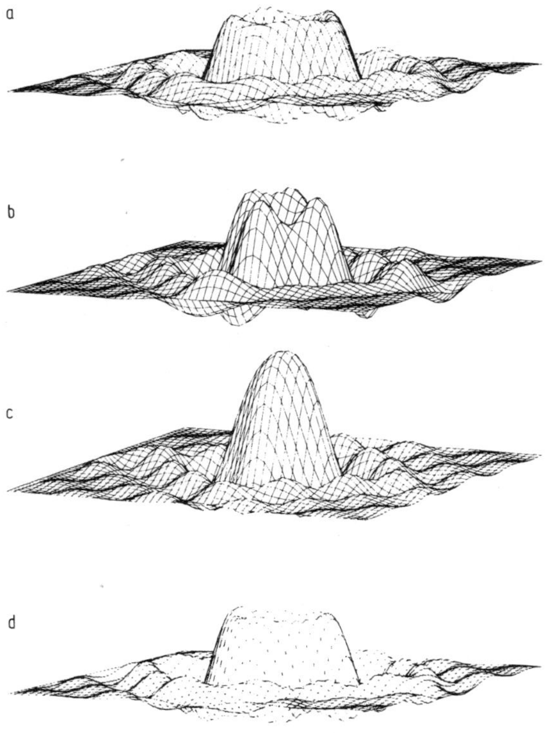


FIG. 4.2-3. Diffraction patterns similar to Fig. 4.2-2 but for an array of $64 \times 64$ hydrophones and an aperture $A = 64d$. The focusing distance equals $R = 5A = 320d$. (a) $R_0 = 6.5A$, (b) $R_0 = 6A$, (c) $R_0 = 4.5A$, (d) $R_0 = 4A$. (Courtesy N. Mohamed, Government of Kuwait.)

while the imaged point is actually at the distance $R_0 = 6.5A$, $6A$, $4.5A$, or $4A$.

The comparison of Figs. 4.2-2 and 4.2-3 shows that the reduction of the relative focusing distance $R/A$ from 10 to 5 makes the patterns more sensitive to an incorrect distance $R$. This sensitivity is well known in photography as the reduction of the depth of field with decreasing distance of the object.

The depth of field is generally defined as the deviation of the object distance $R_0$ from the focusing distance $R$ that can be tolerated without making the image of a point at $R_0$ appear objectionably blurred (Kingslake, 1967; Levi, 1968; Valasek, 1969). The more specific definition varies according to the intended use. In optics one generally uses a definition based on geometric optics. The image of a point *out of* focus is a circular disk according to geometric optics. The image of the same point *in* focus yields a diffraction pattern according to wave optics. The point is within the depth of field if the radius of the disk is no larger than the distance of the first zero crossing of the diffraction pattern from its center. This definition is not suited for acoustic imaging. However, it defines the depth of field essentially by requiring that the diffraction pattern of the out-of-focus image should not be more than twice as wide as the diffraction pattern of the focused image. This definition is usable in acoustic imaging.

Figures 4.2-2 and 4.2-3 show that the diffraction patterns are not rotationally symmetric and that they depend slightly on the number of hydrophones. Hence, we will consider the depth of field only for one dimension and a large number of hydrophones. Substitution of $R_0$ for $R$ in $\psi(i, R, \beta)$ yields the following equations from Eq. (4.1-25):

$$\begin{aligned} v(R, R_0, \gamma, \beta) &= [A^2(R, R_0, \gamma, \beta) + B^2(R, R_0, \gamma, \beta)]^{1/2} \\ A(R, R_0, \gamma, \beta) &= \sum_i \cos[\chi(i, R, \gamma) - \psi(i, R_0, \beta)] \\ B(R, R_0, \gamma, \beta) &= \sum_i \sin[\chi(i, R, \gamma) - \psi(i, R_0, \beta)] \end{aligned} \tag{1}$$

For $R = R_0$ and $\gamma = 0$, we obtain a diffraction pattern $v(R, R, 0, \beta) = v(R, 0, \beta)$ similar to the one shown by the solid line in Fig. 2.1-3. The distance between the first two zero crossings equals $2\lambda/A$. For a different value of $R_0 \neq R$, we obtain a pattern $v(R, R_0, 0, \beta)$ with a larger distance between the zero crossings. The two values $R_0 = R_{max} > R$ and $R_0 = R_{min} < R$, for which the distance between the zero crossings is doubled to $4\lambda/A$, are of interest. The difference $R - R_{min} = \Delta R_F$ is the *front depth of field*, and the difference $R_{max} - R = \Delta R_B$ is the *back depth of field*. The sum $\Delta R_F + \Delta R_B$ is the *depth of field*.

The only practical way to obtain $R_{max}$ and $R_{min}$ is to let a computer determine $v(R, R_0, 0, \beta)$ from Eq. (1) for various values of $R_0$ and check for which values the distance between the first zero crossings is just twice that of $v(R, R, 0, \beta)$. The values so obtained for $\Delta R_F$ and $\Delta R_B$, normalized by

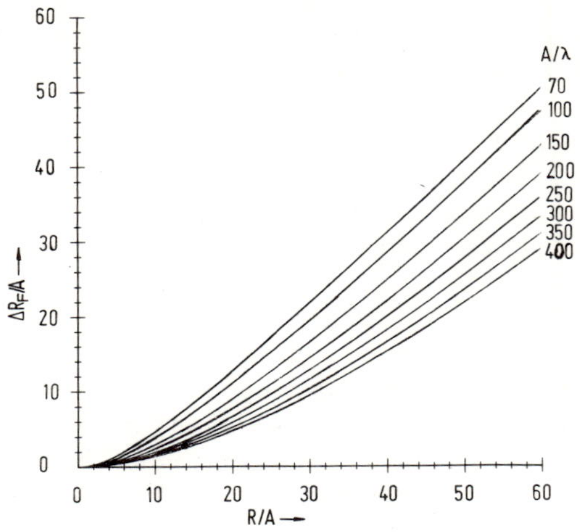


FIG. 4.2-4. Normalized front depth of field $\Delta R_F/A$ as function of the normalized focusing distance $R/A$ for various values of $A/\lambda$. $A$ = aperture, $\lambda$ = wavelength. (Courtesy N. Mohamed, Government of Kuwait.)

the aperture $A$, are plotted versus $R/A$ for various values $\lambda/A$ in Figs. 4.2-4 and 4.2-5. One may readily see that the back depth of field accounts for most of the depth of field and that it increases much more rapidly with $R/A$ than the front depth of field.

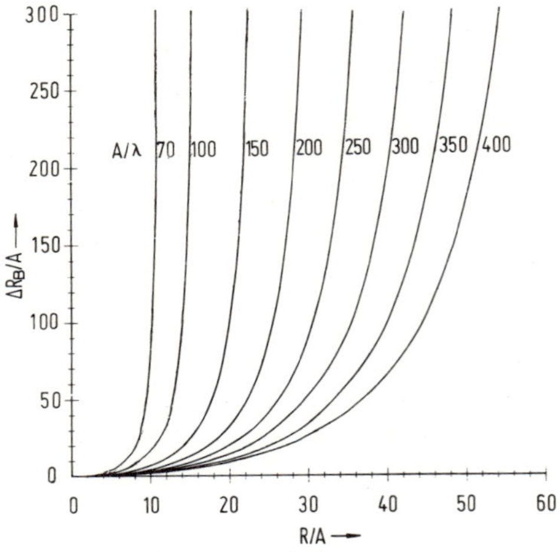


FIG. 4.2-5. Normalized back depth of field $\Delta R_B/A$ as function of the normalized focusing distance $R/A$ for various values of $A/\lambda$. $A$ = aperture, $\lambda$ = wavelength. (Courtesy N. Mohamed, Government of Kuwait.)

## 4.3 Focusing Circuits

In this section we discuss circuits that implement the equations derived in Section 4.1. Consider Eq. (4.1-9). It requires the multiplication of two time variable voltages $v^*(i, R, \beta, t)$ and $v(i, R, \beta, t)$ with two constants $\sin(\pi i^2 d^2/R\lambda)$ and $\cos(\pi i^2 d^2/R\lambda)$. The constants have to be computed every time a new focusing distance $R$ is required. This computation is most readily done by a microprocessor, and the constants are then available as binary numbers. Hence, we need a multiplier that can multiply a rapidly varying voltage with a binary number. The operation of such an analog-digital multiplier will be explained with reference to Fig. 4.3-1.

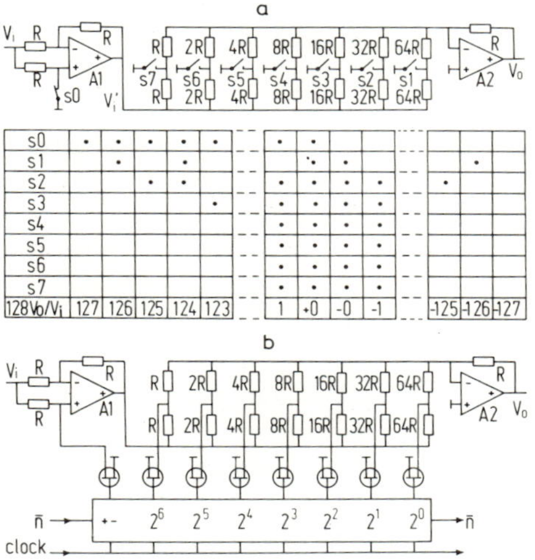


FIG. 4.3-1. Principle of a multiplier with one input digital and the other analog, producing an analog output voltage (a). A practical circuit is shown by (b).

A voltage $V_i$ applied to the input terminal of this circuit will produce the voltage $V_i' = -V_i$ at the output of the operational amplifier A1 if switch s0 is closed, and the voltage $V_i' = +V_i$ if s0 is open. This relationship may be verified by observing that there can be no voltage difference between the reversing and the nonreversing input terminals of an operational amplifier. The network resistors and switches connected to A1 permits a further multiplication by 0, $\frac{1}{128}$, $\frac{2}{128}$, ..., $\frac{127}{128}$. Consider the case shown with all switches s1–s7 open. The following current flows to the reversing input terminal of the operational amplifier A2:

$$i = V_i'\left(\frac{1}{2R} + \frac{1}{4R} + \frac{1}{8R} + \frac{1}{16R} + \frac{1}{32R} + \frac{1}{64R} + \frac{1}{128R}\right) = \frac{127}{128}\frac{V_i'}{R}$$

The output voltage $V_o$ must have the following value to drive an equal current through the resistor $R$:

$$\frac{V_o}{R} = -\frac{127}{128}\frac{V_i'}{R} = \frac{127}{128}\frac{V_i}{R}$$

The value $128 V_o/V_i = 127$ is shown in the second column of the table in Fig. 4.3-1a. The reversal of the sign $V_i' = -V_i$ is due to the closed switch s0.

Let now s0 and s1 be closed. The current flowing to amplifier A2 and the output voltage $V_o$ become

$$i = \frac{V_i'}{R}\left(\frac{1}{2} + \frac{1}{4} + \frac{1}{8} + \frac{1}{16} + \frac{1}{32} + \frac{1}{64}\right) = \frac{126}{128}\frac{V_i'}{R}$$

$$V_o = -\frac{126}{128} V_i' = \frac{126}{128} V_i$$

This value is shown in the third column of the table in Fig. 4.3-1a.

A practical circuit is shown in Fig. 4.3-1b. The switches are implemented by field effect transistors. A shift register with serial input and parallel output supplies a positive voltage to a transistor if the respective digit is 1, and no voltage if the digit is 0. A positive sign of the binary number is represented by 1, a negative sign by 0. The positive voltages make the transistors conducting. Hence, one does not supply the numbers $n$, representing $\cos(\pi i^2 d^2/R\lambda)$ and $\sin(\pi i^2 d^2/R\lambda)$, to the shift register but the complementary numbers $\bar{n}$ ($\overline{1010} = 0101$).

Networks of resistors and switches as shown in Fig. 4.3-1b are commercially available as digital–analog converters. The implementation of the circuit of Fig. 4.3-1 by commercial components is shown by the two multipliers in Fig. 4.3-2. The amplifier A1 in Fig. 4.3-1 is represented by either the amplifier A1 or A2. The switch s0 in Fig. 4.3-1a is replaced by the analog switches[1] AS1 and AS2, and the network of switches and resistors by the digital–analog converters DAC1 and DAC2. Note that the multipliers are four-quadrant multipliers; the time-variable voltages as well as the digital numbers may be either positive or negative.

The amplifier A0 in Fig. 4.3-2 performs the integration of Eq. (4.1-7) if one chooses $(R1 + P1)C1 = \lambda/2\pi v$. The multiplications of $v^*(i, R, \beta, t)$ by $\sin(\pi i^2 d^2/R\lambda)$ and of $v(i, R, \beta, t)$ by $\cos(\pi i^2 d^2/R\lambda)$ is performed by the two multipliers, and the summation of Eq. (4.1-9) is done by the amplifier A3. The reversal of sign of the output voltage compared with Eq. (4.1-9) is of no consequence as long as the voltage for every value of $i$ is reversed.

[1] AH0141 (National Semiconductor Corp.), DAC-IC8BC (Datel Systems, Inc.), SN74164 (Texas Instruments, Inc.).

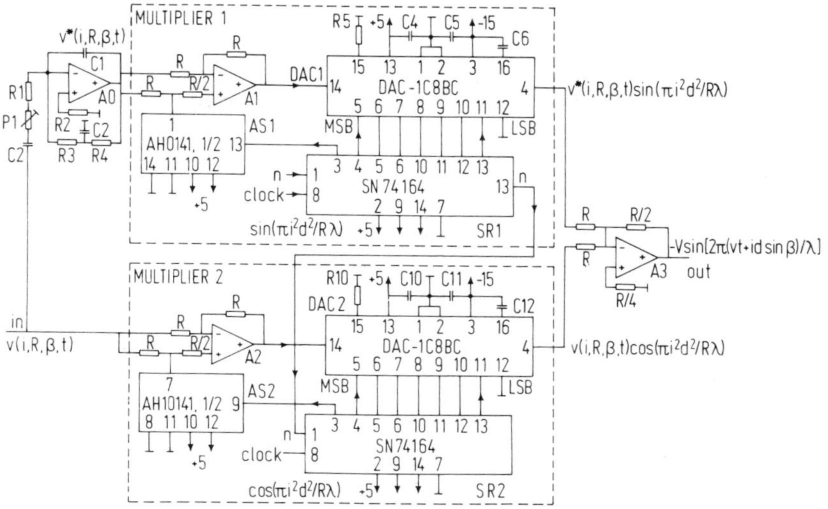


FIG. 4.3-2. Circuit implementing the integration of Eq. (4.1-7) and the multiplications with summation of Eq. (4.1-9). AS, analog switch; DAC, digital–analog converter; SR, series-parallel shift register. Note that the number $n$ rather than its complement $\bar{n}$ as in Fig. 4.3-1 is used due to the construction of the particular digital–analog converter used.

Since $i^2$ appears in $\sin(\pi i^2 d^2/R\lambda)$ and $\cos(\pi i^2 d^2/R\lambda)$ one may combine the circuits for $+i$ and $-i$ to save components. This is shown in Fig. 4.3-3. The notation $+i$ and $-i$ is replaced by $+p$ and $-p$ in order to conform with the notation used for focusing of a two-dimensional array in Figs. 4.1-5 and 4.1-6. Two shift registers are saved compared with doubling the circuit of Fig. 4.3-2, which is not much unless one considers that hundreds to tens of thousands of hydrophones have to be focused. In addition to the saving of components, one also reduces the time required for loading of the constants $\sin(\pi i^2 d^2/R\lambda)$ and $\cos(\pi i^2 d^2/R\lambda)$ to one-half.

In order to implement focusing for an array of $16 \times 16$ hydrophones according to Figs. 4.1-5 and 4.1-6, one has to assemble eight circuits according to Fig. 4.3-3 on a printed circuit card as shown in Fig. 4.3-4. The important point of this illustration is that the string of binary digits $n$, representing $\sin[\pi(p^2 + s^2)d^2/R\lambda]$ and $\cos[\pi(p^2 + s^2)d^2/R\lambda]$ is fed serially through all eight circuits and comes out of the last circuit denoted FOC7.

Sixteen cards according to Fig. 4.3-4 are required. They are shown for $k = 0, 1, \ldots, 15$ in Fig. 4.3-5. Overlayed are the $16 \times 16$ hydrophones of Fig. 4.1-5 with the notation $s = -7, \ldots, +7$ and $p = -7, \ldots, +7$. The numbers $n$ are fed in parallel to the cards $k = 8, 9, \ldots, 15$ and $k = 7, 6, \ldots, 0$ as shown, due to the symmetry between the left and right half plane in Fig. 4.1-5. The numbers $n$ are fed serially through the focusing

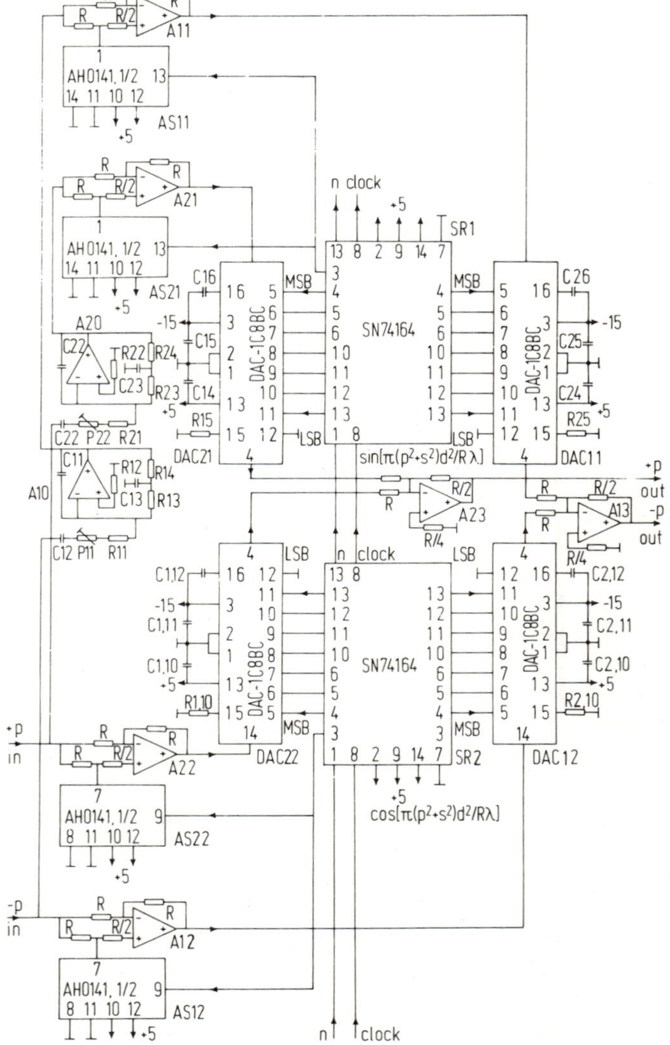


FIG. 4.3-3. Circuit according to Fig. 4.3-2 for pairs of input voltages $v(p, R, \beta, t)$ and $v(-p, R, \beta, t)$ that require multiplication by the same constants $v_s(s, p)$ and $v_c(s, p)$ according to Eqs. (4.1-14) and (4.1-15).

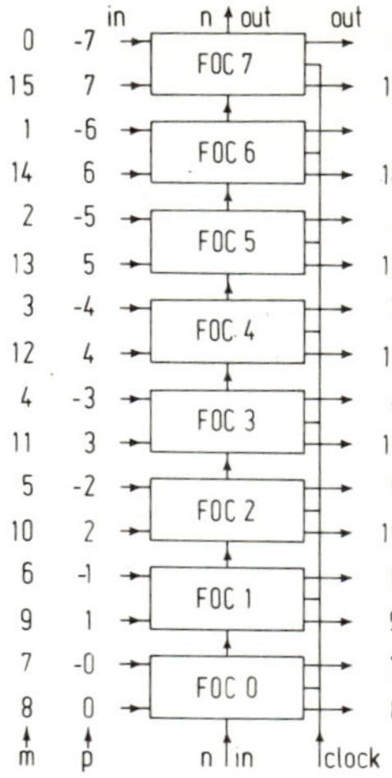


FIG. 4.3-4. Arrangement of 8 circuits according to Fig. 4.3-3 on one printed circuit card to permit focusing of 16 voltages.

circuits for the hydrophones $|p| = 0, 1, \ldots, 7$ on the card $k = 8$, then through the circuits for the hydrophones $|p| = 0, 1, \ldots, 7$ on the card $k = 9$, etc. The result is that the first computed number $n$ must be $\sin[\pi(7^2 + 7^2)d^2/R\lambda]$, the second $\cos[\pi(7^2 + 7^2)d^2/R\lambda]$, the third $\sin[\pi(6^2 + 7^2)d^2/R\lambda]$, the fourth $\cos[\pi(6^2 + 7^2)d^2/R\lambda]$, and so on until $\sin[\pi(0^2 + 0^2)d^2/R\lambda]$ and $\cos[\pi(0^2 + 0^2)d^2/R\lambda]$ are reached.

The numbers $n$ are best produced by a microprocessor.[1] Only the numbers required by Fig. 4.1-6 are needed to focus all the hydrophones in Fig. 4.1-5. The following programming instructions have to be carried out:

1. Permanently store $s' = 7$, $p' = 7$, $d$, $\lambda$.
2. Key in focusing distance $R$ and start program.
3. Compute $\xi = \pi(p^2 + s^2)d^2/R\lambda$ for $p = p'$, $s = s'$.
4. Compute $\sin \xi$.
5. Compute $\cos \xi$.
6. Add 1 to $\sin \xi$ to shift the range from $-1 \leq \sin \xi \leq +1$ to $0 \leq 1 + \sin \xi \leq +2$.
7. Add 1 to $\cos \xi$.
8. Convert $1 + \sin \xi$ and $1 + \cos \xi$ to binary numbers if they are not already represented by binary strings.

[1] A pocket calculator SR52 with a printer PC-100A, both made by Texas Instruments, Inc., were actually used. An interface card was needed since the numbers $n$ are not delivered as a binary string to the output terminals of this calculator. Such details are ignored here due to the rapidly changing state of the art of microprocessors.

9. Shift $n$ serially to the focusing cards ($n$ in Fig. 4.3-3 contains eight digits; hence, eight clock pulses are required for each number $n$).
10. Decrement $p$ by 1 and repeat steps 3 to 9.
11. If $p$ is decremented to less than zero, restore $p' = 7$, decrement $s$ by 1, repeat steps 3 to 10.
12. If $s$ is decremented to less than zero, stop program and return to step 2.

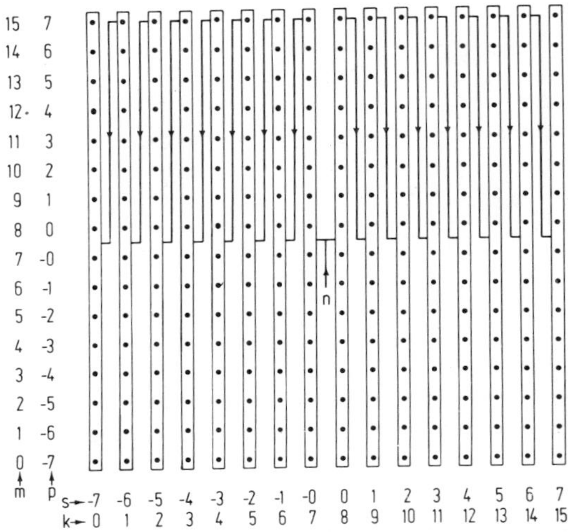


FIG. 4.3-5. Arrangement of 16 printed circuit cards according to Fig. 4.3-4 in two dimensions to permit focusing of 16 × 16 voltages. Note the serial feed of the focusing constants represented by $n$.

This completes the discussion of focusing circuits based on the phase shifting method. A printed circuit card for focusing of 2 × 8 hydrophones with eight circuits essentially equal to the one in Fig. 4.3-3 is shown in Fig. 4.3-6.

We turn to the implementation of the second method of focusing, based on frequency conversion and represented by Eqs. (4.1-16)–(4.1-18). Figure 4.3-7 shows an appropriate circuit for focusing 16 input voltages. The products $v(i, R, \beta, t)e(i, t)$ are produced by the multipliers MOD81, MOD82, ..., MOD01. These multipliers can actually be ring modulators, which explains the choice of the notation MOD. The output of these multipliers is theoretically the product $v(i, R, \beta, t)e(i, t)$, but practically one can design the multipliers so that the terms with the sum frequency $f + f_1$ are suppressed and only the terms $v'(i, R, \beta, t)$ with the difference frequency $f' = f - f_1$ are obtained.

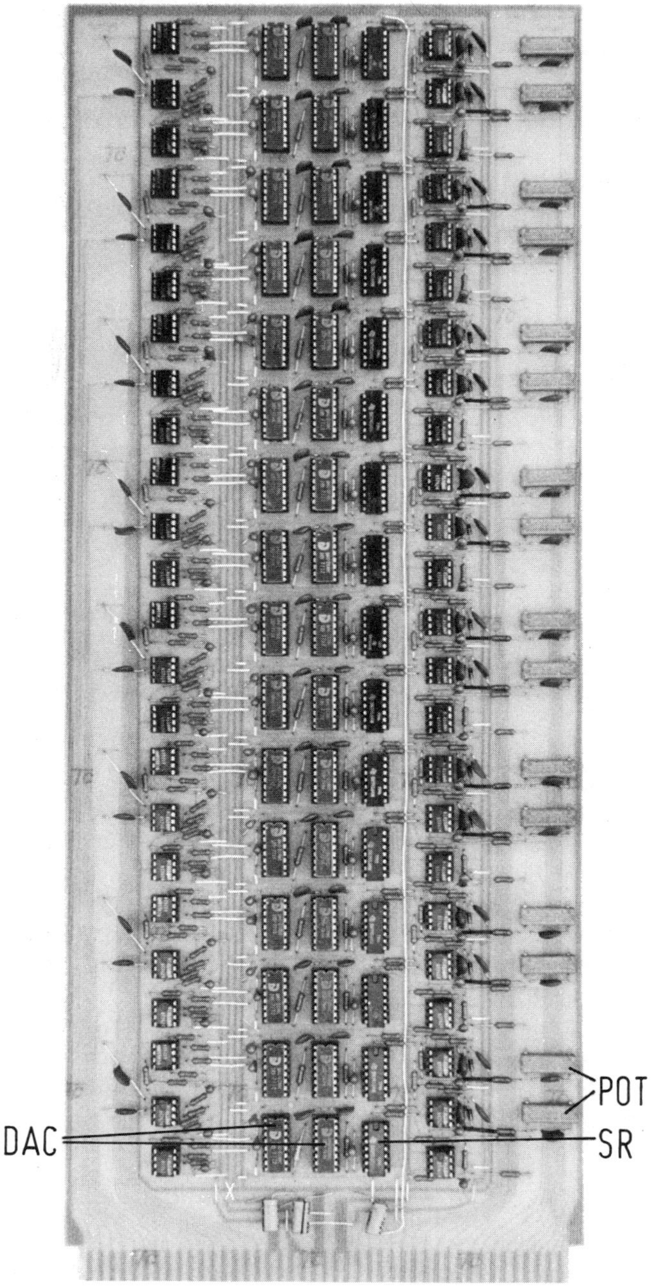


FIG. 4.3-6. Printed circuit card for focusing according to Fig. 4.3-4. SR, shift register; DAC, digital-analog converter; POT, potentiometers P11 and P21 in Fig. 4.3-3.

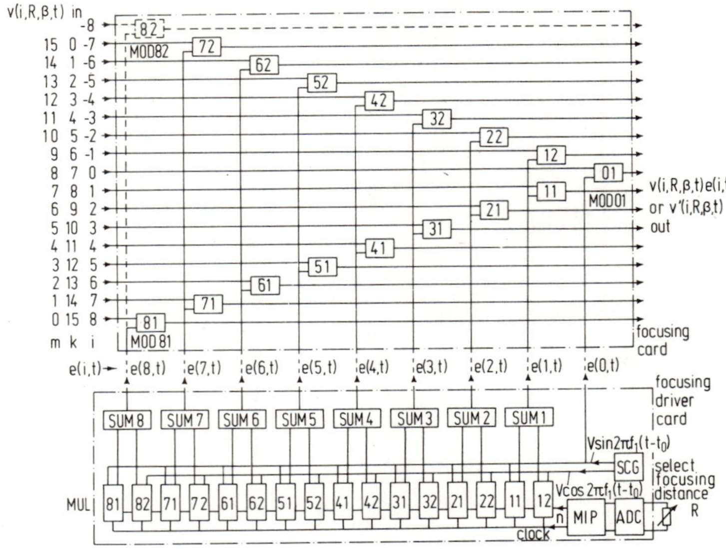

FIG. 4.3-7. Circuit for focusing 16 voltages. One needs 16 focusing cards for an array of 16 × 16 hydrophones, but only one focusing driver card. MOD, multipliers (ring modulators); MUL, multipliers (one input digital, the other input and the output analog); SUM, summing circuit; SCG, generator for a sinusoidal and cosinusoidal voltages; MIP, microprocessor; ADC, analog-to-digital converter.

Let us turn to the generation of the voltages $e(i, t)$ according to Eq. (4.1-17). We need the time-variable voltages $\sin 2\pi f_1(t - t_0)$ and $\cos 2\pi f_1(t - t_0)$, which are produced by the sinusoidal generator SCG in Fig. 4.3-7. The frequency $f_1$ of the voltages produced will not exceed 100 kHz, and thus there is no need to discuss this generator any further. The terms $\cos(\pi i^2 d^2/R\lambda)$ and $\sin(\pi i^2 d^2/R\lambda)$ are time invariant. The only variable in them is the distance $R$, which is only changed when the focusing distance is changed. We can produce a voltage representing $R$ by means of a potentiometer. This voltage is fed in Fig. 4.3-7 via an analog-to-digital converter ADC to a microprocessor MIP, which produces a stream of binary digits $n$ that represent $\cos(\pi i^2 d^2/R\lambda)$ and $\sin(\pi i^2 d^2/R\lambda)$ for $i = 1, \ldots, 8$ and the constants $\pi$, $d^2$, $\lambda$.

The products

$$\cos(\pi i^2 d^2/R\lambda) \sin 2\pi f_1(t - t_0) \qquad \text{and} \qquad \sin(\pi i^2 d^2/R\lambda) \cos 2\pi f_1(t - t_0)$$

of Eq. (4.1-17) are produced in the multipliers MUL12–MUL81 in Fig. 4.3-7, and the difference of the products is produced in the summing circuits SUM1–SUM8. Each summing circuit contains one operational amplifier with differential input and four resistors; these circuits require no further discussion. The multipliers have to multiply a digital number representing $\cos(\pi i^2 d^2/R\lambda)$ or $\sin(\pi i^2 d^2/R\lambda)$ with an analog voltage $V \sin 2\pi f_1(t - t_0)$ or $V \cos 2\pi f_1(t - t_0)$. Appropriate circuits are shown in Figs. 4.3-1 to 4.3-3.

We now turn to the design of circuits for focusing without approximation as represented by Eqs. (4.1-19)–(4.1–23). We need, according to Eq. (4.1-22), binary numbers representing $\cos[\chi(i, R, \gamma)]$ and $\sin[\chi(i, R, \gamma)]$, where $\chi(i, R, \gamma)$ is defined by Eq. (4.1-19). The only variable is the distance $R$. Just as in Fig. 4.3-7, one can produce a voltage proportionate to $R$ by means of a potentiometer and send it via an analog-to-digital converter to a microprocessor. The constants $\lambda$, $\gamma$, $i$, and $d$ are stored in the processor. The processor can thus compute $\chi(i, R, \gamma)$ and then $\cos[\chi(i, R, \gamma)]$ and $\sin[\chi(i, R, \gamma)]$. The computation is more complicated than that of $\cos(\pi i^2 d^2/R\lambda)$ and $\sin(\pi i^2 d^2/R\lambda)$, but it does not exceed the capability of the better programmable pocket calculators.

Next we have to produce the phase-shifted voltage $v^*(i, R, \beta, t)$ of Eq. (4.1-21) from the voltage $v(i, R, \beta, t)$ of Eq. (4.1-5). One can do so by integration:

$$(V/RC) \int \sin 2\pi[vt/\lambda - \psi(i, R, \beta)]\, dt$$

$$= V(\lambda/2\pi RCv) \cos 2\pi[vt/\lambda - \psi(i, R, \beta)]$$

The time constant $RC$ of the integrator may be chosen $RC = \lambda/2\pi v$ to yield $v^*(i, R, \beta, t)$ after integration. However, the circuit would then work for a particular wavelength $\lambda$ only and fail, e.g., if a Doppler shift changed the wavelength. Let $\lambda$ be changed to $\lambda/q$, where $q$ shall be close to 1, say $0.9 < q < 1.1$. The integrator will then still produce the correct phase shift but the amplitude will be wrong. We will get $q^{-1}v^*(i, R, \beta, t)$ and a multiplication by $q$ is required to obtain $v^*(i, R, \beta, t)$. This multiplication is done in Fig. 4.3-8 by multipliers MUL-7 to MUL8, which follow the integrators INT-7 to INT8 on the left.

A practical circuit for these multipliers based on the circuit of Fig. 4.3-1 is shown in Fig. 4.3-9. The factor $q$ may have values between $(9 - 64/64)/9 = 0.89$ and $(9 + 63/64)/9 = 1.095$ in steps $1/64 = 0.0156$. The network of resistors and switches may be replaced by digital–analog converters of the type shown in Fig. 4.3-2.

We have now $v(i, R, \beta, t)$, $v^*(i, R, \beta, t)$, $\cos[\chi(i, R, \gamma)]$, and $\sin[\chi(i, R, \gamma)]$. The products $v(i, R, \beta, t)\cos[\chi(i, R, \gamma)]$ and $v^*(i, R, \beta, t)\sin[\chi(i, R, \gamma)]$ required for Eq. (4.1-22) are produced in the $16 \times 32 - 2 = 510$ multipliers MUL$i$, $l\varepsilon$,c and MUL$i$, $l\varepsilon$,s in Fig. 4.3-8. The letter $i$ indicates the row, and the letters $l\varepsilon$ the column in which the multiplier is found; the letters s and

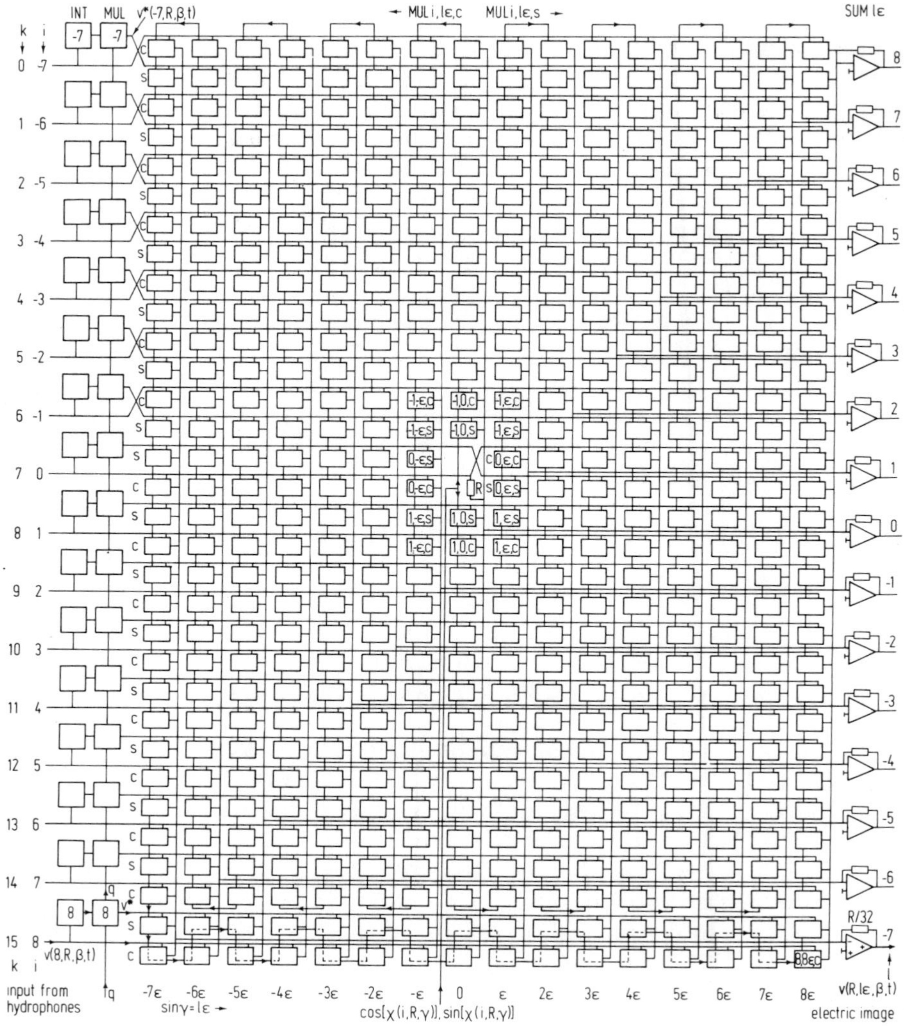


FIG. 4.3-8. Circuit for focusing and image generation without approximation according to Fig. 4.1-1. INT, integrator; MUL, multiplier; SUM, summing circuit.

c are required since there are two multipliers for $v(i, R, \beta, t) = V \sin 2\pi[vt/\lambda - \psi(i, R, \beta)]$ and for $v^*(i, R, \beta, t) = V \cos 2\pi[vt/\lambda - \psi(i, R, \gamma)]$. The multipliers are of the type shown in Fig. 4.3-1, but a resistor $R$ shall be connected to the output terminal so that the voltage $V_0$ is driving a current $V_0/R$ through this resistor. No multiplier MUL0,0,s is required since $\chi(i, R, \gamma)$ equals 0

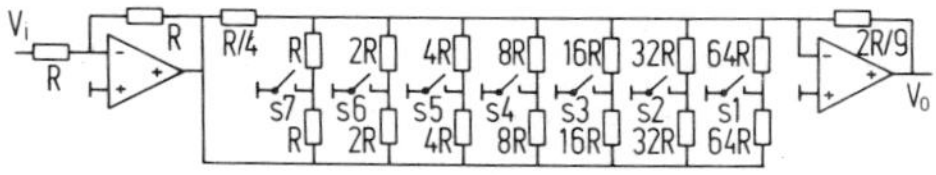


| | | | | | | | | | |
|---|---|---|---|---|---|---|---|---|---|
| s1 | • | | – | • | | • | – | • | • |
| s2 | • | • | – | | | • | – | | |
| s3 | • | • | – | | | • | – | | |
| s4 | • | • | – | | | • | – | | |
| s5 | • | • | – | | | • | – | | |
| s6 | • | • | – | | | • | – | | |
| s7 | • | • | – | • | • | | – | | |
| $9V_0/V_i$ | 9-64/64 | 9-63/64 | – | 9-1/64 | 9 | 9+1/64 | – | 9+62/64 | 9+63/64 |

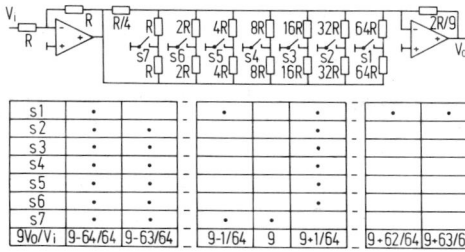

FIG. 4.3-9. Principle of a circuit for the multipliers MUL-7 to MUL8 in Fig. 4.3-8.

for $i = 0$ and $l\varepsilon = 0$, which implies $\gamma = 0$ and $\sin \gamma = 0$. The multiplier MUL0,0,c is not needed because $\cos \chi$ equals 1. The digital numbers representing $\cos[\chi(i, R, \gamma)]$ and $\sin[\chi(i, R, \gamma)]$ are fed as a serial stream of digits to the multipliers MUL1,0,s and MUL-1,0,s, and they advance through the two chains of multipliers as indicated by arrows to MUL8,8ε,s and MUL7,8ε,c.

The output voltages $v(i, R, \beta, t) \cos[\chi(i, R, \gamma)]$ and $v^*(i, R, \beta, t) \sin[\chi(i, R, \gamma)]$ are summed for $i = -7, \ldots, +8$ by the summers SUM$l\varepsilon$. This means that the two summations of Eqs. (4.1-22) and (4.1-23) are produced in one step. Note that this summation of 32 voltages in one step is done to simplify the circuit diagram of Fig. 4.3-8; it is for the same reason that the summing resistors were assumed to be included in the multipliers. Good engineering practice would call for the summation of $4 \times 8$ voltages in a first step and of the resulting 4 voltages in a second step; furthermore, the summing resistors must be mounted with the summing amplifiers rather than the multipliers to reduce cross talk.

The circuit of Fig. 4.3-8 replaces the two circuits of Figs. 3.4-2 and 4.3-4 or 4.3-7. Nevertheless, this circuit is much more expensive than the two it replaces. Even though it provides superior focusing it is not likely to be used without a better reason. There is such a better reason. The hydrophones do not have to be mounted along a straight line as in Fig. 4.1-1 or on a plane as in Fig. 4.1-4, but may be mounted on a curved surface such as the bow of a ship as long as the curvature is not too violent. There is no need to discuss the advantage of not having to push a large flat hydrophone array through the water. One can readily replace the straight line array in Fig. 4.1-1 by hydrophones located on some curve. The distance between a point $W\beta$ and a hydrophone $i$ will have a value different from the one given in Eq. (4.1-2), and in general one cannot express it by a formula but by numerical values only. The numerical values of $\chi(i, R, \gamma)$ are harder to calculate, but the circuit of Fig. 4.3-8 is not affected. There is also no need to have a plane of focus in Fig. 4.1-1. One can focus on a curved surface and mount the hydrophones on a curved surface without creating more problems than a more complicated computation of $\chi(i, R, \gamma)$. The only thing to watch out for is that the two-dimensional transform can be separated into two one-dimensional transforms for simplicity, but the separation does not have to be done in cartesian coordinates.

In Section 2.3 we had investigated image generation by means of sampled storage circuits, and we will briefly discuss focusing for this method. The delay time due to the curvature of a spherical wave for the arrival at a hydrophone $s$, $p$ in Fig. 4.1-6 is defined by Eq. (4.1-11):

$$t_{s,p} = (s^2 + p^2)d^2/2Rv$$

These delay times have to be added to the times at which the switches of the two-dimensional sampling filter in Fig. 2.3-4 are closed. The sampling filter and the storage and distribution circuit in Fig. 2.3-4 remain thus unchanged, but the timing circuits become more complicated.

# 5 Practical Equipment

## 5.1 Principles of Array Design

The observation angle $\gamma$ of Eq. (2.2-6) may be written in the form

$$\sin\gamma = l\lambda/A, \qquad l = 0, \pm 1, \pm 2, \ldots \tag{1}$$

The parameter $l$ may assume $2^n$ values if there are $2^n \times 2^n$ receptors in a square array. A typical choice for $l$ is $-2^n/2, \ldots, 0, \ldots, +2^n/2 - 1$ or $-2^n/2 + 1, \ldots, 0, \ldots, +2^n/2$. The *viewing angle* or *field of view angle* is defined by inserting the smallest and largest value of $l$ into Eq. (1):

$$\begin{aligned} \sin\alpha_- &= -2^n\lambda/2A & \text{or} \quad \sin\alpha_- &= (-2^n/2 + 1)\lambda/A \\ \sin\alpha_+ &= (2^n/2 - 1)\lambda/A & \text{or} \quad \sin\alpha_+ &= 2^n\lambda/2A \end{aligned} \tag{2}$$

For large values of $n$ one does not need to distinguish between $2^n/2$ and $2^n/2 - 1$. One obtains the following simpler formula:

$$\sin\frac{\alpha}{2} = 2^n\lambda/2A = \lambda/2d$$

$$A = 2^n d, \qquad \alpha \doteqdot 2\alpha_- \doteqdot 2\alpha_+ = \text{viewing angle} \tag{3}$$

The *field of view* at a distance $L$ from the array equals $D \times D$, where $D$ is defined by the equation

$$D = 2L\tan(\alpha/2) \tag{4}$$

The distance of resolvable points at the distance $L$ is defined by $D/2^n$. Let us point out that one could space the curves in Fig. 2.1-3 differently. For instance, the maximum of the curve $\gamma = \lambda/A$ could be shifted from the zero crossing of the curve $\gamma = 0$ at $\beta = \lambda/A$ to the minimum of the curve $\gamma = 0$ further to the right. One would obtain different values for the viewing angle $\alpha$ and the field of view $D \times D$. The spacing used in Fig. 2.1-3 has the advantage that all curves are zero when one has its maximum.

Table 5.1-1 shows some representative values derived from Eqs. (3) and (4) for acoustic imaging underwater. The three operating frequencies of 10 kHz, 100 kHz, and 1 MHz are assumed. The corresponding wavelengths for sound in water are 15 cm, 1.5 cm, and 1.5 mm. Furthermore, hydrophone arrays with $16 \times 16$, $64 \times 64$, and $256 \times 256$ hydrophones are assumed.

TABLE 5.1-1

REPRESENTATIVE FIGURES FOR ACOUSTIC IMAGING IN WATER BY MEANS OF TWO-DIMENSIONAL ELECTRIC FILTERS

| | | | | | | | | | |
|---|---|---|---|---|---|---|---|---|---|
| Frequency of insonifying wave | | 10 kHz | | | 100 kHz | | | 1 MHz | |
| Wavelength | | 15 cm | | | 1.5 cm | | | 1.5 mm | |
| Hydrophone array | 16 × 16 | 64 × 64 | 256 × 256 | 16 × 16 | 64 × 64 | 256 × 256 | 16 × 16 | 64 × 64 | 256 × 256 |
| Number of hydrophones | 256 | 4096 | 65,536 | 256 | 4096 | 65,536 | 256 | 4096 | 65,536 |
| $\alpha = 30°$ | | | | | | | | | |
| $d = \lambda/2 \sin(\alpha/2)$ | | 29 cm | | | 2.9 cm | | | 2.9 mm | |
| Size of hydrophone array | $4.6^2$ m² | $18.5^2$ m² | $74^2$ m² | $0.46^2$ m² | $1.85^2$ m² | $7.4^2$ m² | $4.6^2$ cm² | $18.5^2$ cm² | $74^2$ cm² |
| $D = 2L \tan(\alpha/2)$ | | | | | | | | | |
| $D \times D$ = field of view at $L = 200$ m | | | | | $106^2$ m² | | | | |
| Distance of resolvable points at $L = 200$ m | 6.6 m | 1.66 m | 41.4 cm | 6.6 m | 1.66 m | 41.4 cm | 6.6 m | 1.66 m | 41.4 cm |
| Field of view for $L = 20$ m | | | | | $10.6^2$ m² | | | | |

| | | | | | | | | | |
|---|---|---|---|---|---|---|---|---|---|
| Distance of resolvable points at $L = 20$ m | 66 cm | — | — | 66 cm | 16.6 cm | 4.2 cm | 66 cm | 16.6 cm | 4.2 cm |
| $\alpha = 10°$ | | | | | | | | | |
| $d = \lambda/2 \sin(\alpha/2)$ | | 86 cm | | | 8.6 cm | | | 8.6 mm | |
| Size of hydrophone array $D = 2L \tan(\alpha/2)$ $D \times D$ = field of view at $L = 200$ m | $13.8^2$ m$^2$ | $55^2$ m$^2$ | $220^2$ m$^2$ | $1.4^2$ m$^2$ | $5.5^2$ m$^2$ $35^2$ m$^2$ | $22^2$ m$^2$ | $14^2$ cm$^2$ | $55^2$ cm$^2$ | $2.2^2$ m$^2$ |
| Distance of resolvable points at $L = 200$ m | 2.2 m | 55 cm | — | 2.2 m | 55 cm | 13.7 cm | 2.2 m | 55 cm | 13.7 cm |
| Field of view for $L = 20$ m | | | | | $3.5^2$ m$^2$ | | | | |
| Distance of resolvable points at $L = 20$ m | 22 cm | — | — | 22 cm | 5.5 cm | 1.37 cm | 22 cm | 5.5 cm | 1.37 cm |

The two viewing angles $\alpha = 30°$ and $\alpha = 10°$ are considered. The distance $d$ between adjacent hydrophones is defined by Eq. (3), the size of the array by $2^n d \times 2^n d$, the field of view $D \times D$ by Eq. (4), the distance of resolvable points by $D/2^n$.

Let us note that hydrophone arrays with thousands of hydrophones are well within the state of the art. The fabrication of such arrays was described by Prokhorov (1972).

For a frequency of 10 kHz one obtains dimensions that are too large for a movable array. A frequency of 1 MHz, on the other hand, implies absorption losses that are very high. The frequency of 100 kHz is representative for practical equipment. An array with a viewing angle of $\alpha = 10°$ and $16 \times 16$ hydrophones has been built. The size of $1.4 \times 1.4$ m$^2$ appears to be about the upper limit for a practical movable array. An array with $64 \times 64$ hydrophones and a viewing angle $\alpha = 10°$ calls for an array of $5.5 \times 5.5$ m$^2$ according to Table 5.1-1. This is too large. An increase of $\alpha$ to 30° yields an array size of $1.85 \times 1.85$ m$^2$. This is practical, particularly if one increases the frequency from 100 to about 150 kHz. For an array of $256 \times 256$ hydrophones one obtains the array size of $7.4 \times 7.4$ m$^2$ even for $\alpha = 30°$. This is too large. An increase of the frequency to about 750 kHz would be required to reduce the array size to a practical value, but the absorption losses[1] would increase from 0.02 dB/m at 100 kHz to 0.2 dB/m at 750 kHz. Hence, a distance of 50 m to the object is all one can expect to achieve if one wants a resolution of $256 \times 256$ points. For larger distances, one must operate at lower frequencies, and this implies fewer resolved points for a given size of the array. The synthetic aperture techniques discussed in Chapter 9 promise to overcome this limitation, but they are currently beyond our technological capabilities.

Let us turn to acoustic arrays for medical diagnosis. In optical imaging we distinguish between images produced by a telescope and by a microscope. Although it is sometimes difficult to distinguish between the two, the telescope is generally used to look at objects that are at a distance large compared with the focal length of the objective, while this is not so for the microscope. Acoustic imaging has been implemented so far for the telescope case, but medical diagnosis calls for the implementation of the microscope. For a better explanation, refer to Fig. 5.1-1.

An array with 16 hydrophones is shown on the right, spaced at distances $d$ equal to the wavelength $\lambda$. The resolution angle $\varepsilon$ equals $\lambda/16d$. If we use a focusing method that simulates the focusing by a lens, we can focus the points located on the circles of focus with radius $R$ or spheres of focus

[1] The attenuation of a sinusoidal sound wave in seawater is usually approximated by $2.75 \times 10^{-9} f^2$ dB/km for frequencies $f$ below about 100 kHz. The attenuation at higher frequencies is considerably less than suggested by this formula (Albers, 1965, p. 31).

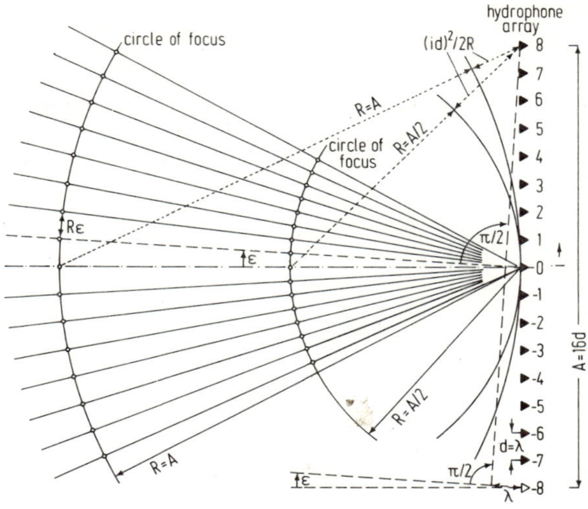


FIG. 5.1-1. Geometric relations for an acoustic array used as a microscope.

with radius $R$. Two such circles with $R = A/2$ and $R = A$ are shown in Fig. 5.1-1. For $R = \infty$ one obtains the limit case of a planar wavefront. The distance between two resolved points on the circle $R = A$ equals essentially the distance $d$ between two hydrophones; the distance between two resolved points is larger than $d$ for $R > A$ and smaller than $d$ for $R < A$. Instead of $R > A$ and $R < A$ one would obtain $R > nA$ and $R < nA$ if the distance between the hydrophones had been chosen $d = n\lambda$. Hence, in terms of array imaging, the distance between two resolved points is larger than the distance between adjacent hydrophones for a telescope and smaller for a microscope. Electronic processing permits us to replace the circles or spheres of focus by planes, which the lens cannot do, but the distinction between telescope and microscope is not affected.

Figure 5.1-2 shows how active or passive beam forming works for finite distances. The black triangles denoted $i = -7, \ldots, +8$ are sound projectors for active beam forming, and hydrophones for passive beam forming.[1] The sinusoidal oscillations produced by the 16 sound projectors are timed so that the waves are in phase at the 16 points of the circle of focus. This is shown in detail for the point $\gamma = 8\varepsilon$. If the hydrophone array were infinitely long, one would obtain a perfect concentration of the sound waves in this point. The finite size of the array makes the amplitudes of the sum

[1] The discussion is in terms of active beam forming because it makes it easier to visualize the diffraction patterns.

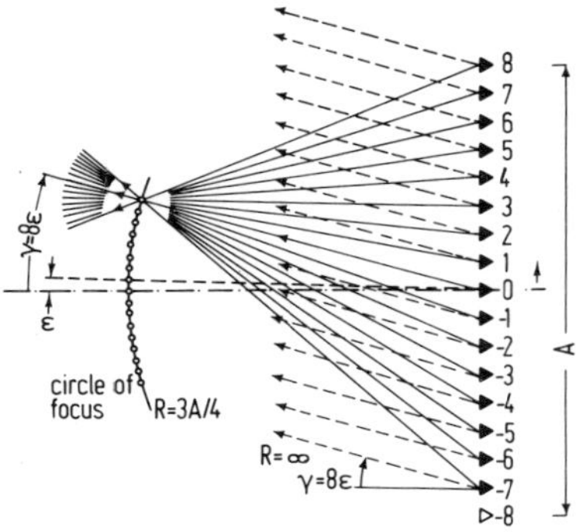


FIG. 5.1-2. Generation of an image of an area that is smaller than the hydrophone array.

of the sound waves very similar to the diffraction pattern of Fig. 4.2-1 if the projector array in Fig. 5.1-2 is two-dimensional with 16 × 16 projectors rather than one-dimensional. The rays in Fig. 5.1-2 diverge in front and behind the point $\gamma = 8\varepsilon$ on the circle of focus. One may readily infer that as a result the depth of field decreases with the distance $R$. For $R \to \infty$ the rays through the point $\gamma = 8\varepsilon$ become the parallel rays shown. The terms beam and beam forming are usually applied to the case $R = \infty$, since most radar and acoustic arrays operate over distances that are large compared with the length of the array. The development of electronic focusing circuits is only very recent, and we still have to develop a generally accepted vocabulary for it.

Figure 5.1-3 shows a possible mechanical implementation of an acoustic microscope using electronic processing. The object under observation—typically some part of the human body—is on the left. We assume that the maximum depth of penetration of interest is 150 mm, since few people have a diameter of more than 300 mm. Furthermore, we assume a hydrophone array[2] with 32 × 32 hydrophones with a spacing of 8.6 mm. An operating frequency of 1 MHz yields the wavelength $\lambda = 1.5$ mm. If we want the distance of two resolvable points in the object plane to be 1.5 mm, we obtain an observable area of $(1.5 \times 32)^2 = 48^2\ \text{mm}^2$ in the object plane. The other numerical values in Fig. 5.1-3 follow from these assumptions.

[2] Hydrophones may be used either for the reception or the radiation of sound. It is tacitly assumed here that the hydrophones serve alternately as radiators and receptors, but a more detailed discussion is postponed until Section 6.2, where the use of this double function for sidelobe reduction is shown.

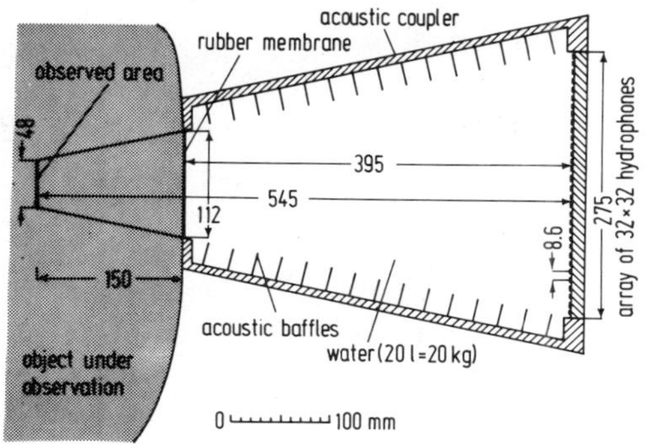


FIG. 5.1-3. Mechanical diagram showing the coupling of a hydrophone array to a human body. All distances are in millimeters.

The acoustic coupler in Fig. 5.1-3 connects the hydrophone array to the object under observation. The rubber membrane on the left insures good acoustic coupling without a layer of air. The acoustic baffles reduce returns from grating lobes and sidelobes of the array. The hydrophones are used alternately to insonify the object plane and to receive the scattered waves. The water in the acoustic coupler weighs 20 kg, which is a reasonable value.

The distance between resolvable points equals 1.5 mm, that between the centers of the hydrophones 8.6 mm. This yields an enlargement of 8.6/1.5 = 5.7. Note that the enlargement is important for the practical construction of the hydrophone array, even though it means theoretically no more than an enlargement—as contrasted to resolution—in optics.

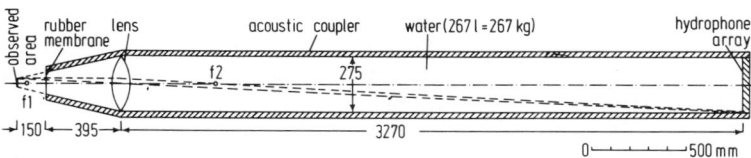

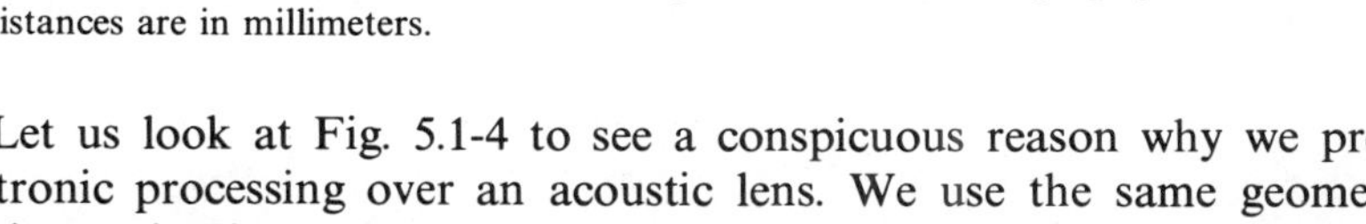


FIG. 5.1-4. Representative mechanical diagram for acoustic imaging by means of a lens. All distances are in millimeters.

Let us look at Fig. 5.1-4 to see a conspicuous reason why we prefer electronic processing over an acoustic lens. We use the same geometric relations as in Fig. 5.1-3, but replace the hydrophone array with a rectangular lens of the same size. Furthermore, we round the enlargement from 5.7 to 6. As a result, we need a square tube with cross section $275^2$ $mm^2$ and a length of 3.27 m to couple the lens to the hydrophones on the right. The coupling between the lens and the object is done by the same acoustic coupler as in Fig. 5.1-3. One now needs 267 liters of water for the coupling, which means a weight of 267 kg. Considering that the container for such a load must be mechanically strong and thus heavy, one must allow a total weight of at least half a ton for the construction of Fig. 5.1-4, which is about 20 times as much for Fig. 5.1-3. Size and weight favor electronic processing, even though one can improve the design of Fig. 5.1-4, e.g., by using a system of lenses.

One may readily infer that the example given here does *not* exaggerate the problems caused by the distance required between the lens and the hydrophone array. The barrel of a microscope or a telescope is always long compared with the diameter of the lenses. A camera without a highly corrected lens system needs a ratio (distance between lens and film)/(diameter of lens) of 8 or more. In optics, this is usually no problem since the wavelength of light is so short and we operate in air; only the large astronomical telescopes have the size of acoustic imaging systems using lenses, and they offer comparable mechanical problems.

Consider the required flatness of an array. In principle, the hydrophones must not deviate by more than a fraction of the wavelength from the nominal reception plane, or a phase error will be introduced. A phase error of 1° translates into a distance of $15/360 = 0.042$ mm at 100 kHz, and into 4.2 $\mu$m at 1 MHz. The actual required flatness is not as perfect as suggested by these figures if the hydrophones deviate randomly in both directions from the nominal plane.

A major concern in the past were deviations from the flatness caused by a curvature of the array rather than local random deviations. The development of electronic focusing brought a welcome change. Consider an array that has a slightly spherical curvature.[1] A reflection of Figs. 4.1-2 and 4.1-3 shows that a spherical curvature of the array can be corrected by the same focusing circuits that change a spherical wavefront to a planar one.

The general case for the correction of the imperfect flatness of an array will be discussed with reference to Fig. 5.1-5. First, observe that we are not satisfied with correcting the mechanical deviations of the array, but we want to correct the phase of the electric voltages delivered by the hydrophones. These phases may be wrong due to a mechanical curvature of the array, variations of the hydrophones, phase shifts in the cables between the array and the processor, and phase shifts in the preamplifiers between the hydrophones and the processor. The way to correct all these possible errors at once is to use a sound source that radiates a spherical wave.

[1] This case is important because a slightly spherical curvature can be obtained by grinding *two* nominally flat surfaces against each other. To obtain a flat surface, one must alternately grind *three* nominally flat surfaces against each other.

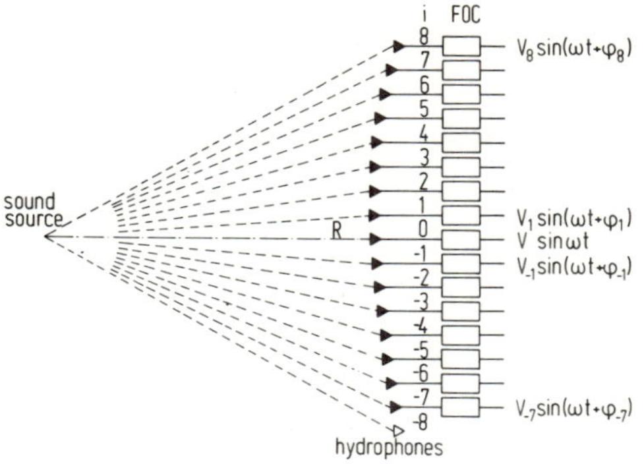


FIG. 5.1-5. Correction of deviations from the acoustic–electric flatness of a hydrophone array. FOC, focusing circuits.

The focusing circuits FOC in Fig. 5.1-5 are set for the nominal distance $R$ of the sound source. If everything is correct, all the output voltages will have the same phase and the same amplitude. The focusing circuits FOC must contain a circuit according to Fig. 4.3-2 rather than the one of Fig. 4.3-3, unless we know that the deviations from the flatness are symmetric to the axis of the array. The phase of the voltage $V_i \sin(\omega t + \varphi_i)$ can be made equal to that of the reference voltage $V \sin \omega t$ from the hydrophone $i = 0$ by changing $R$ a little in the constants $\cos(\pi i^2 d^2/R\lambda)$ and $\sin(\pi i^2 d^2/R\lambda)$ which are fed into the circuit of Fig. 4.3-2. Denote this new value of $R$ by $R + \Delta R_i$. The correction $\Delta R_i$ is a characteristic value for the hydrophone $i$ that has to be added to the focusing distance whenever the constants $\cos(\pi i^2 d^2/R\lambda)$ and $\sin(\pi i^2 d^2/R\lambda)$ are calculated in the future.[1]

Figure 5.1-5 shows that not only the phases of the output voltages can be wrong but also their amplitudes $V_i$. This is primarily due to variations between the hydrophones. To equalize $V_i$ with the reference amplitude $V$ of the hydrophone $i = 0$, we must multiply both $\cos(\pi i^2 d^2/R\lambda)$ and $\sin(\pi i^2 d^2/R\lambda)$ in Fig. 4.3-2 with a factor $q_i$ that is close to 1. Hence, the two characteristic values $\Delta R_i$ and $q_i$ have to be determined for every hydrophone, and the corrected constants $q_i \cos[\pi i^2 d^2/(R + \Delta R_i)\lambda]$ and $q_i \sin[\pi i^2 d^2/(R + \Delta R_i)\lambda]$ have to be used for focusing. As one says these days, a software solution for a hardware problem was found.

The one thing we cannot correct electronically are time variations of the hydrophone array. The array must be stiff and rigid. The array shown in Fig. 1.2-9 uses a solid aluminum plate to meet this requirement, but

[1] This is correct only for small viewing angles. For very wide ones, one must move the sound source in Fig. 5.1-5 to various observation angles $\gamma \neq 0$ and use the circuit of Fig. 4.3-8.

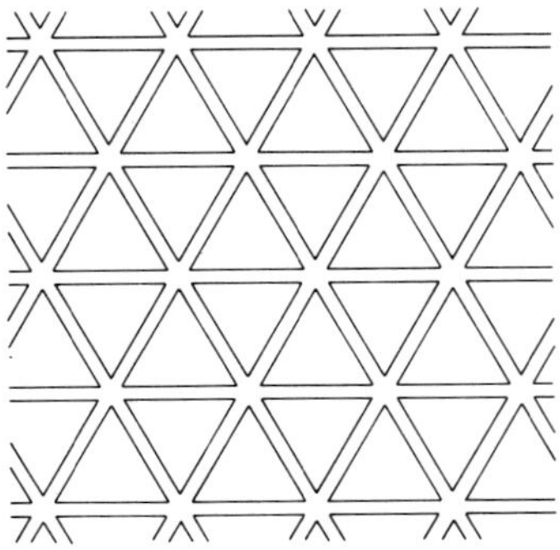

FIG. 5.1-6. Lightweight design for a rigid back plate of a hydrophone array.

this is not a good design since the array is too heavy. A triangular grid as shown in Fig. 5.1-6 is just as rigid and much lighter.

One may readily infer from Fig. 5.1-5 that this correction method for deviations from acoustic–electric flatness also works, if the hydrophones are mounted on a surface that is purposely rounded to reduce drag in water.

An important consideration in array design is the filling factor, which states the fraction of the array area actually occupied by the areas of the hydrophones. Figure 5.1-7a shows a completely filled array. The hydrophones

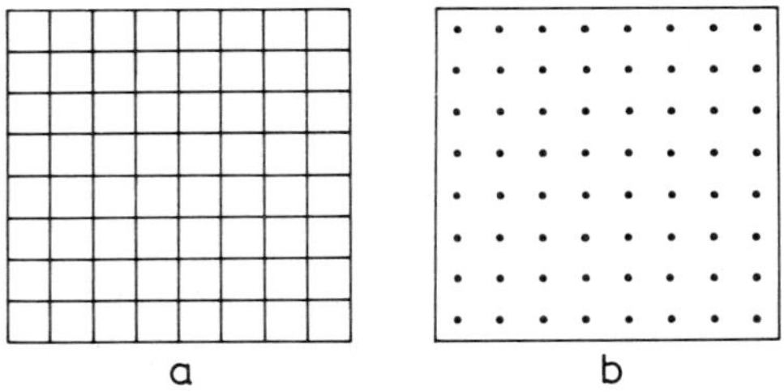


FIG. 5.1-7. Two limit cases of array design. (a) Completely filled array, (b) completely empty array.

are assumed to be square, and they cover the whole area of the array. The other extreme is the completely empty array of Fig. 5.1-7b, which assumes pointlike hydrophones. The difference between these two cases is not only the different sensitivity of the hydrophones, but the different beam patterns. Table 5.1-1 shows that the distance between the centers of the hydrophones equals about two wavelengths for a viewing angle $\alpha = 30°$, and about six wavelengths for $\alpha = 10°$. The array of Fig. 5.1-7b will thus yield grating lobes at multiples of the viewing angle, while the array of Fig. 5.1-7a will not have grating lobes due to the beam pattern of the individual hydrophones. It will be shown in Chapter 6 on multiplexing and sidelobe reduction that these two limit cases have important practical applications. How close the two limits can be approached depends on the state of the art of hydrophone design. The small hydrophones required for the completely empty array have necessarily a low sensitivity. On the other hand, it is not easy to produce hydrophones with the large area required for the completely filled array.[1]

## 5.2 Circuit Architecture

The basic organization of an acoustic imaging system with electronic circuits was shown in Fig. 1.2-7. A more detailed arrangement of the circuits is shown in Fig. 5.2-1 for an imaging system that resolves 16 × 16 points. Four stacks of printed circuit cards are shown. The first stack performs the inverse transform for the $x$-direction, that is, for the wavefronts $W(\beta_x)$ or

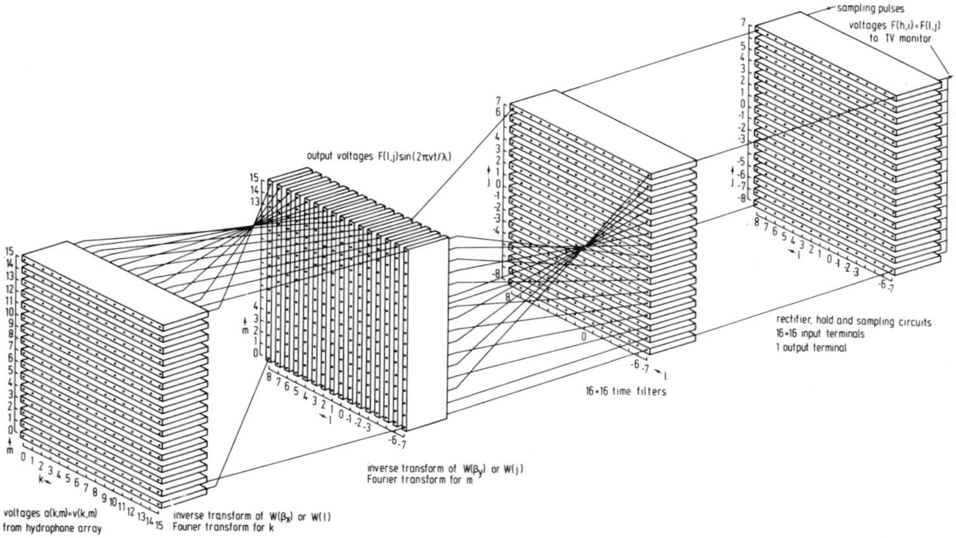


FIG. 5.2-1. The four stacks of printed circuit cards required by a basic processor for acoustic imaging. The two stacks of cards on the left produce 16 × 16 two-dimensional beams, the third improves the signal-to-noise ratio, and the fourth converts the electric image into a form suitable for display on a TV monitor.

[1] For a more detailed discussion of sound radiation by arrays see the chapter "Sound Radiation of Arrays and Membranes" in the book by Skudrzyk (1971).

$W(l) = W(\pm l\varepsilon)$. The circuit of the cards is shown by Fig. 3.4-2 for $2^n = 16$. The second stack contains cards with the same circuit, and it performs the inverse transform for the $y$-direction.

The input terminals of the circuit in Fig. 3.4-2 for the $x$-direction are denoted $k = 0, 1, \ldots, 2^n - 1 = 15$, and the output terminals are denoted $l = 0, +1, -1, \ldots, +7, -7, \pm 8$. Hence, the input terminals are ordered, but the output terminals do not have the order $l = -7, -6, \ldots, +8$ required for the stack of cards performing the inverse transform for the $y$-direction in Fig. 5.2-1. A certain output terminal $l$ must be connected to the card at location $l$. This is shown for the top row of terminals only, but the connections are the same for all rows.

The cards performing the inverse transform for the $y$-direction in Fig. 5.2-1 have the input terminals denoted $m = 0, 1, \ldots, 15$ and the output terminals $j = -8, -7, \ldots, +7$. This notation is also shown in Fig. 3.4-2.

The output voltages of the second stack of cards in Fig. 5.2-1 are denoted $F(l, j)\sin(2\pi vt/\lambda)$, where $F(l, j)$ is defined by Eqs. (2.1-17) and (2.2-6):

$$F(l, j) \doteq V(A/d)^2 v(\gamma_x, \gamma_y, \beta_x, \beta_y) \tag{1}$$
$$\sin\gamma_x = l(\lambda/A), \qquad \sin\gamma_y = j(\lambda/A)$$

We call $F(l, j)$ the electric image. The time variable voltages $F(l, j)\sin(2\pi vt/\lambda)$ are ordered for the variable $l$ but not for $j$. The third stack of cards in Fig. 5.2-1, denoted "16 × 16 time filters," has the input terminals ordered for $l$ and $j$. The output terminals $j$ of the second stack of cards must be connected to the input terminals $j$ of the time filters as shown for the card at location $l = -7$.

We have only used the space variables $x$ and $y$, or $k$ and $m$, so far. The time variable $t$ of the voltages $F(l, j)\sin(2\pi vt/\lambda)$ has not yet been exploited. There are a variety of uses for $t$. One is to increase the signal-to-noise ratio by passing each voltage $F(l, j)\sin(2\pi vt/\lambda)$ through a narrow filter tuned to the frequency $f = v/\lambda$. This is the purpose of the 16 × 16 time filters in Fig. 5.2-1. The filters can be inserted after the beam forming is done without any detrimental effect from the phase shifts caused by narrow filters. If the time filters were inserted in front of the cards performing the inverse spatial transform, one would have to control the frequency response of their phase shifts very carefully. The time filters will be discussed in more detail in the next section.

The output voltages of the time filters in Fig. 5.2-1 are fed to a stack of printed circuit cards containing rectifier, hold, and sampling circuits. The purpose of these circuits is to transform the 16 × 16 input voltages, having the time variation of pulsed sinusoidal functions, into voltages of a form suitable for display on a TV monitor.

Sixteen printed circuit cards as shown in Fig. 4.3-5 have to be placed in front of the stacks of cards in Fig. 5.2-1 if one wants to focus for a finite distance of the object plane. Furthermore, preamplifiers are needed between the hydrophones and the focusing cards in order to lift the voltages to the level of about 1 V before processing.

The interconnection of so many stacks of printed circuit cards is a typical and severe problem introduced by two-dimensional spatial processing.[1] One may implement the structure of Fig. 5.2-1 by means of the usual printed circuit hangers, wired together according to the illustration. This was actually done in the first experimental equipment. However, this is no easy task, as one can verify by counting the number of wire ends to be soldered or wrapped in Fig. 5.2-1 and calculating their number for a resolution of more than $16 \times 16$ points. A more promising way is to build two-dimensional spatial processors with the arrangement suggested by Fig. 5.2-1, but to place the individual stacks of cards vertically above each other rather than horizontally behind each other.

To do so, one has to design the printed circuit cards of Fig. 3.2-6 and Fig. 4.3-6 not with the input and output terminals at the bottom, but with the input terminals on the left side and the output terminals on the right side, in close correspondence to the circuit diagrams of Figs. 3.2-3, 3.4-2, and 4.3-4. This is a considerable advantage since the leads are shortened, and the input terminals are better separated from the output terminals. Furthermore, the number of terminals on one edge of the card is effectively reduced to one-half, and they are located on the long rather than the short edges as in Figs. 3.2-6 and 4.3-6. The crisscrossing of the wires between the cards can be avoided by rearranging the output terminals on each card so that they are ordered like the input terminals on the next stack of cards.

One problem remains. In any practical equipment using printed circuit cards one must be able to extract readily any one card. This is no problem if all contacts are on one edge as in Figs. 3.2-6 and 4.3-6, but a very grave problem if contacts are located on more than one edge. A practical solution is shown in Fig. 5.2-2. A set of input sockets IS1 to IS16 is rigidly mounted on brackets BR1 and BR2. A set of output sockets OS1 to OS16 is connected via flexible wires to the input sockets. Hence, the stacks of cards in Fig. 5.2-1, in vertical arrangement, can be inserted into their rigidly mounted input sockets, and the flexibly mounted output sockets can then be pushed on. There is no known practical way to mount card guides for this structure. However, if the output sockets OS1 to OS16 are

[1] Acoustic imaging is one example of two-dimensional spatial processing. The other important example is image processing with hardware, which means primarily the classification of aerial photographs.

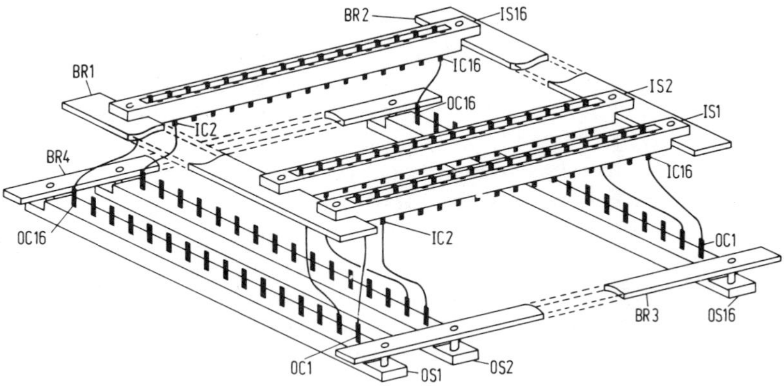


FIG. 5.2-2. Crossed sockets for the connection of stacks of printed circuit cards according to Fig. 5.2-1. OS1–OS16, output connectors; IS1–IS16, input connectors; OC1–OC16, output contacts; IC1–IC16, input contacts; BR1–BR4, brackets.

made rigid after insertion of the cards, either by screwing or by clipping them to brackets BR3 and BR4, one can hold the cards on two opposite edges with sockets to obtain a very rigid structure.

There are many practical ways to implement the crossed sockets of Fig. 5.2-2. Certain brands of sockets permit one to solder the tails of the contacts before the contacts are inserted into the body of the socket. Hence, one can do all the soldering first and push the contacts in afterward.

With some care, one can make printed circuit cards sufficiently thin so that they can be mounted with 1.25 cm distance from center to center. The cards of Figs. 3.2-6 and 4.3-6 permit this spacing, even though the integrated circuits are not soldered to the cards but sit on top of sockets that are soldered to the card. One can thus mount 32 cards side by side in the usual telephone rack with a width of 48 cm. Beyond a resolution of 32 × 32 points, one runs into a space problem. One can push the printed circuit card technology to 64 × 64 points by making the cards somewhat thinner and using an oversized rack. Beyond 64 × 64 points, we can turn to the multiplexed use of the processor, as will be discussed in Section 6.1, or one can turn to thick and thin film technology.

## 5.3 TIME FILTERS FOR NOISE REJECTION

The 16 × 16 time filters in Fig. 5.2-1 are narrow-band filters that improve primarily the signal-to-noise ratio. They can also be used to provide a Doppler resolution and to perform other tasks to be discussed later.

A typical frequency for the acoustic wave used is 100 kHz. The hydrophones and the preamplifiers connected to the hydrophones ahead of the processing equipment have a very large frequency bandwidth. This is necessary to prevent any phase shifts between the voltages fed to the processing equipment caused by any time filtering in the preamplifiers. Such phase shifts would, of course, interfere with the beam forming. Once the beam forming is done and the electric image $F(l, j)\sin(2\pi vt/\lambda)$ is obtained, one is no longer interested in the phase. Hence, one may use time filters with very narrow bandwidth. A typical filter for a center frequency of 100 kHz and a bandwidth of about 1 kHz is shown in Fig. 5.3-1. This is a very primitive filter, but the large number of filters required makes small size and low cost mandatory.

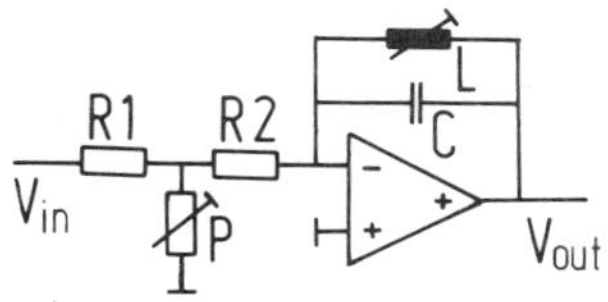


FIG. 5.3-1. Low-cost, small-size filter for a center frequency of 100 kHz and a bandwidth of about 1 kHz.

To produce the sensation of a moving image one must generate about 25 images per second. This calls for a bandwidth of 12.5 Hz according to the sampling theorem. It is theoretically possible to produce filters with this bandwidth at a center frequency of 100 kHz. The practical way to do it is to use frequency conversion to, say, 1 kHz, and to build a filter with about 12.5 Hz bandwidth. However, a filter operating at 1 kHz poses a problem both in size and cost.

We derived in Chapter 3 circuits that performed the Fourier transform of space signals in a very satisfactory way. Let us convert this circuit into one that transforms time signals, and see how it can be used for filtering.

Figure 5.3-2 shows the circuit of Fig. 3.1-3 with the required modifications to transform time signals. A voltage $Vf(t)$ shall be fed to the input terminal of the stage 0 of the analog shift register ASR. The clock pulse used for shifting $Vf(t)$ through the register shall have the period $\tau$. Hence the voltage[1] $Vf(t + k\tau)$, $k = 0, \ldots, 15$, is applied to the input terminal of the stage $k$ of the shift register at the time $t$. This voltage replaces the function

[1] Note that a voltage applied to the input terminal 1 had to arrive at the input terminal 0 earlier. Hence, $t + k\tau$ rather than $t - k\tau$ is correct.

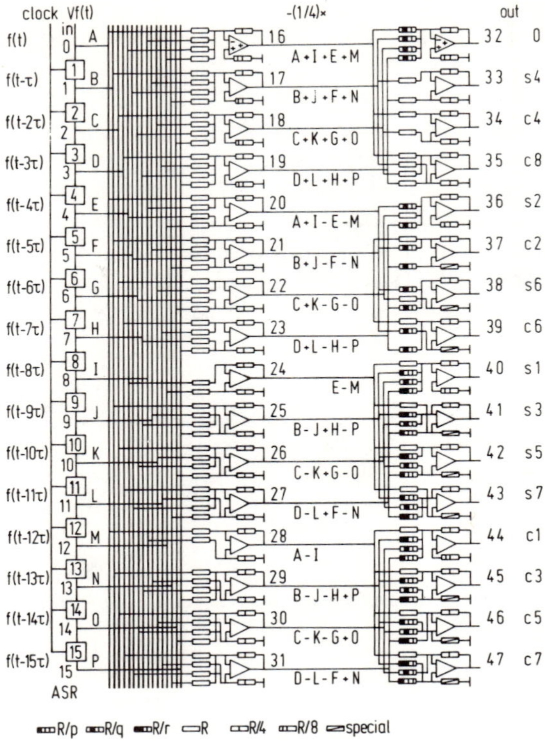


FIG. 5.3-2. Filter for time-variable signals based on the fast Fourier transform circuit of Fig. 3.1-3. ASR, analog shift register. The heavy lines indicate the part of the transform circuit required if one wants only the voltage $v(3, f, s, t)$ at the output terminal s3. $p = \sin \pi/8$, $q = \sin \pi/4 = \sqrt{2}/2$, $r = \sin 3\pi/8$.

$F(k)$ on top of Fig. 3.1-1. For reasons soon to be seen, we use the variable $i = k - 2^n/2 + 1$ instead of $k$, $i = -2^n/2 + 1, \ldots, 2^n/2$, and denote the voltage samples in the analog shift register of Fig. 5.3-2 by $Vf(t + i\tau)$.

The sinusoidal function $\sin 2\pi ft$ is substituted for $f(t)$:

$$Vf(t + i\tau) = V \sin 2\pi(ft + if\tau) = v(i, t) \tag{1}$$

Equations (1) and (2.2-1) become equal if one substitutes

$$v/\lambda = f, \qquad (d/\lambda) \sin \beta = f\tau \tag{2}$$

The unknown variable for the spatial transformation was $\sin \beta$, while the unknown variable for the time transformation is the frequency $f$; the clock period $\tau$ is known.[1] Hence, the following substitution should be made in Eq. (2.2-2):

$$(d/\lambda)\sin\gamma = f_l\tau \tag{3}$$

The frequency $f_l$ will have certain values determined by the design of the filter just as $\sin\gamma$ had certain values determined by the equipment. Equation (2.2-6) assumes the following form if Eq. (3) is substituted:

$$\begin{aligned} 2^n f_l\tau &= l && \text{for } f_l\tau > 0 \\ &= -l && \text{for } f_l\tau < 0 \qquad l = 1, 2, \ldots \qquad 2^n\tau = T \\ &= 0 && \text{for } f_l\tau = 0 \end{aligned} \tag{4}$$

The meaning of positive and negative angles $\beta$ and $\gamma$ in Eqs. (2.2-3)–(2.2-6) was clear, while the meaning of negative values for $f\tau$ in Eq. (2) and $f_l\tau$ in Eq. (4) is not. For explanation, let $\tau$ change its sign. The signal is then shifted in Fig. 5.3-2 from the last stage of the shift register ASR to the input stage rather than in the opposite direction. Hence, no mysterious negative frequencies are encountered.

The number $2^n$ in Eq. (4) represents the number of stages of the shift register ASR in Fig. 5.3-2, as one can readily see from the summation of the terms $i = -2^n/2 + 1, \ldots, 2^n/2$ [Eqs. (2.2-3)–(2.2-5)]. Hence, one must use the following substitutions when interpreting Eqs. (2.2-7)–(2.2-9) for time transforms:

$$(A/\lambda)\sin\beta = 2^n(d/\lambda)\sin\beta = 2^n f\tau = fT \tag{5}$$

$$(A/\lambda)\sin\gamma = 2^n f_l\tau = f_l T \tag{6}$$

$2^n\tau = T =$ time required to shift one sample through the shift register

Equations (2.2-7) to (2.2-9) thus assume the following form:

$$v(0, \beta, t) \to v(0, f, t) = \tfrac{1}{2}\sqrt{2}\,2^n V \frac{\sin(\pi f T)}{\pi f T}\sin(2\pi f t) \tag{7}$$

$$v(\gamma, \beta, s, t) \to v(l, f, s, t) = \tfrac{1}{2}2^n V\left(\frac{\sin[\pi(f - f_l)T]}{\pi(f - f_l)T} - \frac{\sin[\pi(f + f_l)T]}{\pi(f + f_l)T}\right)\cos(2\pi f t) \tag{8}$$

$$v(\gamma, \beta, c, t) \to v(l, f, c, t) = \tfrac{1}{2}2^n V\left(\frac{\sin[\pi(f - f_l)T]}{\pi(f - f_l)T} + \frac{\sin[\pi(f + f_l)T]}{\pi(f + f_l)T}\right)\sin(2\pi f t) \tag{9}$$

[1] One could just as well determine an unknown clock period $\tau$ from a known frequency $f$, but this possibility is currently of no practical interest.

The values of $f_l$ follow from Eq. (4)

$$f_l = l/2^n\tau = l/T, \qquad l = 0, \pm 1, \pm 2, \ldots \tag{10}$$

The voltages $v(0, f, t)$, $v(l, f, s, t)$, and $v(l, f, c, t)$ of Eqs. (7)–(9) are obtained at the output terminals 0, s$l$, and c$l$, $l = 1, \ldots, 8$, of the circuit in Fig. 5.3-2. The normalized amplitudes of these voltages are the frequency response of the amplitudes, denoted $v(0, f)$, $v(l, f, s)$ and $v(l, f, c)$:

$$v(0, f) = \sqrt{2}\,\frac{\sin(\pi f T)}{\pi f T} \tag{11}$$

$$v(l, f, s) = \frac{\sin[\pi(f - f_l)T]}{\pi(f - f_l)T} - \frac{\sin[\pi(f + f_l)T]}{\pi(f - f_l)T} \tag{12}$$

$$v(l, f, c) = \frac{\sin[\pi(f - f_l)T]}{\pi(f - f_l)T} + \frac{\sin[\pi(f + f_l)T]}{\pi(f - f_l)T} \tag{13}$$

The frequency response of the phases is constant since the arguments of $\sin 2\pi ft$ and $\cos 2\pi ft$ in Eqs. (7)–(9) contain no phase term $ft_0(f) = \varphi(f)/2\pi$.

The sixteen curves $v(0, f)$, $v(l, f, s)$, $l = 1, \ldots, 7$, and $v(l, f, c)$, $l = 1, \ldots, 8$, for the circuit of Fig. 5.3-2 are shown in Fig. 5.3-3. The curves $v(l, f, c)$ and $v(l, f, s)$ become more and more equal with increasing values of $l$. In essence, a filter of this type with $2^n - 1$ shift register stages divides a frequency band $0 < f < 1/2\tau$ into $2^n/2$ channels.

In order to implement a filter with similar properties as the one shown in Fig. 5.3-1, one does not need the whole circuit of Fig. 5.3-2, but only the part feeding to one output terminal. This is shown for the terminal s3 by the section of the transform circuit drawn with heavier lines. Although the circuit has several good features—no inductivity, constant frequency response of the phase, tuning by the clock pulse period—it is currently too expensive in this form. However, we will come back to it in Section 6.1, where banks of stable filters tuned to different frequencies are required.

In order to improve the frequency response of the amplitudes plotted in Fig. 5.3-3, one can connect several of the circuits of Fig. 5.3-2 in series. Consider several circuits of the part of Fig. 5.3-2 shown with heavy lines. The output terminal s3 of the first circuit is connected to the input terminal of the second circuit, the output terminal s3 of the second circuit to the input terminal of the third circuit, etc. The frequency response of the amplitude at the output terminal s3 of the first circuit is $v(3, f, s)$, at the second circuit it is $v^2(3, f, s)$, at the third circuit $v^3(3, f, s)$, etc. Curves for one, two, three, and four circuits are plotted in Fig. 5.3-4. The frequency response of the power, $v^{2n}(3, f, s)$, is plotted for $n = 1, 2, 3, 4$ circuits in series on a logarithmic scale, $10 \log v^{2n}(3, f, s)$, in Fig. 5.3-5.

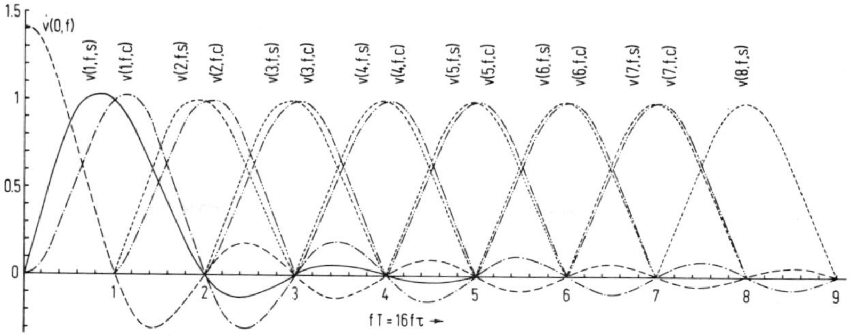


FIG. 5.3-3. Frequency response of the amplitude for the 16 output terminals 0, s1, c1, ..., c8 of Fig. 5.3-2. The frequency response of the phase is constant.

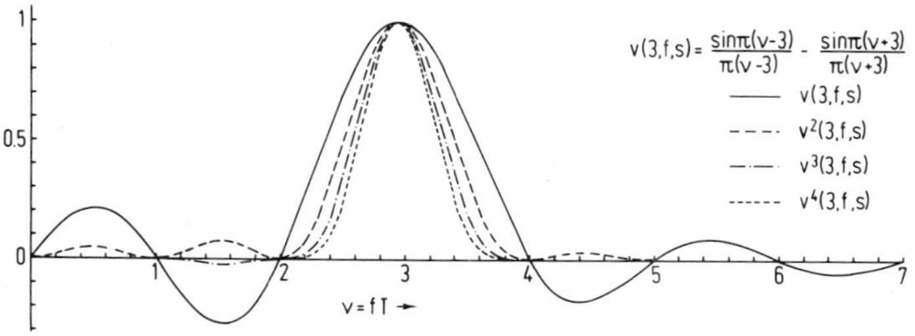


FIG. 5.3-4. Frequency response of the amplitude at the output terminal s3 in Fig. 5.3-2 if one, two, three, or four circuits are connected in series. The curve $v(3, f, s)$ is the same as in Fig. 5.3-3.

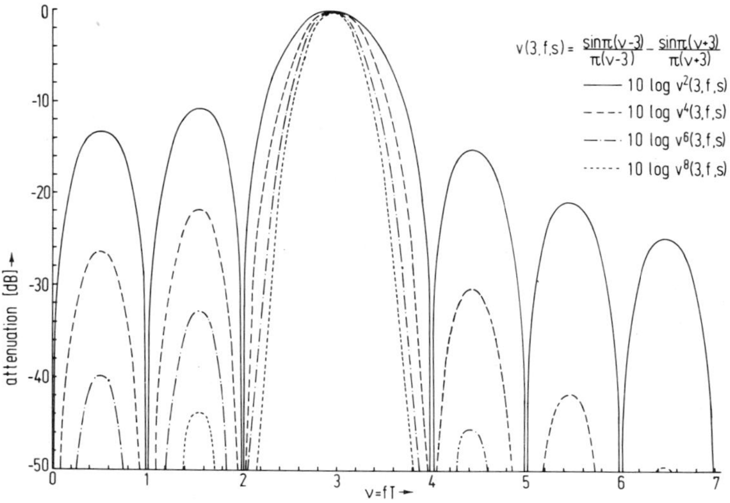


FIG. 5.3-5. Frequency response of the power at the output terminal s3 of Fig. 5.3-2 if one, two, three, or four circuits are connected in series. The curves are the logarithm of the square of the curves in Fig. 5.3-4.

## 5.4 Image Display and Test Results

The output voltages of the time filters in Fig. 5.2-1 have to be brought into a form suitable for electrooptic display. A light-emitting diode array was used in early work. In essence, one may connect an emitter follower to the output terminal of each time filter to convert from voltage to current source and drive a light-emitting diode. This display was discarded because the control of brightness and contrast was too costly, and the light-emitting diodes were not sufficiently uniform to display halftones. The use of a TV monitor solved these problems and at the same time made it possible to display an image at several places. Figure 5.2-1 is thus drawn for a TV monitor.

The output voltages of the time filters are fed to a stack of printed circuit cards containing rectifier, hold, and sampling circuits. Details of these circuits are shown in Fig. 5.4-1. The rectifier is a standard absolute value circuit

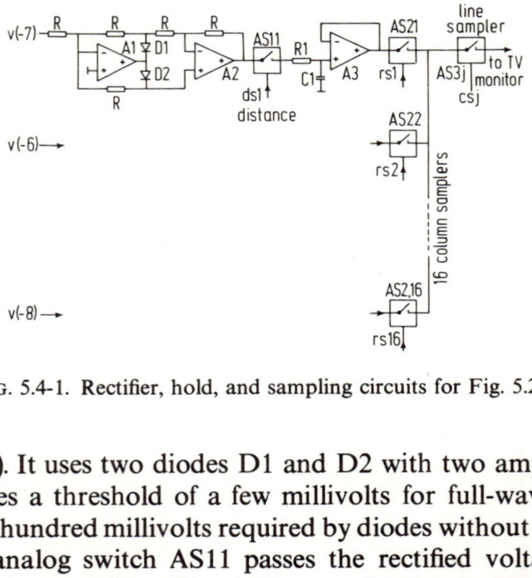


FIG. 5.4-1. Rectifier, hold, and sampling circuits for Fig. 5.2-1.

(Graeme, 1973). It uses two diodes D1 and D2 with two amplifiers A1 and A2. This assures a threshold of a few millivolts for full-wave rectification instead of a few hundred millivolts required by diodes without an operational amplifier. An analog switch AS11 passes the rectified voltage through a resistor R1 to the capacitor C1, which smooths the rectified voltage. The analog switch is required, since a practical system uses pulsed sinusoidal waves and not continuous ones. The analog switch AS11 is closed at the time when the pulse arrives from a certain selected distance; this image selection or range gating will be discussed in more detail in Section 8.1. When the switch AS11 is open, the charge in the capacitor C1 cannot be discharged through the amplifier A2, and the brightness of the displayed image will not "flash," which means it will not change rapidly with time depending on whether the pulsed sinusoidal wave is received or not. A simpler and not so good method is to use a diode, bridged by a resistor with a resistance large compared with that of R1, instead of the switch AS11.

The amplifier A3 has a high input impedance and a low output impedance, and thus holds the charge in capacitor C1. The output of A3 is sampled and fed to the TV monitor. The sampling in Fig. 5.2-1 requires 16 switches AS21 to AS2,16 per card for the column sampling, and one switch AS3$j$ per card for the line sampling. This method of sampling requires one scale 16 counter for the rows and a second scale 16 counter for the columns. The voltages for the analog switches AS21–AS2,16 and AS3$j$ in Fig. 5.4-1 can readily be derived from these counters by standard logic circuits. Digital–analog converters connected to the two counters provide the horizontal and vertical scanning voltages with proper synchronization for a TV monitor.

Figures 5.4-2 and 5.4-3 show typical images obtained by experimental equipment. The photographs were copied from a 16-mm movie produced by J. Dierks of the Applied Research Laboratories of the University of Texas at Austin.[1] Consider first the six photographs of Fig. 5.4-2. A model of the target is shown in the lower left corners in the positions of the actual target. The spheres of the model represent the sound sources of the actual target, which are seen by the imaging equipment. The images produced and displayed on a TV screen are shown by the photographs a to f. The resolution of 16 × 16 points is very poor, but the photographs clearly show the target in various positions rotating around a vertical axis. The much

[1] The experimental equipment was designed and built at The Catholic University of America, except for the work done by the three subcontractors listed below. The team consisted of J. Kamal, a postdoctoral fellow from India; S. S. R. Murty, a student from India; Y. K. Hong, a student from Korea; N. Mohamed, a student from Kuwait; T. Frank, an American student; and the author. The printed circuit cards were fabricated at ITT Electro-Physics Laboratory of Columbia, Maryland, and at ITE, Inc. of Beltsville, Maryland. The acoustic array with a projector, shown in Fig. 1.2-9, was fabricated at Bendix Electrodynamics Division of Sylmar, California, under the leadership of G. Zilinskas and B. Lee.

The tests were conducted at the Lake Travis Test Station of the University of Texas, Applied Research Laboratories, Austin, under the leadership of J. Dierks. A considerable amount of equipment development and modification had to be done during the tests. For instance, the available equipment for the measurement of beam patterns could not resolve the beams of about 30′ width of the imaging equipment; a remote controlled movable target had to be installed in Lake Travis; the light-emitting diode display was replaced by a TV monitor; etc. The movie, from which Figs. 5.4-2 and 5.4-3 were copied, was made by J. Dierks during the summer of 1976.

Financial support of about $150,000 was required for design and construction, about one-third of it for the hydrophone array. This financial support was provided by the Sensor Technology Group of the Office of Naval Research, Washington, D.C.

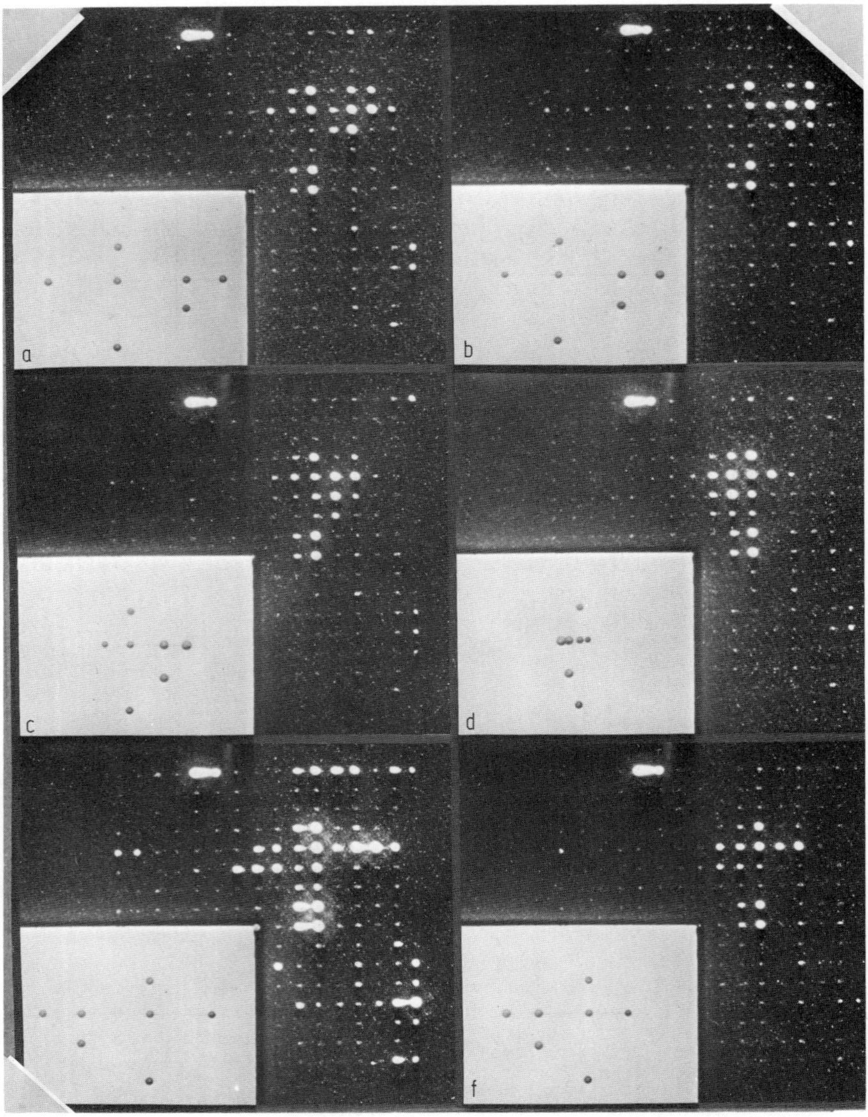


FIG. 5.4-2. Rotation of a target around a vertical axis as shown by the acoustic imaging equipment and optically by a model of the target copied into the lower left corner of each photograph. The usual bright round dots of a TV monitor are used for display.

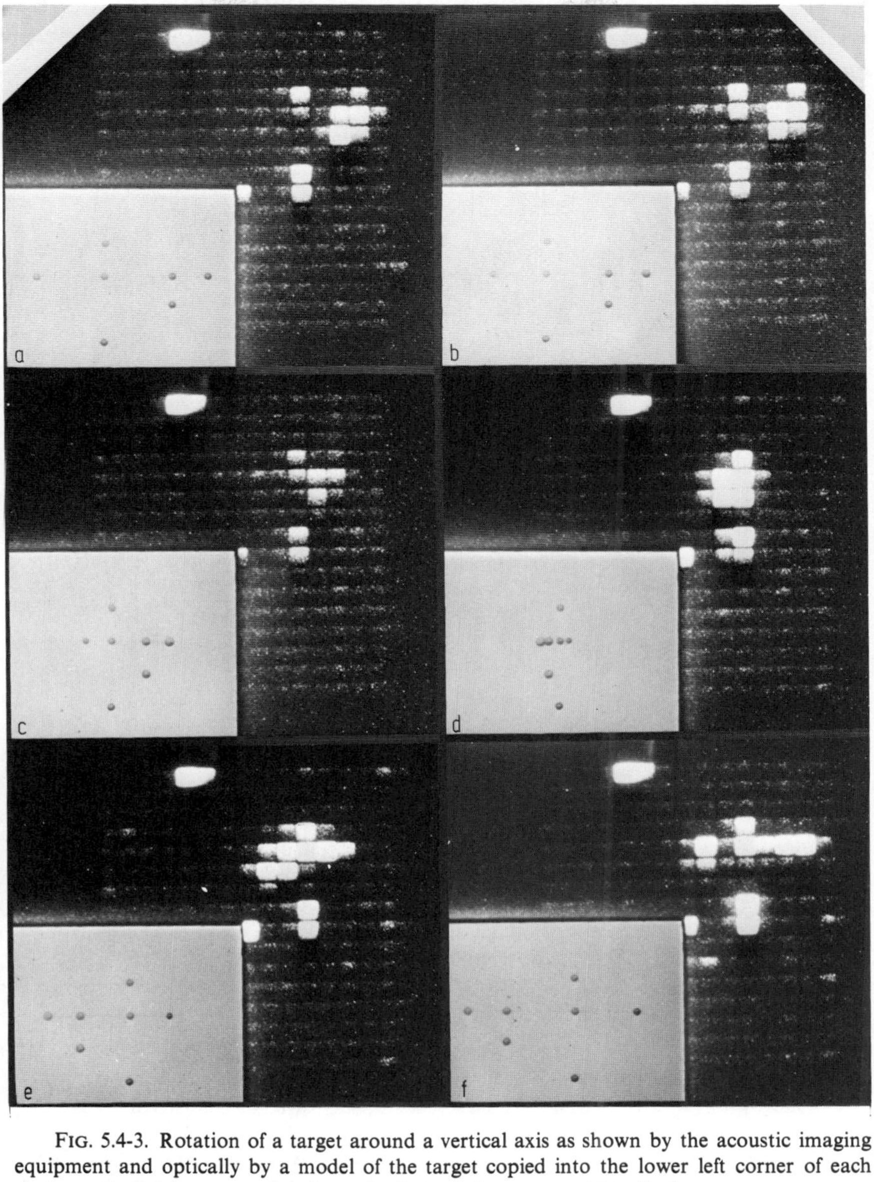


FIG. 5.4-3. Rotation of a target around a vertical axis as shown by the acoustic imaging equipment and optically by a model of the target copied into the lower left corner of each photograph. Bright square dots instead of round dots are used for display.

more impressive motion shown by the original movie, unfortunately, cannot be represented in print.[1]

Figure 5.4-3 shows essentially the same as Fig. 5.4-2, but the display has been modified so that the small round dots of light on the TV display become squares that almost touch each other. This is one of many modifications tried to make it easier for the observer to recognize the shape of the imaged object.

[1] Copies of the movie are in the possession of J. Dierks, Applied Research Laboratories, University of Texas, P. O. Box 8029, Austin, Texas 78712; M. Blizard, Sensor Technology Group, Office of Naval Research, Department of the Navy, Washington, D.C. 20360; H. F. Harmuth, Department of Electrical Engineering, Catholic University, Washington, D.C. 20064.

# 6 Multiplexing and Sidelobe Reduction

## 6.1 MULTIPLEXING HYDROPHONE ARRAY AND PROCESSOR

One needs about 25 images per second to create the sensation of movement. The circuits discussed are capable of generating about 100,000 images per second, using low-cost integrated-circuit operational amplifiers. Whenever there is such an enormous difference in required and actual speed, one will try to save equipment cost by the multiplex use of the equipment.

To recognize the principle of multiplexing let us study the beam pattern of a low-resolution hydrophone array with 16 × 16 hydrophones and a viewing angle of 10° × 10° in cartesian coordinates. The array shall be close to the completely empty type shown in Fig. 5.1-7b. The axes of the beams formed in one dimension are shown in Fig. 6.1-1. The main group[1] of beams—or group 0—has the 16 beam axes denoted $W(-7\varepsilon)$ to $W(+8\varepsilon)$.

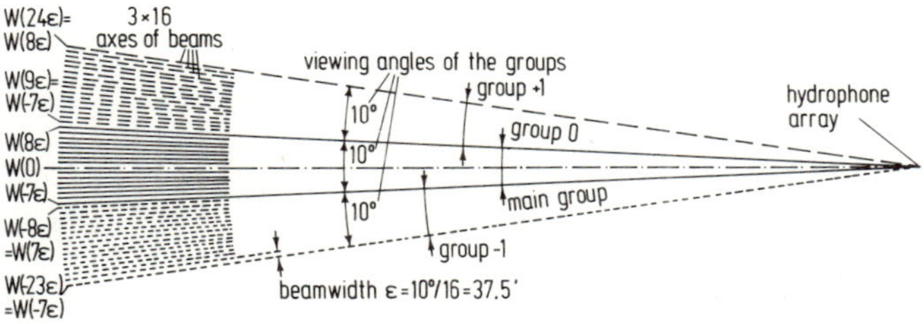


FIG. 6.1-1. The beam pattern of an array of 16 hydrophones with a viewing angle of 10° and a beamwidth of 10°/16 = 37.5′. The pattern of the main group of beams—or group 0—is repeated periodically for positive and negative multiples of 10°.

The beams of the main group are repeated periodically as so-called *grating beams* or *grating lobes* for positive and negative multiples of 10°, within the limits of the approximation $\sin l\varepsilon \doteq l\varepsilon$. So far we have been interested in the main group, since only the viewing angle of the main group was insonified. Obviously, one could also produce images by means of the grating lobes in the other groups.

[1] In telephony multiplexing one uses the terms channel, group (of channels), supergroup, and mastergroup. We use the same terminology, except that channel is replaced by beam. Many of the concepts of telephony multiplexing are applicable here, and the use of the same terms is intended to help build a bridge to telephony.

Let the viewing angle of the main group in Fig. 6.1-1 be insonified by a wave with a frequency of 100 kHz, the viewing angle of the group +1 with 100 kHz + 100 Hz, and the viewing angle of the group −1 with 100 kHz − 100 Hz. The difference between the frequencies is so small that the beams and the characteristics of the hydrophones will be essentially the same for all three frequencies, but we can now resolve 3 × 16 angles or points even though we have only 16 hydrophones.

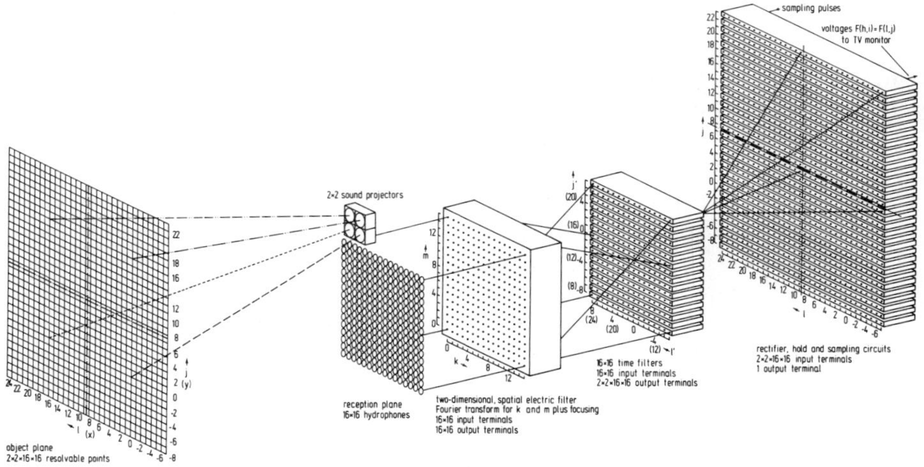


FIG. 6.1-2. Principle of multiplexing the hydrophone array and the two-dimensional spatial electric filter to resolve more points in the object plane. The illustration holds for any kind of multiplexing, even though frequency multiplexing is assumed in the discussion.

For an extension of this principle to two dimensions and the equipment required refer to Fig. 6.1-2. Four sections of the object plane with 16 × 16 points each are shown on the left. Each section is insonified with a sound wave of slightly different frequency. These sound waves come from 2 × 2 projectors. One may direct each projector mechanically to insonify an assigned section of the object plane, but the better technique is to use electronic beam steering by feeding voltages with all four required frequencies and proper phase shifts to all four projectors.

The scattered waves from the four sections of the object plane are received by the 16 × 16 hydrophones in the reception plane. Their output voltages are processed by the two-dimensional spatial electric filter. The output voltages of the filter represent four electric images, one for each section of the object plane. These four electric images can be separated, since each image is represented by a voltage with different frequency. The separation is done by the 16 × 16 time filters. Each one of these filters consists in essence of four resonant filters tuned to, e.g., 100 kHz − 100 Hz, 100 kHz, 100 kHz + 100 Hz, and 100 kHz + 200 Hz. The four filters have a common input terminal but four separate output terminals. Each of the four output terminals is connected to the respective section of the rectifier, hold, and sampling circuits. The four images received and processed by common equipment are thus separated by the time filters. The hydrophone array and the two-dimensional spatial electric filter are used in frequency multiplex.

Frequency multiplex is not the only multiplexing method that can be used. Time multiplexing could be used if it were not for the too long propagation times of sound in water. Other methods of orthogonal multiplexing can be used; the decision among them is one of technology and economy. Frequency multiplexing lends itself readily to explanation since we have discussed everything so far in terms of sinusoidal sound waves. The practical way to produce sound waves that differ by 100 or 200 Hz from a center frequency of 100 kHz is to amplitude modulate a sinusoidal carrier with a frequency of 100 kHz by sinusoidal baseband signals having a frequency of 100 or 200 Hz. Single-sideband modulation would have to be used to conform to our discussion, but it is not difficult to modify everything for double-sideband modulation. Using modulation, one can replace the sinusoidal baseband signals by more general ones.

Let us discuss some circuits. The Fourier transform circuits of Figs. 3.1-3, 3.2-3, and 3.2-5 are essentially independent of the frequency of the sound wave. The same holds for the circuit of Fig. 4.3-7. All are wide-band circuits. Figures 4.3-2, 4.3-3, and 4.3-8, on the other hand, are typically narrow-band circuits, since proper phase shift and correct amplitude of the voltages phase shifted 90° can be achieved for one frequency only. However, 1% of 100 kHz is 1 kHz, and the tolerances of the components make differences of frequencies of less than 1% generally insignificant. A change of the frequency by 1% will change $\lambda/A$ and thus $\varepsilon = \lambda/A$ also by 1%. The direction of a beam $\gamma = l\varepsilon$ will be changed by $l$%.

A possible circuit for the 16 × 16 time filters in Fig. 6.1-2 is shown in Fig. 6.1-3. It is not practical to build narrow filters for nominal frequencies of 100 kHz − 100 Hz, 100 kHz, 100 kHz + 100 Hz, and 100 kHz + 200 Hz. Hence, one must use an auxiliary carrier $V \sin \omega_b t$ to shift the signals from a band around 100 kHz to a band below 10 kHz. For instance, an auxiliary carrier with a frequency of 95 kHz will produce the four shifted frequencies 4900, 5000, 5100, and 5200 Hz. These frequencies can be separated by the four resonant filters in Fig. 6.1-3.

The circuit of Fig. 6.1-3 is technically possible but not good. The coils and capacitors required for resonant frequencies around 5 kHz are bulky. Furthermore, the resonant frequency and the amplification of each filter

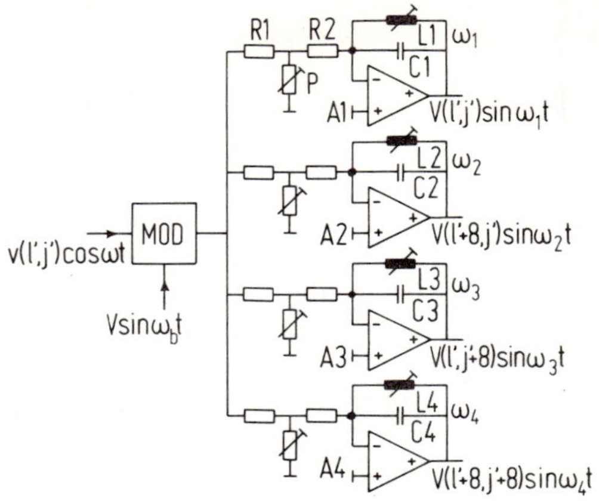


FIG. 6.1-3. Principle of a time filter that separates the images of the four sections of the object plane in Fig. 6.1-2. One needs 16 × 16 such filters. MOD, amplitude modulator (multiplier).

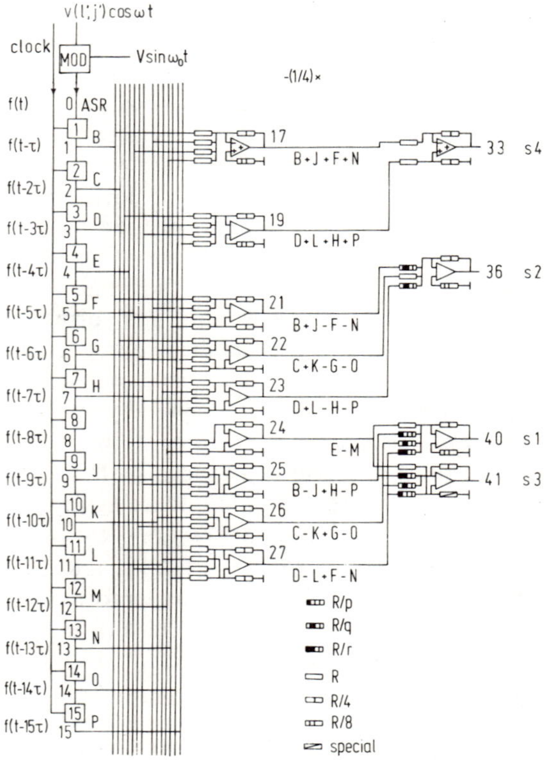


FIG. 6.1-4. Time filter for the separation of signals from four sections of the object plane. No coils and capacitors are used, and no tuning is required. MOD, amplitude modulator; ASR, analog shift register with 16 stages.

must be tuned. The only good features of the circuit are the modulator MOD used for frequency shifting, and the ease with which it can be explained.

A more promising circuit has been derived in Section 5.3. The circuit shown there in Fig. 5.3-2 is shown again with some modifications in Fig. 6.1-4. The modulator MOD is again used to convert the received signal to a fixed lower frequency. The low-frequency signal is then fed to the analog shift register ASR. Let us assume that the frequency of the auxiliary carrier fed to the modulator is 99,800 Hz. Sinusoidal signals with the frequencies 100, 200, 300, and 400 Hz are then delivered from the modulator MOD to the shift register ASR. Let the clock deliver $16 \times 100 = 1600$ pulses per second to the shift register. This means the value of $T$ in Fig. 5.3-3 equals $T = 16\tau = 16/1600 = 0.01$ sec. Hence, $fT = 1, 2, 3, 4$ in Fig. 5.3-3 implies $f = 100, 200, 300, 400$ Hz. If the part of the circuit of Fig. 5.3-2, feeding either the output terminals s1, s2, s3, s4 or c1, c2, c3, c4, is connected to the analog shift register ASR in Fig. 6.1-4, one can separate the sinusoidal signals with a frequency of 100, 200, 300, or 400 Hz. The circuit of Fig. 6.1-4 shows the part of the circuit of Fig. 5.3-2 that is required. The separation of the signals is quite good since the frequency response of the amplitude for one terminal in Fig. 5.3-3 has essentially its maximum when the other terminals yield no output voltage.

Consider the contours where the four sections of the object plane in Fig. 6.1-2 are joined to each other. One cannot realistically expect to insonify the points in the section with $l = -7, \ldots, +8$ and $j = -8, \ldots, +7$ with a sine wave of frequency $f_1$ but not the points on the adjacent lines $l = +9$ and $j = +8$. The weakly insonified points on the line $l = +9$ in the object plane will produce *alias* points on the line $l = -7$ of the display, while line $j = +8$ will produce alias points in the line $j = -8$ in the display. There is a simple way to combat this *contour effect*.

Let us insonify, receive, and process image sections with $16 \times 16$ points but display the $14 \times 14$ inner points only as indicated by the shaded areas in Fig. 6.1-5 for the sections H0V0 in the lower left corner to H1V1 in the upper right corner. The properly assembled sections of the optical display are shown in Fig. 6.1-6. The full display has $4 \times 14 \times 14 = 28 \times 28$ points rather than $32 \times 32$ points as in Fig. 6.1-2. Hence, by using only the inner points of each image section one can avoid the contour effect but obtains a smaller total image. These considerations may readily be extended to the use of only $13 \times 13$ or $12 \times 12$ inner points.

Experience has shown that many people find it difficult to comprehend Figs. 6.1-5 and 6.1-6. The problem disappears immediately if one makes a model. Copy Fig. 6.1-5 on transparent material, such as the foil used for viewgraphs, and cut out the four squares. Overlay them as indicated by the

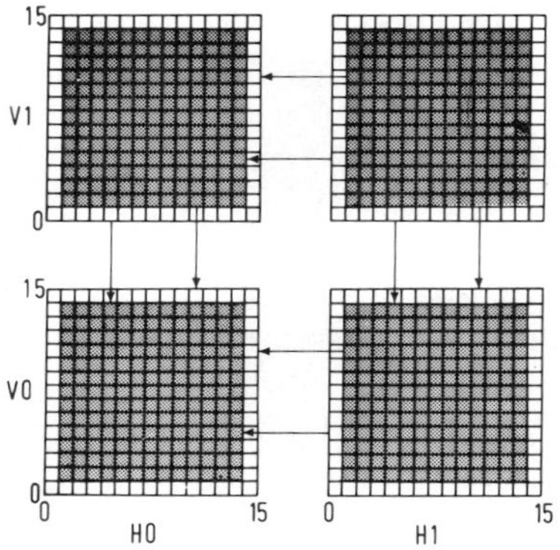


FIG. 6.1-5. Use of 14 × 14 inner points of sections with 16 × 16 points to avoid the contour effect. Only the points in the shaded areas are displayed.

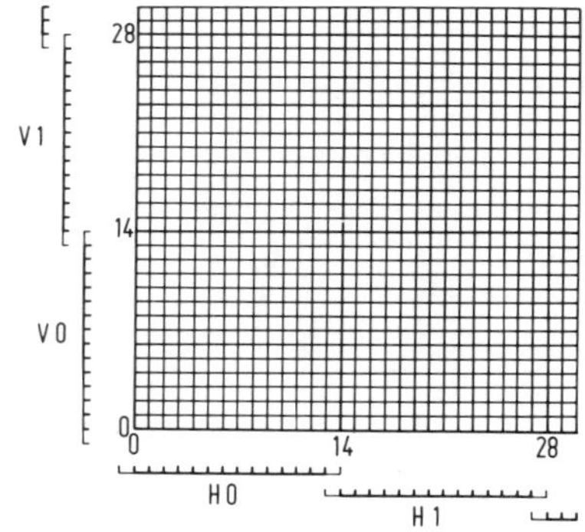


FIG. 6.1-6. Properly connected image sections with 14 × 14 points according to Fig. 6.1-5.

arrows in Fig. 6.1-5. Only the shaded areas will remain visible, and Fig. 6.1-6 is obtained.

We have so far ignored the bandwidth required for image generation. According to the sampling theorem one needs a bandwidth of 12.5 Hz to produce 25 images per second. If we divide an object plane with 256 × 256 points into image sections with 16 × 16 points, we must frequency multiplex $256^2/16^2 = 256$ image sections. This calls for a bandwidth of $12.5 \times 256 = 3200$ Hz. The circuits for the Fourier transform of Fig. 3.1-3 and for focusing according to Fig. 4.3-3 will readily operate over this bandwidth. However, using an array of 16 × 16 hydrophones for a resolution of 256 × 256 points requires an array of 16 × 16 projectors, which are more expensive than hydrophones. An array with 32 × 32 hydrophones requires 8 × 8 projectors and a bandwidth of 12.5 × 64 = 800 Hz, while an array with 64 × 64 hydrophones calls for 4 × 4 projectors and a bandwidth of 12.5 × 16 = 200 Hz. The practical array for the implementation of an acoustic imaging system with a resolution of 256 × 256 points appears thus to be one with either 32 × 32 or 64 × 64 hydrophones.

TABLE 6.1-1

REPRESENTATIVE NUMBERS FOR MOVING ACOUSTIC IMAGES WITH A RESOLUTION OF NOMINAL 256 × 256 POINTS USING HYDROPHONE ARRAYS WITH 32 × 32 OR 64 × 64 HYDROPHONES

| | | | |
|---|---|---|---|
| 1 | Hydrophones | 32 × 32 | 64 × 64 |
| 2 | Image sections | 9 × 9 | 5 × 5 |
| 3 | Used inner points per image section according to Fig. 6.1-5 | 28 × 28 | 52 × 52 |
| 4 | Guard lines around image sections not displayed to avoid contour effect according to Fig. 6.1-5 | 2 | 6 |
| 5 | Projectors and orthogonal signals for insonification | 9 × 9 | 5 × 5 |
| 6 | Bandwidth required for display of 25 images per second | 1012.5 Hz | 312.5 Hz |
| 7 | Printed circuit cards according to Figs. 3.4-2 and 3.2-3 or 3.2-5 with 32 input and output terminals (a) or with 64 input and output terminals (b) | 2 × 32(a) | 2 × 64(b) |
| 8 | Time filters with 81 output terminals (a) or with 25 output terminals (b) according to Figs. 6.1-3 or 6.1-4 | 28 × 28(a) | 52 × 52(b) |
| 9 | Actually resolved points | 252 × 252 | 250 × 250 |

Table 6.1-1 gives some representative numbers for imaging systems with 32 × 32 or 64 × 64 hydrophones. Arrays of such size are well within our technology. The arrays of 9 × 9 or 5 × 5 projectors listed in line 5 of the table are also not too difficult to implement. The bandwidths listed in line 6 are quite acceptable if one considers that the nominal operating frequency is about 100 kHz. The printed circuit cards in line 7 are manageable. The real problem is created by the time filters. One can readily see why the filter of Fig. 6.1-3 was called impractical if one visualizes 28 × 28 such filters with 81 rather than 4 output terminals each. The time filters are shown separated from the rectifier, hold, and sampling circuits in Fig. 6.1-2, but they must practically be combined on printed circuit cards if one wants to avoid 252 × 252 = 63,504 or 250 × 250 = 62,000 interconnecting wires.

## 6.2 Increased Dynamic Range by Sidelobe Reduction

The diffraction pattern or—more precisely—the amplitude pattern produced by an electronic processor with one-dimensional array or "one-dimensional lens" was derived in Eq. (2.1-8). For a beam in the direction of the array axis, $\gamma = 0$, one obtains

$$v(0, \beta) = \frac{\sin(\pi A\beta/\lambda)}{\pi A\beta/\lambda} \tag{1}$$

where $A$ is the length of the array, $\lambda$ the wavelength, and $\beta$ the angle of incidence. A plot of $v(0, \beta)$ for $\beta > 0$ is shown in Fig. 6.2-1 by the

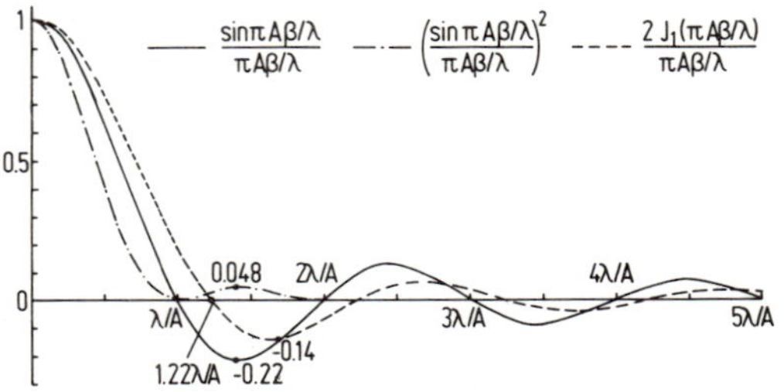


FIG. 6.2-1. Amplitude reception pattern for a line array for omnidirectional insonification (solid line) and for directional insonification (dashed-dotted line). The dashed line shows the pattern for a circular lens with a diameter equal to the length $A$ of the line array.

solid line. A wavefront with angle of incidence $\beta = 0$ produces the relative amplitude 1 at the output terminal for $\gamma = 0$ of the processor, while a wavefront with angle of incidence $\beta = 1.4\lambda/A$ produces the relative amplitude $-0.22$. If both wavefronts have the same power, the ratio of the magnitudes of the amplitudes of the processor output voltages will be $1/0.22 \doteqdot 4.5$, and the power ratio will be $(1/0.22)^2 \doteqdot 21$ or $10 \log 21 = 13$ dB. This is the dynamic range, since a wavefront coming from the correct direction $\beta = 0$ will produce the same output voltage as a wavefront coming from the wrong direction $\beta = 1.4\lambda/A$ but having 13 dB more power. Putting it differently, a small scatterer in direction $\beta = 0$ will produce the same output voltage as a larger scatterer returning 13 dB more power in direction $\beta = 1.4\lambda/A$, if both are equally insonified.

If the line array is replaced by a circular lens with diameter $A$, one obtains the amplitude pattern

$$v_c(0, \beta) = \frac{2J_1(\pi A\beta/\lambda)}{\pi A\beta/\lambda} \tag{2}$$

instead of Eq. (1), where $J_1$ is a Bessel function. A plot of $v_c(0, \beta)$ is shown by the dashed line in Fig. 6.2-1. The peak of the largest sidelobe is now only $-0.14$, which implies an amplitude ratio of $1/0.14 \doteqdot 7.15$ and a power ratio of $(1/0.14)^2 \doteqdot 50$ or 18.5 dB. This appears to be a significant improvement over Eq. (1), but the price for it is that the main lobe has been widened. The first zero has shifted from $\lambda/A$ for the solid line in Fig. 6.2-1 to $1.22\lambda/A$ for the dashed line. We know that the sidelobes can be reduced at the price of a wider main lobe by means of a linear transformation. This process is usually referred to as shading or tapering. However, the price is too high and we prefer to search for a better method to increase the dynamic range.

Such a better method is to insonify *unequally*. Consider Fig. 2.1-2, and let hydrophone $\gamma$ there be used as a sound projector. Most of the power will be radiated in the direction of $\gamma$, but there will be an amplitude transmission pattern around this angle that is equal to the amplitude reception pattern of Eq. (1) for a "one-dimensional lens," except that the maximum will be in the direction $\gamma$ rather than $\gamma = 0$. If we now stop the radiation and use the hydrophone for reception, the amplitude of the output voltage will be proportionate to the product of the amplitude transmission pattern and the amplitude reception pattern. Hence, the *round trip amplitude pattern* $v_r(\gamma, \beta)$ equals the square of Eq. (1). It is shown for $\gamma = 0$ by the dashed-dotted line in Fig. 6.2-1. The peak amplitude of the first sidelobe has dropped from $-0.22$ to $(-0.22)^2 \doteqdot 0.048$, which implies an amplitude ratio of $(1/0.22)^2 \doteqdot 21$ and a power ratio of $(1/0.22)^4 \doteqdot 427$ or 26 dB. This ratio is sufficiently good for halftone images.

In optics, we are usually not too worried about the dynamic range. The resolution of a photographic camera is rarely limited by the diffraction pattern of the lens. The film, the positioning of the film, and the chemical development limit the resolution long before the lens does, and as a result, the dynamic range is not determined by the first sidelobe of the diffraction pattern but by the tenth or higher numbered sidelobe. The sidelobe problem is overcome by using a large lens. Cost and mechanical problems prevent us from using larger hydrophone arrays in acoustics.

The principle of improving the dynamic range by means of the round trip amplitude pattern is used in radar, when the same parabolic dish is used first for beam forming for radiation and then for reception. The problem is to transfer the principle from one beam to many beams. A scanning radar solves this problem by forming many beams in a time sequence, or by *time multiplexing*. We cannot do so, since we want to produce eventually several hundred thousand beams. Doing this in a time sequence would take so long that no moving images could be produced, since a moving image requires that all beams are produced about 25 times per second. However, the term time multiplexing suggests the use of some other form of multiplexing, and we will try frequency multiplexing.

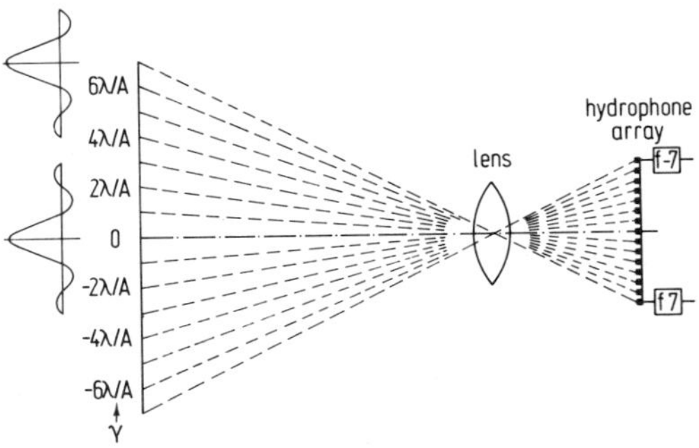


FIG. 6.2-2. Frequency multiplexing of round trip beam patterns.

Figure 6.2-2 shows an array of hydrophones used as projectors that produce beams in the directions $\gamma = -7\lambda/A, \ldots, +7\lambda/A$. Let each one be produced by a sine wave of slightly different frequency. The frequencies shall differ so little that we can ignore the change of the wavelength and its influence on the beam pattern, but they shall differ enough to make them separable by filters upon reception. Even though the beam pattern shown centered at $\gamma = 0$ will be produced with centers at $\gamma = -7\lambda/A$, $-6\lambda/A, \ldots, +7\lambda/A$, only one of them will be produced by a wave with correct frequency; all the others become in essence invisible due to suppression by the filters upon reception. Only the two filters for the frequencies

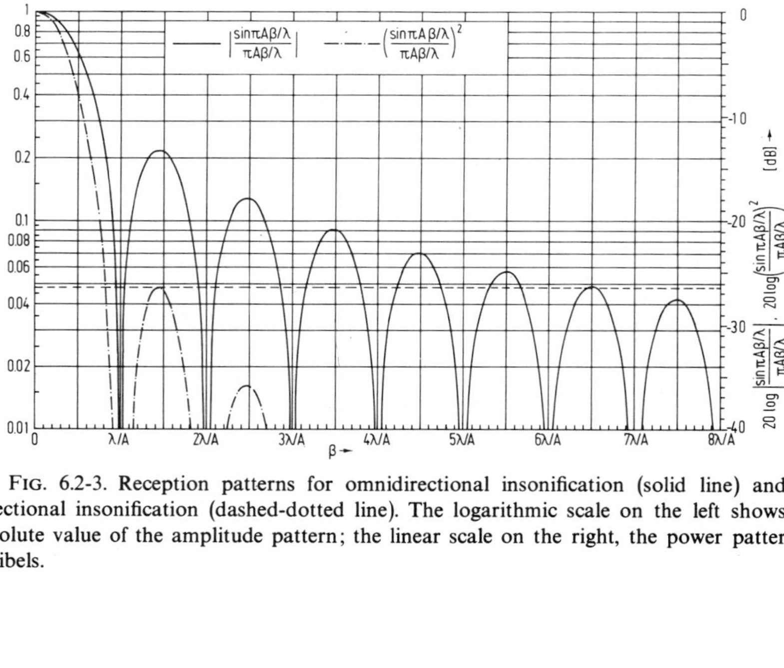


FIG. 6.2-3. Reception patterns for omnidirectional insonification (solid line) and for directional insonification (dashed-dotted line). The logarithmic scale on the left shows the absolute value of the amplitude pattern; the linear scale on the right, the power pattern in decibels.

$f-7$ and $f+7$ are shown in Fig. 6.2-2, but each hydrophone is connected to a filter. We have thus solved the problem of producing simultaneous, frequency multiplexed, round trip beam patterns.

A new problem arises. We cannot use a different frequency for each beam for a number of reasons, but we must find a way to make do with a relatively small number of frequencies. To see how this can be done we plot the curves $(\sin x)/x$ and $[(\sin x)/x]^2$ of Fig. 6.2-1 again in Fig. 6.2-3, but on a semilogarithmic scale. The problem of negative values of $(\sin \pi A\beta/\lambda)/(\pi A\beta/\lambda)$ is solved by plotting the absolute value. One can see that the peak of the sixth sidelobe of $(\sin \pi A\beta/\lambda)/(\pi A\beta/\lambda)$ has about the same magnitude, 0.048, as the peak of the first sidelobe of $[(\sin \pi A\beta/\lambda)/(\pi A\beta/\lambda)]^2$. As a result we may use the same frequency for the beams in the directions $\gamma = 0$, $\pm 7\lambda/A$, $\pm 14\lambda/A$, ... in Fig. 6.2-2 without reducing the dynamic range appreciably. A safer choice is to space the beams with equal frequency by multiples of eight, particularly since eight is a better number that seven for semiconductor technology.

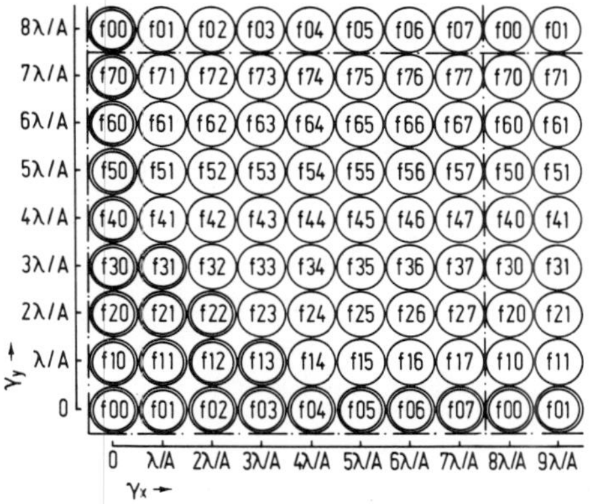


FIG. 6.2-4. Frequency allocations for image frames with 8 × 8 beams using 64 different frequencies $f00$ to $f77$ or $f_i$, $i = 1, \ldots, 64$.

Figure 6.2-4 shows a possible choice of frequencies if eight different frequencies are used in one dimension and thus 8 × 8 in two dimensions. Let us note that actually only the 21 frequencies in double circles are needed; either one of the frequencies $f14$, $f23$, $f32$, or $f41$ could be replaced by $f00$. Although there is great practical interest in this reduction of frequencies, no acceptable assignment better than the one of Fig. 6.2-4 has yet been worked out.

The principle of using frequency multiplexing of round trip beam patterns was explained in Fig. 6.2-2 with the help of a lens. It can be done just as well with an electronic circuit. The lens can be used for beam forming for radiation as well as reception; the electronic circuits for beam forming in both directions are also very similar, as was discussed in Sections 3.4 and 3.6. The electronic implementation of sidelobe reduction will be studied in more detail in the following section.

## 6.3 Circuits for Sidelobe Reduction

The architecture of circuits for sidelobe reduction as discussed in the preceding section is shown in Fig. 6.3-1. On the right we have 2 × 16 active beam-forming cards for $\gamma_x$ and $\gamma_y$. The circuits on these cards are shown by

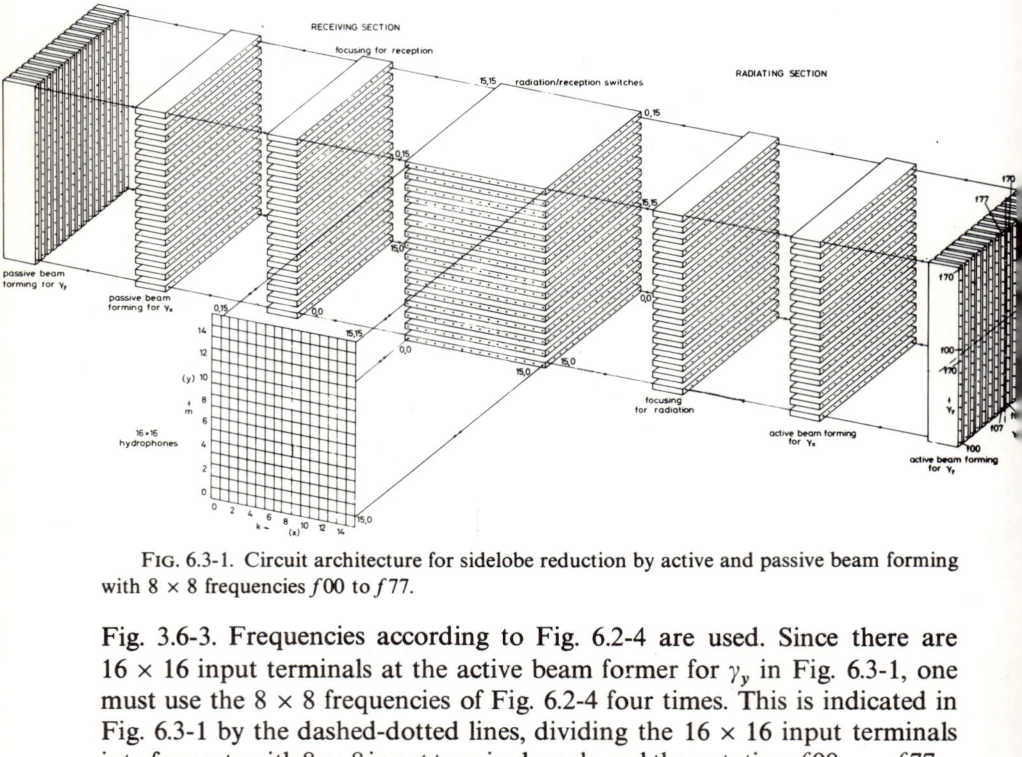


FIG. 6.3-1. Circuit architecture for sidelobe reduction by active and passive beam forming with 8 × 8 frequencies $f00$ to $f77$.

Fig. 3.6-3. Frequencies according to Fig. 6.2-4 are used. Since there are 16 × 16 input terminals at the active beam former for $\gamma_y$ in Fig. 6.3-1, one must use the 8 × 8 frequencies of Fig. 6.2-4 four times. This is indicated in Fig. 6.3-1 by the dashed-dotted lines, dividing the 16 × 16 input terminals into four sets with 8 × 8 input terminals each, and the notation $f00, \ldots, f77$.

The two sets of active beam-forming cards in Fig. 6.3-1 produce nominally planar wavefronts. These wavefronts will "focus" points at infinity. In order to focus for a finite distance $R$, one must phase shift the voltages at the hydrophones so that the radiated sound wave has equal phase on the circle with radius $R$ in Fig. 4.1-2 rather than on the surface of the flat hydrophone array. This calls for exactly the same focusing circuits as for reception.

Hence, the stack of cards in Fig. 6.3-1 denoted "focusing for radiation" contains circuits according to Fig. 4.3-4.

The following set of cards, "radiation/reception switches," connects the hydrophones to the radiating section during radiation, and to the receiving section during reception. These switches must also prevent the direct transfer of the power from the radiating to the receiving section.

Received signals pass from the hydrophones via the radiation/reception switches to a stack of cards denoted "focusing for reception," and to the passive beam-forming cards with circuits according to Fig. 3.4-2. Beyond the passive beam-forming cards come the time filters, rectifier, hold, and sampling circuits as shown in Fig. 5.2-1.

A block diagram of all the electronic circuits required for acoustic imaging with sidelobe reduction is shown in Fig. 6.3-2. A hydrophone

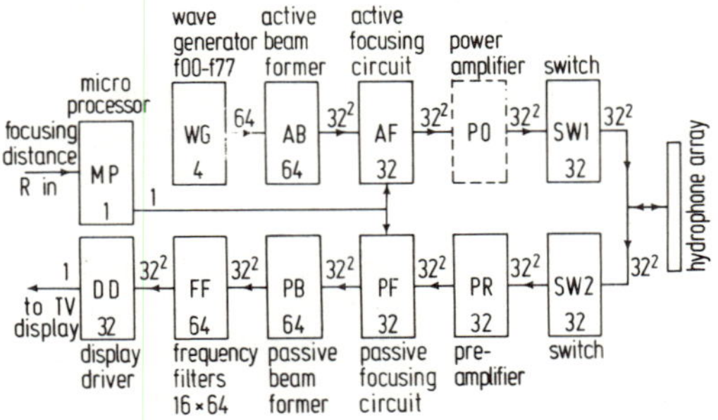


FIG. 6.3-2. Block diagram of electronic circuits for sidelobe reduction using an array with 32 × 32 hydrophones. The figures in the blocks show the number of printed circuit cards, while the figures between the blocks show the number of connecting wires.

array with 32 × 32 hydrophones is assumed. A wave generator WG produces 64 sinusoidal oscillations with the frequencies $f00 = 300$ Hz, $f01 = 600$ Hz, $f10 = 900$ Hz, ..., $f77 = 19{,}200$ Hz. They are shifted by suppressed carrier, double-sideband modulation to the center frequency of 1 MHz. The resulting frequencies are 1 MHz ± 300 Hz to 1 MHz ± 19,200 Hz. The relative deviation $19200/10^6 = 0.0192$ is less than 2% of the center frequency and is thus generally negligible. A further compensation is achieved by using the double-sideband modulation. For every frequency 1 MHz + $f_i$ that is too high there is another frequency 1 MHz − $f_i$ that is too low. Since the sum of the signals with both frequencies is used, one obtains a diffraction pattern equal to the average pattern of the signals with too high and too low a frequency.

The 64 sinusoidal functions with frequencies $f00$ to $f77$ are produced by digital circuits to guarantee stability; 64 PROMs[1] with 512 × 8 bits storing 512 samples of sin $\omega t$ to sin $64\omega t$ are a practical way to implement

[1] Programmable Read Only Memories.

these circuits. By deriving all functions from one clock rate one can change the frequencies proportionately by changing the clock rate. This permits experimentation with frequency allocations different from the multiple of 300 as discussed here. The shifting by a carrier with frequency 1 MHz can be done with a ring modulator.

The $8 \times 8$ carriers modulated with the 64 sinusoidal voltages with frequency $f00$ to $f77$ are fed to the $32 \times 32$ input terminals of the active beam former AB in Fig. 6.3-2 according to the pattern of Fig. 6.2-4. Hence, each voltage is fed to $4 \times 4 = 16$ input terminals.

The beam former produces the necessary voltages for $32 \times 32$ beams for planar wavefronts. The active focusing circuit AF produces the phase shifts for a finite focusing distance $R$. The focusing is achieved by feeding binary digital numbers to the circuit, which are then stored in shift registers. A set of these numbers has to be produced whenever a new distance $R$ is chosen.

The power amplifier PO in Fig. 6.3-2 is not always needed. Since we have $32 \times 32 = 1024$ hydrophones/projectors we need only 10 mW per projector to produce 10 W total power. The 10 mW can be produced by operational amplifiers. A power of 10 W is about right for medical diagnosis equipment, but not sufficient for acoustic imaging underwater.

The radiation/reception switch in Fig. 6.3-1 is replaced by two switches SW1 and SW2 in Fig. 6.3-2. The switches SW1 are closed during radiation and open otherwise. The switches SW2 may be closed whenever the switches SW1 are open, or they may be closed for a shorter time only. The second option makes it possible to produce the image of a plane at a certain distance only, as will be discussed more fully in Section 8.1.

The received signals pass through the preamplifier PR and are lifted to a voltage level of about 1 V; this calls for a gain of 60 to 80 dB. The passive focusing circuit PF and the passive beam former PB need no further comment.

The frequency filters FF contain a demodulator that removes the carrier with frequency 1 MHz, separates the 64 frequencies $f00$ to $f77$, and performs some more operations. This circuit will be discussed in more detail below.

The display driver DD in Fig. 6.3-2 contains sampling circuits that sample the $32^2$ voltages arriving simultaneously from the frequency filters FF according to the lines of a TV display, and feeds the resulting serial voltage samples to a TV monitor. This operation calls for $32 + 1$ sampling circuits with 32 input terminals each, plus 2 five-stage binary counters. Such circuits are commercially available as integrated circuits.

We return now to the frequency filters FF. For an explanation of the required operations, consider the 64 frequencies $f00$ to $f77$, but let them be denoted $f_i$, $i = 1, \ldots, 64$. Let furthermore $f_0$ denote a carrier frequency.

We modulate the carrier $\sin(2\pi f_0 t + \varphi)$, having an arbitrary phase, by one of the voltages $\sin 2\pi f_i t$:

$$\sin(2\pi f_i t)\sin(2\pi f_0 + \varphi) = [\cos 2\pi(f_0 - f_i)t - \cos 2\pi(f_0 + f_i)t]\cos\varphi$$
$$-[\sin 2\pi(f_0 - f_i)t - \sin 2\pi(f_0 + f_i)t]\sin\varphi$$

A signal with the same time variation will also be received, but the phase angle $\varphi$ will be changed to $\varphi'$ due to the propagation delays. Let us multiply this returned signal with the original carrier $\sin(2\pi f_0 t + \varphi)$:

$$[\sin(2\pi f_i t)\sin(2\pi f_0 + \varphi')]\sin(2\pi f_0 + \varphi)$$
$$= \tfrac{1}{2}[\cos(\varphi - \varphi')(1 - \cos 4\pi f_0 t) + \sin(\varphi + \varphi')\sin 4\pi f_0 t)]\sin 2\pi f_i t$$

The terms with frequency $2f_0$ can easily be suppressed. However, the resulting low-frequency term $\cos(\varphi - \varphi')\sin 2\pi f_i t$ has an amplitude that can be very small if $\varphi - \varphi'$ is close to $\pm\pi/2$.

Let us demodulate the received signal with $\cos(2\pi f_0 t + \varphi)$:

$$[\sin(2\pi f_i t)\sin(2\pi f_0 t + \varphi')]\cos(2\pi f_0 t + \varphi)$$
$$= \tfrac{1}{2}[\sin(\varphi' - \varphi)(1 - \cos 4\pi f_0 t) + \sin(\varphi + \varphi')\cos 4\pi f_0 t]\sin 2\pi f_i t$$

Again, the terms with frequency $2f_0$ can be suppressed. The two remaining terms

$$\tfrac{1}{2}\cos(\varphi - \varphi')\sin 2\pi f_i t, \qquad \tfrac{1}{2}\sin(\varphi' - \varphi)\cos 2\pi f_i t$$

yield after rectification

$$\tfrac{1}{2}\cos(\varphi - \varphi'), \qquad \tfrac{1}{2}\sin(\varphi - \varphi')$$

and one obtains by squaring and adding

$$\tfrac{1}{4}[\cos^2(\varphi - \varphi') + \sin^2(\varphi - \varphi')] = \tfrac{1}{4}$$

The phase angles are eliminated. Figure 6.3-3 shows the block diagram of a circuit that performs these operations.

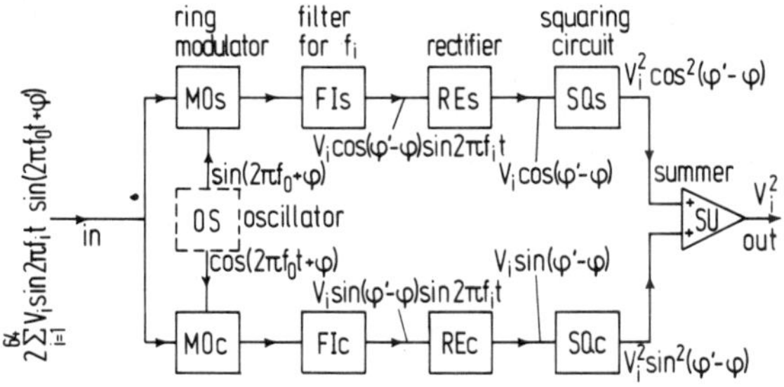


FIG. 6.3-3. Demodulation and separation of the oscillations with different frequencies $f_i = f00, \ldots, f77$. The oscillator OS is shared by the circuits for all $32^2$ beams.

# 7 Implementation of Imaging by Digital Circuits

## 7.1 Digital Delay Circuits

Let the output voltages of the hydrophones in Fig. 1.2-7 be fed to $16 \times 16$ analog-to-digital converters that convert at a sufficiently high sampling rate. We may then use digital circuits for processing instead of the analog circuits used so far. The delay circuits may be replaced by digital shift registers. Each delay circuit has to be replaced by $r$ shift registers if the analog-to-digital converter produces $r$ binary digits per sample. The analog circuits had a dynamic range for the voltage of about 1000 : 1. Digital circuits would require $r = 10$ binary digits to obtain the same dynamic range ($2^{10} = 1024$). Hence, the circuit of Fig. 2.1-4 calls for $10 \times 16 = 160$ shift registers, while the two-dimensional circuit of Fig. 2.1-5 requires $160 \times 16 \times 2 = 5120$ shift registers.

To get a rough estimate of the number of stages in the shift registers, let us calculate the delay time in Fig. 2.1-4 between the taps of the delay circuit connected to hydrophone 6 and leading to the summing amplifiers denoted $-7\lambda/A$ and $-6\lambda/A$:

$$
\begin{aligned}
t_0 + (6 - 2^4/2 + 1)(d/v)\sin(-7\lambda/A) \\
- [t_0 + (6 - 2^4/2 + 1)\sin(-6\lambda/A)] \\
= (d/v)[\sin(7\lambda/A) - \sin(6\lambda/A)] \doteq d\lambda/vA
\end{aligned}
$$

This is the smallest incremental delay required, which means that each stage of the shift registers should produce that much delay. The total delay of a delay circuit divided by $d\lambda/vA$ yields the number of stages required. The total delay of the circuit connected to hydrophone zero can be expressed in two ways. Counting the number of taps, one sees that it must be equal to $15(t_0/8)$. The delay can also be expressed by $t_0 + (0 - 4^4/2 + 1)(d/v)\sin(-7\lambda/A)$. Hence, one can calculate $t_0$:

$$
\begin{aligned}
15(t_0/8) = t_0 + (0 - 4^4/2 + 1)(d/v)\sin(-7\lambda/A) \\
t_0 = 8(d/v)\sin(7\lambda/A) \doteq 56d\lambda/vA
\end{aligned}
$$

All delay lines in Fig. 2.1-4 require a delay between $t_0$ and a little more than $2t_0$. The average delay is thus $1.5t_0$, and the average number of stages per shift register equals $1.5 \times 56 = 84$. The total number of shift register stages in Fig. 2.1-4 is thus about $160 \times 84 = 13{,}440$ and about

5120 × 84 = 430,080 in Fig. 2.1-5. These numbers are very large. The problem of cost becomes even more impressive when one considers the 16 × 16 = 256 analog-to-digital converters in front of the delay circuits in Fig. 2.1-5. As a result, digital delay circuits are currently important for one-dimensional arrays with a low dynamic range only. Typically, the value $r = 1$ is used, which means a positive voltage is converted into the digit 1 and a negative voltage into the digit 0. The analog-to-digital converter becomes in this case a simple threshold detector.

Let us see whether the implementation of the inverse Fourier transform by digital circuits is more promising.

## 7.2 Digital Circuits for the Fourier Transform

Figure 7.2-1 shows the digital version of the circuit of Fig. 3.1-3. The major difference between the two circuits is in the summers. Analog circuits can readily sum four voltages at a time, but digital circuits typically sum

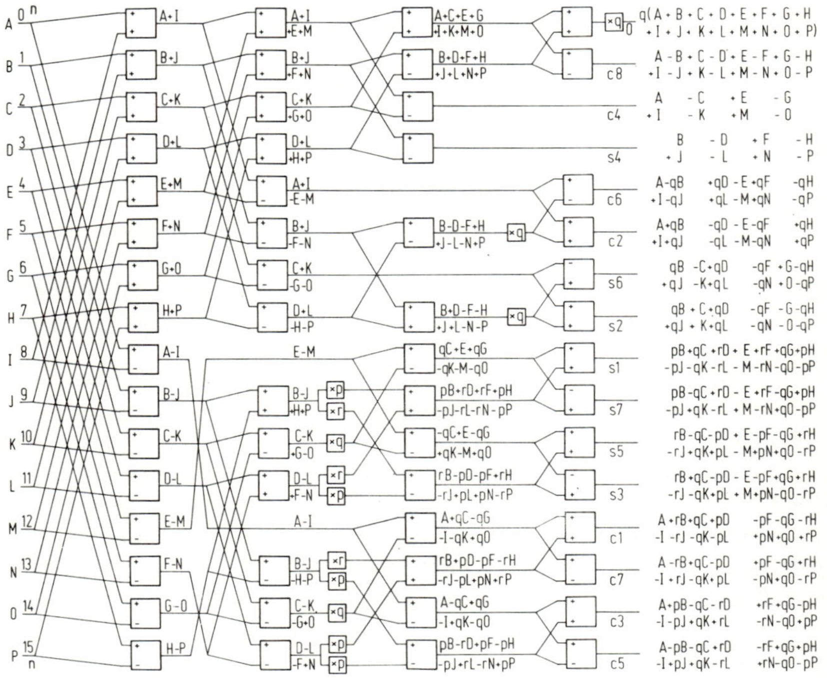


FIG. 7.2-1. Digital circuit performing the fast Fourier transform of 16 input voltages. The circuit blocks with two positive signs (+ +) are adders; those with a positive and a negative sign (+ −), subtracters; those with a multiplication sign ($\times p$, $\times q$, $\times r$), multipliers. $p = \sin \pi/8$, $q = \sin \pi/4$, $r = \sin 3\pi/8$.

only two numbers at a time; hence the need for almost twice as many adders (+ +) or subtracters (+ −) as there are operational amplifiers in Fig. 3.1-3. The multiplications performed by resistors in Fig. 3.1-3 require digital multipliers, represented by the blocks with the notation $\times p$, $\times q$, and $\times r$. Digital multipliers are rather expensive, and the cost of the circuit of Fig. 7.2-1 is determined by the few multipliers, not by the many adders and subtracters.

At the present, the circuit of Fig. 7.2-1 is too expensive to be acceptable. However, it is perfectly suited for large-scale integration, and it is quite possible that it will be one day[1] more practical than the analog circuit of Fig. 3.1-3.

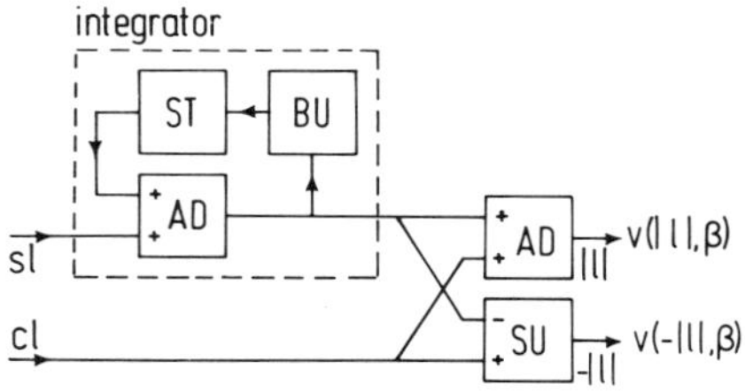


FIG. 7.2-2. Digital version of the circuit in Fig. 3.4-1c. AD, adder; SU, subtracter; ST, storage; BU, buffer storage.

A digital implementation of one of the circuits in Fig. 3.4-1c is shown in Fig. 7.2-2. The integration is performed by adding an incoming number and a number stored in storage ST by the adder AD. The sum of the two numbers is fed back via a buffer storage BU to the storage ST. Hence, the storage ST contains the sum of all received numbers.

One can use time sharing to make the digital circuits of Figs. 7.2-1 and 7.2-2 less expensive. This is what is done if the Fourier transform is performed by computer processing. A new problem is encountered. If the acoustic wave used has a frequency of 100 kHz one should take upward of 200,000 samples per second and convert them into digital numbers. This is a very fast rate for a digital circuit using multipliers, and the rate is prohibitive if one wants to time share the multipliers. Synchronous demodulation permits operation at much slower rates.

[1] Large-scale integrated (LSI) circuits are common today. Very large scale integrated (VLSI) circuits are about to enter mass production. The circuit of Fig. 7.2-1 is either a complicated VLSI circuit or a simple ULSI (ultra large-scale integrated) circuit. In ten years, it may be a routine component like the microprocessor is now.

## 7.3 Synchronous Demodulation as the Key to Practical Circuits

The output voltage $v(i, t)$ of a hydrophone of an array according to Eq. (2.2-1),

$$v(i, t) = V \sin[2\pi(vt + id \sin \beta)/\lambda] = V \sin 2\pi[ft + i(d/\lambda) \sin \beta] \quad (1)$$

is multiplied by a voltage $e(t)$ with amplitude 2:

$$e(t) = 2 \sin 2\pi f_1(t - t_0) \quad (2)$$

One obtains:

$$\begin{aligned} v(i, t)e(t) = V\{&\cos 2\pi[(f - f_1)t + f_1 t_0 + i(d/\lambda) \sin \beta] \\ &- \cos 2\pi[(f + f_1)t - f_1 t_0 + i(d/\lambda) \sin \beta]\} \end{aligned} \quad (3)$$

Let the frequencies $f$ and $f_1$ be about equal. The difference frequency $f - f_1 = f'$ is then very small, while the sum frequency $f + f_1$ equals about $2f$. One may suppress the term with the sum frequency in Eq. (3) by a simple filter. Stray capacitances of the circuits will usually do this filtering. The remaining low-frequency term is denoted $v'(i, t)$:

$$\begin{aligned} v'(i, t) &= V \cos 2\pi[f'(t + t_0') + i(d/\lambda) \sin \beta] \\ &= V \sin[2\pi(v't' + id \sin \beta)/\lambda] \\ f' &= f - f_1, \qquad t_0' = t_0 f_1/f', \qquad v' = \lambda f' = v(1 - \lambda/\lambda_1) \\ t' &= t + t_0' + 1/4f', \qquad \lambda f = v, \qquad \lambda_1 = v/f_1 \end{aligned} \quad (4)$$

A comparison of Eq. (4) with Eq. (1) shows that we have the same voltage except that the frequency $f$ is replaced by $f'$, the velocity $v$ by $v'$, and the time variable $t$ by $t'$. The results of Eqs. (2.2-1)–(2.2-12) remain unchanged except for the replacement of the time-variable terms $\sin(2\pi vt/\lambda)$ and $\cos(2\pi vt/\lambda)$ by $\sin(2\pi v't'/\lambda)$ and $\cos(2\pi v't'/\lambda)$.

The reduction of the operating frequency $f$ in the water to the lower processing frequency $f'$ in the equipment has obvious advantages. One can choose $f$ to be 100 kHz to keep the hydrophone array small, but convert $f$ to a frequency $f'$ on the order of a few hundred or a few thousand hertz to avoid the need for shielded wiring and to improve the performance of the operational amplifiers. However, one must be somewhat careful not to choose the processing frequency $f'$ too low, lest the integrator in Fig. 3.4-1c, as well as certain filters, become too bulky.

When one uses digital circuits one does not have to be concerned about bulkiness caused by a low processing frequency. Hence, one may reasonably choose $f_1 = f$ and convert the sinusoidal output voltages of the hydrophones into dc voltages. Let us consider $v'(k)$ in the form shown by Eq. (3) for $f_1 \to f$:

$$v'(i, t) = V\{\cos 2\pi(f - f_1)t \cdot \cos 2\pi[f_1 t_0 + i(d/\lambda) \sin \beta] \\ - \sin 2\pi(f - f_1)t \cdot \sin 2\pi[f_1 t_0 + i(d/\lambda) \sin \beta]\} \quad (5)$$

For $f - f_1 = 0$, the first term becomes

$$v'(i, t) = V \cos 2\pi[f_1 t_0 + i(d/\lambda) \sin \beta] \quad (6)$$

but the second term vanishes and with it the value of $\sin 2\pi[f_1 t_0 + i(d/\lambda) \sin \beta]$. One cannot reconstruct $v'(i, t)$ of Eq. (5) without the knowledge of $\sin 2\pi[f_1 t_0 + i(d/\lambda) \sin \beta]$. Hence, Eq. (6) contains less information than Eq. (5). To regain the lost term we multiply Eq. (1) with the voltage $e'(t)$

$$e'(t) = 2 \cos 2\pi f_1(t - t_0) \quad (7)$$

We obtain

$$v'(i, t)e'(t) = V\{\sin 2\pi[(f - f_1)t + f_1 t_0 + i(d/\lambda) \sin \beta] \\ + \sin 2\pi[(f + f_1)t - f_1 t_0 + i(d/\lambda) \sin \beta]\} \quad (8)$$

The low-frequency term is denoted $v''(i, t)$:

$$v''(i, t) = V\{\sin 2\pi(f - f_1)t \cdot \cos 2\pi[f_1 t_0 + i(d/\lambda) \sin \beta] \\ + \cos 2\pi(f - f_1)t \cdot \sin 2\pi[f_1 t_0 + i(d/\lambda) \sin \beta]\} \quad (9)$$

Now the first term vanishes for $f - f_1 = 0$, and we obtain the term lost by Eq. (6):

$$v''(i, t) = V \sin 2\pi[f_1 t_0 + i(d/\lambda) \sin \beta] \quad (10)$$

In the present context the term $f_1 t_0$ in Eqs. (6) and (10) is undesirable. One may eliminate it by using synchronized voltages $e(t)$ and $e'(t)$ in Eqs. (2) and (7). For instance, one may use $v(i, t)$ for $i = 0$ in Eq. (1) for $e(t)$, and produce $e'(t)$ by a phase shift of $v(i, t)$. We will accept without further discussion that $v'(i, t)$ and $v''(i, t)$ can be produced without the term $f_1 t_0$:

$$v'(i, t) = V \cos[2\pi i(d/\lambda) \sin \beta], \qquad v''(i, t) = V \sin[2\pi i(d/\lambda) \sin \beta]$$

We may now multiply $v'(i, t)$ by $\sqrt{2}/2$ and $\cos[2\pi i(d/\lambda) \sin \gamma]$ from Eq. (2.2-2), while $v''(i, t)$ is multiplied by $\sin[2\pi i(d/\lambda) \sin \gamma]$, and produce the sums of Eqs. (2.2-7)–(2.2-9). However, one may just as well add $v'(i, t)$ and $v''(i, t)$, multiply and sum, since the sums of the terms

$$V \cos[2\pi i(d/\lambda) \sin \beta] \sin[2\pi i(d/\lambda) \sin \gamma], \\ V \sin[2\pi i(d/\lambda) \sin \beta] \cos[2\pi i(d/\lambda) \sin \gamma]$$

and $(\sqrt{2}/2)V \sin[2\pi i(d/\lambda) \sin \gamma]$ vanish. One obtains

$$
\begin{aligned}
v'(0, \beta, t) &= \frac{1}{2}\sqrt{2} \sum_{i=-2^n/2+1}^{2^n/2} [v'(i, t) + v''(i, t)] \\
&= \frac{1}{2}\sqrt{2} \sum_{i=-2^n/2+1}^{2^n/2} \cos(2id\lambda^{-1} \sin \gamma) \\
v'(\gamma, \beta, s, t) &= \sum_{i=-2^n/2+1}^{2^n/2} [v'(i, t) + v''(i, t)] \sin(2\pi id\lambda^{-1} \sin \gamma) \\
&= V \sum_{i=-2^n/2+1}^{2^n/2} \sin(2\pi id\lambda^{-1} \sin \beta) \sin(2\pi id\lambda^{-1} \sin \gamma) \\
v'(\gamma, \beta, c, t) &= \sum_{i=-2^n/2+1}^{2^n/2} [v'(i, t) + v''(i, t)] \cos(2\pi id\lambda^{-1} \sin \beta) \\
&= V \sum_{i=-2^n/2+1}^{2^n/2} \cos(2\pi id\lambda^{-1} \sin \beta) \cos(2\pi id\lambda^{-1} \sin \gamma)
\end{aligned} \tag{11}
$$

A comparison with Eqs. (2.2-3)–(2.2-5) shows that $v'(0, \beta, t)$, $v'(\gamma, \beta, s, t)$, and $v'(\gamma, \beta, c, t)$ are identical with $v(0, \beta, t)$, $v(\gamma, \beta, s, t)$, and $v(\gamma, \beta, c, t)$, except that the time-variable terms $\sin(2\pi vt/\lambda)$ and $\cos(2\pi vt/\lambda)$ have to be replaced by 1. As a result, the integration of Eq. (2.2-10) is not needed to obtain the voltages $v(l, \beta, t)$ and $v(-l, \beta, t)$ of Eqs. (2.2-11) and (2.2-12), but without the time-variable term $\sin(2\pi vt/\lambda)$:

$$
\begin{aligned}
v'(0, \beta, t) &= \frac{1}{2}\sqrt{2}V \frac{A}{d} \frac{\sin(\pi A\lambda^{-1} \sin \beta)}{\pi A\lambda^{-1} \sin \beta} \\
v'(l, \beta, t) &= v'(\gamma, \beta, c, t) + v'(\gamma, \beta, s, t) = V \frac{A}{d} \frac{\sin \pi(A\lambda^{-1} \sin \beta - l)}{\pi(A\lambda^{-1} \sin \beta - l)} \\
v'(-l, \beta, t) &= v'(\gamma, \beta, c, t) - v'(\gamma, \beta, s, t) = V \frac{A}{d} \frac{\sin \pi(A\lambda^{-1} \sin \beta + l)}{\pi(A\lambda^{-1} \sin \beta + l)} \\
l &= A\lambda^{-1} \sin \gamma \qquad \text{for} \quad \gamma > 0 \\
-l &= A\lambda^{-1} \sin \gamma \qquad \text{for} \quad \gamma < 0
\end{aligned} \tag{12}
$$

Let us discuss the practical difference between the conversion of the operating frequency $f$ to a lower processing frequency $f - f_1$ and the synchronous demodulation of $f$ to dc. The one multiplier per hydrophone required for frequency conversion is shown in Fig. 7.3-1a. No filter is needed to suppress the term with the frequency $f + f_1$. The output voltage $v'(i, t)$ is fed to the input terminals in Fig. 3.1-3 or—after analog-to-digital conversion—to the input terminals in Fig. 7.2-1. The integrators in Figs. 3.4-1 and 7.2-2 are required.

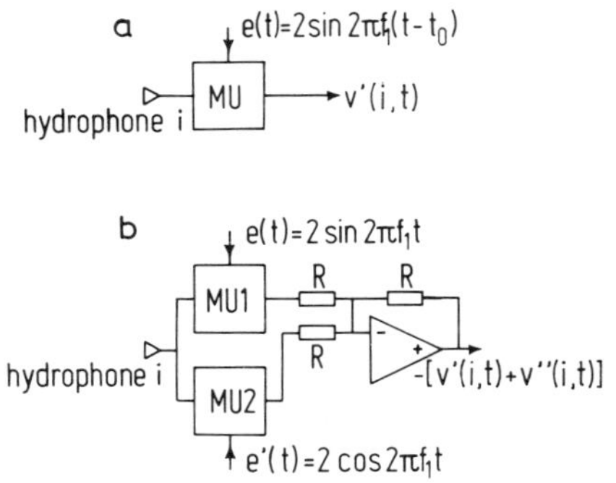


FIG. 7.3-1. (a) Conversion of the operating frequency to a lower processing frequency, (b) synchronous demodulation to dc. MU, multiplier.

Synchronous demodulation requires the two multipliers in Fig. 7.3-1b and the summer. In return, the integrators in Figs. 3.4-1 and 7.2-2 are eliminated. In essence, one trades a multiplier for an integrating capacitor. The main objection to synchronous demodulation is the need to provide synchronized voltages $e(t) = 2 \sin \pi f_1 t$ and $e'(t) = 2 \cos 2\pi f_1 t$. There is little incentive to use dc instead of ac with a frequency of a few hundred to a few thousand hertz, if the Fourier transform is done with the analog circuits of Chapter 3. The same holds true if a parallel digital circuit as in Fig. 7.2-1 is used. However, if this circuit is used with time-shared multipliers, adders, and subtracters, a factor between a few hundred and a few thousand translates directly into an increase in processing time by the same factor. In this case, one will definitely prefer synchronous demodulation to frequency conversion.

The term synchronous demodulation has been used in electrical engineering for upward of half a century. Optical holography is in essence an implementation of synchronous demodulation at the frequencies of coherent light. The word holography was then transferred from optics to acoustics. Holography in acoustics may be implemented by electric voltages as discussed here under the name synchronous demodulation. This is done, e.g., in the equipment developed by Booth and Sutton (1974) or by Marom *et al.* (1971). Let us observe that square waves can and often are used instead of the sinusoidal voltages with the same period in Eqs. (2) and (7), since the harmonics are readily suppressed by stray capacitances. There are, however, other ways to implement acoustical holography for which the term synchronous demodulation would not mean much. These methods typically use interference patterns produced by the sound waves on the interface between two materials, and a laser beam. Examples of this technique are discussed in papers by Mueller and Keating (1969) and Fritzler *et al.* (1969).

# 8 Special Effects Producible by Electronic Circuits

## 8.1 Range Images and Three-Dimensional Images

Photography by means of a lens is mainly useful for light waves. The physical dimensions of lenses have restricted their use in acoustics or radar. The lens produces transformations that are almost exclusively based on delay and summation. Multiplication is used only in the simple forms provided by masks and wavelength-selective filter glasses. More general linear and nonlinear operations are currently being developed under the name optical processing and should eventually become applicable for image generation and processing. Time-variable devices that can be changed in times short compared with the period of the used light are beyond the foreseeable development of our technology.

Holography is a means to obtain and store the amplitudes and phases of wavefronts. In the case of light waves one may use this stored information to generate an image or one may use it as input for optical processing. There is no equivalent to optical processing in acoustics, since acoustic lenses are so cumbersome compared with optical lenses.

The main feature of electrical processing of acoustic waves is the possibility of using time-variable devices that can be changed in a time that is short compared with the period of the wave. A secondary feature, compared with optical processing, is the advanced technology of electronics. Amplification, variation of the frequency, time filtering, amplitude clipping, and so forth can be done more readily and more accurately in electronics than in optics. Of course, electronic components do not work as fast as optical devices, but this is of little concern in acoustics.

An application that makes use of rapid time variation of a circuit is the generation of range images or three-dimensional images. According to Fig. 8.1-1 a projector radiates a sinusoidal *pulse* with $n$ periods of duration $T$ rather than a continuous wave. The projector is connected via a switch to the power oscillator at the time $t_0$ and disconnected at the time $t_0 + nT$. The pulse travels to a scatterer at a distance $R$ and returns. The front of the pulse arrives at the time $t_0 + 2R/v$, the end of the pulse arrives at the time $t_0 + 2R/v + nT$. Let the hydrophones be connected during the time $t_0 + 2R/v < t < t_0 + 2R/v + nT$ to the processing circuit. The waves scattered on a surface of a sphere of radius $R$ will be processed as if there had

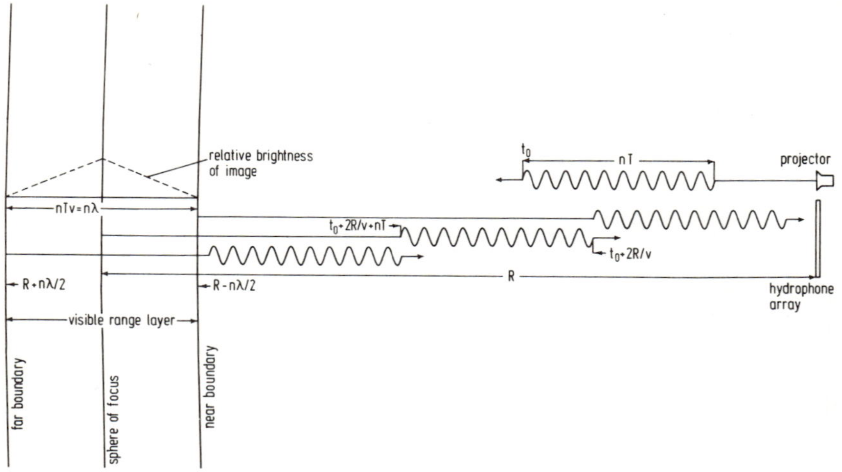


FIG. 8.1-1. Generation of the image of a spherical layer of thickness $n\lambda$ by means of pulsed sinusoidal waves with $n$ cycles per pules.

been no switches. Waves scattered at a distance $r < R$ will have their "heads" cut off by the switches, and pulses scattered at a distance $r > R$ will have their "tails" cut off. Waves scattered at distances $r < R - n\lambda/2$ or $r > R + n\lambda/2$ will be suppressed completely. These distances are denoted *near boundary* and *far boundary* in Fig. 8.1-1. The fraction of the wave returned from the region between these boundaries will vary like the triangle denoted *relative brightness.*

Let the pulse have $n = 100$ periods and let the period be $T = 10$ $\mu$sec, corresponding to a frequency of 100 kHz for the continuous sinusoidal wave. The thickness of the *visible range layer* will then be $nTv = n\lambda = 1.5$ m. The advantage of producing only an image of this layer is the suppression of returns from other scatterers, such as air bubbles, at different distances that would make the image appear "foggy." The optical equivalent of a range image would permit us to see through fog. For instance, shutters in front of the headlights of a car synchronized with shutters in front of the eyes of the driver would permit the driver to see only the light scattered at a certain distance, but suppress the light scattered by the fog at closer or larger distances.

A slight modification permits one to obtain three-dimensional images from range images. Let the projector in Fig. 8.1-1 again send out pulses with $n$ periods, but let no switches be used at the output terminals of the hydrophone array. Instead, the optical display—shown as a light-emitting diode display in Fig. 8.1-2—is projected onto a moving ground-glass plate.

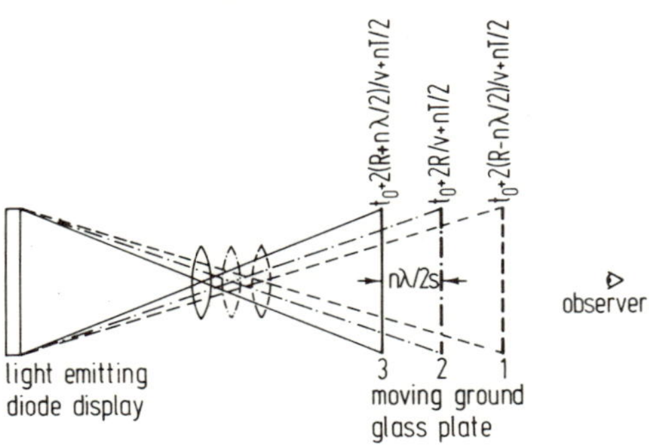


FIG. 8.1-2. Generation of a genuine three-dimensional image by means of pulsed sinusoidal waves and a display on a moving ground-glass plate.

The return from the near boundary in Fig. 8.1-1 produces an image during the time interval $t_0 + 2(R - n\lambda/2)/v < t < t_0 + 2(R - n\lambda/2) + nT$. The peak brightness will occur at the time $t_0 + 2(R - n\lambda/2)/v + nT/2$. The moving ground-glass plate in Fig. 8.1-2 is at that time in position 1. At the time $t_0 + 2R/v + nT/2$ the image of the sphere of focus in Fig. 8.1-1 will appear with peak brightness; the moving ground-glass plate in Fig. 8.1-2 will be in position 2. At the time $t_0 + 2(R + n\lambda/2)/v + nT/2$ the image of the far boundary in Fig. 8.1-1 will appear with peak brightness; the moving ground-glass plate in Fig. 8.1-2 will be in position 3. Hence, images of objects at different distances from the hydrophone array will be seen on the ground-glass plate at different distances from the observer.

The position of the ground-glass plate in Fig. 8.1-2 must be properly scaled to the position of the imaged layers. A linear reduction of $1/s$ means that the position of the ground-glass plate in Fig. 8.1-2 must change $n\lambda/2s$ if the distance of the imaged layer in Fig. 8.1-1 is changed by $n\lambda/2$. If the ground-glass plate is moved physically, its velocity must be $v/s$ if $v$ is the velocity of sound in water.

The displayed image in Fig. 8.1-2 is a genuine three-dimensional image. The observer can move relative to the display to see the three-dimensional image from various angles. This is different from the stereoscopic pictures in photography that have a fixed viewing angle, and which require that one eye see one two-dimensional image and the other eye another two-dimensional image. Optical holography, on the other hand, also yields genuine three-dimensional images, but by a different process.

The movement of the ground-glass plate in Fig. 8.1-2 does not have to be made mechanically. One can substitute many thin layers of a nematic liquid crystal. One layer at a time can be made translucent by applying a voltage across it; the other layers can be made transparent by applying no voltage or only a very small voltage. This layer then acts like a ground-glass plate. The mechanical movement is thus replaced by an electrical movement. The movement of the lens can be avoided by making the ground-glass plate much larger than the light-emitting diode display and by using a mask with a small focal opening. This is basically the same problem as increasing the depth of field of a camera. A fixed lens would make an image on the ground-glass plate in position 1 in Fig. 8.1-2 appear larger than in position 2 or 3 and thus produce a perspective image.

## 8.2 Doppler Images

Let an object approach with the velocity $v_r$ relative to a projector and a hydrophone array. A sinusoidal wave with frequency $f$ produced by the projector will be returned to the hydrophones with a frequency $f(1 + 2v_r/v)$, where $v$ is the velocity of sound in water. For $f = 100$ kHz the frequency shift $\Delta f = 2fv_r/v$ equals 1 kHz for $v_r = 7.5$ m/sec = 14.6 knots. A resonance filter for 100 kHz with a bandwidth of about 1 kHz can be built fairly easily. Let $16 \times 16$ time filters be inserted between the beam former and the driving circuits for the electrooptic display as shown in Fig. 5.2-1. Phase distortions caused by the filters are of no concern if they are inserted at this place. Let the projector radiate a wave with frequency 99 kHz. The image produced by a stationary object $v_r = 0$ will be blocked by the filters and will not be displayed. An object approaching with a velocity $v_r = 7.5$ m/sec will return a wave with frequency 100 kHz. The image of this object will not be blocked by the filters. Hence, by changing the frequency of the insonifying wave one can make objects with certain velocities visible and suppress others.

What is the limit for the Doppler resolution? We have seen in the section on synchronous demodulation that a sinusoidal voltage with a frequency of 100 kHz can readily be converted to a sinusoidal voltage with a much lower frequency. The frequency $\Delta f$ of the Doppler shift remains unchanged during such a conversion. Hence, instead of building a resonance filter at 100 kHz with an approximate bandwidth of 1 kHz, one can convert from 100 kHz to 10 kHz. A resonance filter with a bandwidth of 100 Hz can be built at 10 kHz with about as much effort as one with a bandwidth of 1 kHz at 100 kHz. A Doppler shift of 100 Hz is produced by a relative velocity $v_r = 0.75$ m/sec = 1.46 knots. There is no theoretical limit for the Doppler resolution, since one can convert the frequency of 100 kHz to 1 kHz, 100 Hz, etc. However, building good, small, and inexpensive resonance filters at such low frequencies is a major problem. There is a theoretical limit for the Doppler resolution if one wants to display moving images.

The display of 25 images per second calls for a bandwidth of 12.5 Hz according to the sampling theorem. Hence, the minimum Doppler shift must be about 12.5 Hz, which translates into a relative velocity $v_r = 0.06$ m/sec = 0.18 knots.

Range and Doppler imaging can be combined to yield images of objects at a certain distance with a certain velocity. The practical significance of such images is the elimination of the images of air bubbles, of boundary layers in water or of the sea bottom, which obscure the image of a wanted object. We have explained range images in terms of optics by means of shutters that permit one to look through fog. An optical equivalent of a range–Doppler image is to use in addition the Doppler effect, e.g., to make visible only cars that move with excessive velocity through fog. To do so with light waves is beyond our current technology, but there is no technological barrier to the use of range–Doppler images to look through fog or tropical foliage by means of sound waves.

## 8.3 Telelens Effect

The resolution angle $\varepsilon = \lambda/A$ is a function of the wavelength $\lambda$ for an array with aperture $A$. Figure 8.3-1 shows five representative beam patterns for the wavelength $\lambda$ in the direction of integer multiples of $\pm\varepsilon$. Let a

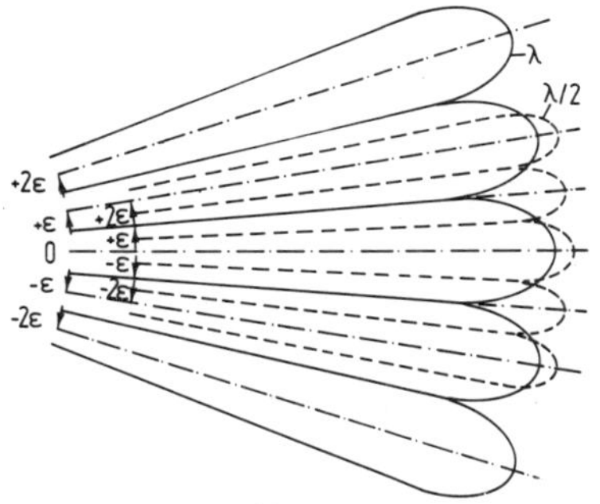


FIG. 8.3-1. Telelens effect for a fixed array of hydrophones and a variable wavelength of the insonificating waves.

wave with the wavelength $\lambda/2$ be used, but let the array remain unchanged. The resolution angle is now $\varepsilon = \lambda/2A$ and the beam patterns shown by dashed lines apply. The change of the wavelength has the same effect as a change of the focal length of a lens in photography.

Let us see what influence a change of wavelength has on the equipment. Hydrophones can work over a fairly wide frequency range, particularly since the sensitivity does not have to be constant. The Fourier transform circuits of Chapter 3 were frequency independent, but the passive and active beam formers in Sections 3.4 and 3.6 were not. The focusing circuit of Fig. 4.3-2 is also frequency dependent. To overcome this frequency dependence, one may use frequency conversion to operate in the processing circuits with a fixed frequency while radiating and receiving sound waves with different frequencies. Alternately, one may make the frequency-dependent integrators in Figs. 3.4-1 and 4.3-2 frequency independent by replacing the fixed input resistors of the integrators by variable resistors. The practical way of doing so is to use digital–analog converters that use a resistor network with switches as discussed in connection with Figs. 4.3-1 and 4.3-9.

One may derive from Fig. 8.3-1 two curious effects of imaging of *active* sound sources. Two sources at essentially the same location radiating waves with wavelengths $\lambda$ and $\lambda/2$ will appear at different locations on the electro-optic display. Furthermore, due to the frequency dependence of the circuits in Fig. 3.4-1, at least one of the active sources will appear in all four quadrants of the image. Of course, both phenomena disappear if one uses narrow-band time filters between the beam former and the electrooptic display as shown in Fig. 5.2-1.

## 8.4 Pseudocolor Images

We have so far assumed insonification by a sine wave with a certain frequency. In terms of photography this would mean illumination with monochromatic light. Such illumination would produce black-and-white pictures, although the shades of gray would differ from the ones obtained with the usual nonmonochromatic light sources. To obtain color pictures in photography, we can illuminate either with white light or we can use three monochromatic light sources emitting red, green, and blue light, respectively. Substituting three monochromatic light sources for one source of white light is not practical in photography, but it is the proper way to produce color images by means of optical holography or spatial electric filters.

The equivalent of three monochromatic light sources in underwater acoustics is three projectors radiating sound waves with three different frequencies. One can choose these frequencies within the technical limitations of the equipment. This gives one more freedom than one has in photography or optical holography, where the light sources must emit the three colors for which the eye is sensitive. Only infrared photography provides independence between the frequency of the light used to produce a picture and the frequencies to which the eye is sensitive.

Using three acoustic waves with different frequencies will provide us with information about the frequency dependence of the scattering of the observed object and the frequency dependence of the absorption in the water. The absorption in seawater changes drastically with frequency, and one will primarily get a range effect. The wave with the highest frequency will produce images primarily of near objects, while the waves with lower frequency will show this effect less strongly. We have discussed previously how images of objects at a certain distance can be produced by gating the projector and the hydrophones. Combining the use of three frequencies with gating eliminates the influence of the frequency dependence of absorption in the water on the final color of the image. The brightness of the image will still depend on the frequency, but this can be compensated by proper amplification, since the distance to the object and thus the absorption in the water is known.

Let us now see how color can be used to discriminate between different materials of scatterers. Rayleigh derived the following expression for the relative amplitude[1] $\sigma_i$ of a wave returned from an incompressible spherical scatterer of diameter $D$ if $D$ is small compared with the wavelength $\lambda$:

$$\sigma_i = (D/\lambda)(4\pi/9)(4C_0^2 + 3C_1^2) \tag{1}$$

The constants $C_0$ and $C_1$ depend on the density and elasticity of the material of the sphere. The relative amplitude $\sigma_c$ of a wave returned from a highly compressible air bubble of diameter $D$ was also derived by Rayleigh:[2]

$$\sigma_c = (D/\lambda)^4 4\pi^4 v^4 (\rho/3\gamma p_0)^{1/2} \tag{2}$$

$v$ = sound velocity in water, $\rho$ = density of water, $p_0$ = pressure of the incident wave, $\gamma$ = ratio of specific heat at constant pressure to specific heat at constant volume (Albers, 1965).

[1] The term "scattering cross section" or "effective target area" is used for $(D^2\pi/4)\sigma_i$. This terminology would be misleading in image generation, since a pointlike scatterer with large effective target area is represented on the display by a very bright point, not by a large bright area.

[2] Rayleigh's law of scattering uses $(D^2\pi/4)\sigma_c$ instead of $\sigma_c$.

Let us assume that a wavelength $\lambda$ is used that yields the same value for $\sigma_i$ and $\sigma_c$. An incompressible and a highly compressible scatterer yield in this case the same brightness and cannot be distinguished. Let a second wave with shorter wavelength $\lambda' < \lambda$ be used. $\sigma_c$ will increase much faster than $\sigma_i$ with decreasing wavelength. Let the image produced by the wave with length $\lambda$ be displayed in red and the image produced by the wave with length $\lambda'$ in green. The image of the incompressible scatterer will then appear yellow, while that of the compressible scatterer will be green-yellow. Hence, pseudocolor can be used to discriminate between scatterers of different compressibility. As a practical example one may think of a skin diver with a foam-rubber suit of high compressibility against a background of clay, sand, or rock with low compressibility. Military skin divers or swimmers, using highly incompressible insulating suits, can still be detected by the compressibility of the air in their lungs.

In addition to the intrinsic color of objects, we have in optics the color of thin films. What is an equivalent for acoustic waves? Consider a ship bottom with a layer of marine growth 1 cm thick. The ship bottom will appear rough to a wave with a wavelength significantly shorter than 1 cm, but smooth to one with a wavelength significantly longer than 1 cm. Hence, pseudocolor can be used to show the surface structure of scatterers.

How can one produce pseudocolor images practically? To have three projectors, three hydrophone arrays, three processors, and one color TV tube would not be practical. Let us instead have three projectors radiating waves with frequencies $f_r$, $f_g$, and $f_b$, but time share the hydrophone array and the processor. To do so one may insert multipliers between the output terminals of the hydrophone array and the input terminals of the processor. Somewhere in the processor one needs narrow-band filters with center frequency $f_0$. During a first time interval of about 100-msec duration one feeds the voltage $V_r \sin 2\pi(f_r - f_0)t$ to the multipliers. The resulting voltages with frequency $f_r - (f_r - f_0) = f_0$ will pass through the filters, while the voltages $f_g \pm (f_r - f_0)$, $f_b \pm (f_r - f_0)$, and $f_r + (f_r - f_0)$ will be blocked. These voltages will be represented by a red image. One then applies the voltages $V_g \sin 2\pi(f_g - f_0)t$ for 100 msec to obtain a green image, and $V_b \sin 2\pi(f_b - f_0)t$ to obtain a blue image. The amplitudes $V_r$, $V_g$, and $V_b$ are to be chosen so as to compensate the different attenuations of the three waves.

If the processor is using the delay principle according to Fig. 2.1-4 or the sampling principle according to Fig. 2.3-1, one will see all scatterers at the same location in the red, green, and blue images. This is not so if one uses the Fourier transform. We have seen from Fig. 8.3-1 that a telelens effect is produced by this circuit if the frequency of the wave is changed.

## 8.5 Spread Spectrum Insonification

There are many military applications for which the insonification required for acoustic imaging poses no problem, but there are others where the user does not want to radiate. Submariners are generally reluctant to radiate in order not to reveal their location. A similar situation existed in radio communications, but was widely overcome by the development of spread spectrum transmission over the last twenty years. The same principles are applicable to acoustic imaging.

To create the sensation of a moving image we have to produce about 25 images per second. A bandwidth of 12.5 Hz is thus required according to the sampling theorem. If we spread the signal power over 1250 Hz, we reduce the spectral power density of the signals by 10 log (12.5/1250) = −20 dB. A bandwidth of 1250 Hz at a nominal frequency of 100 kHz causes no problem with the frequency selective circuits, since the deviation from the nominal frequency is less than 1%. At a nominal frequency of 300 kHz one can spread the signal over a band of 12.5 kHz to achieve a reduction of the spectral power density by −30 dB, and deviate from the nominal frequency by little more than 2%.

The high attenuation in seawater makes spread spectrum techniques for imaging particularly attractive. Let the insonified target have the distance $R$. The returned power will decrease like $1/R^4$ due to the geometric spreading; the attenuation will be proportional to $-40 \log R$ [dB]. The attenuation due to losses in the seawater will equal $-2R\alpha$ [dB], where $\alpha \doteq 2.75 \times 10^{-12} f^2$ [dB/m]. The total attenuation due to geometric spreading and losses equals $-[40 \log(R/R_0) + 2R\alpha]$. Consider now a receiver at a distance $2R$ from the sound source. The received power will decrease like $1/(2R)^2$ due to geometric spreading; the attenuation will be proportional to $-20 \log 2R$ [dB]. The attenuation due to losses in the seawater will equal $-2R\alpha$ [dB] as before. The total attenuation equals now $-[20 \log(2R/R_0) + 2R\alpha]$. If $\alpha$ is very large, the term $2R\alpha$ will dominate either the term $20 \log(2R/R_0)$ or the term $40 \log(R/R_0)$, and the difference between the laws for geometric spreading—$1/R^4$ and $1/(2R)^2$—will become unimportant. Hence, spread spectrum insonification underwater is much more attractive than spread spectrum radiation in radar.

Frequency hopping is a simple method for the implementation of spread spectrum insonification. Every sinusoidal pulse sent out has a different nominal frequency. Since transmitter and receiver are at the same location, one can select the frequencies randomly. Upon reception, the frequencies have to be shifted by multiplication with an auxiliary carrier to a fixed frequency. This is all easy to do. The only practical difficulty is that the time filters in Fig. 5.2-1 must have a bandwidth of close to 12.5 Hz to make full use of spread spectrum insonification.

Frequency hopping is particularly attractive when combined with multiplexing and sidelobe reduction, as discussed in Chapter 6, since one can interchange the frequencies used for different beams. We will not go any further into this subject since the principle is readily understandable, but no equipment using spread spectrum insonification for imaging has been built yet.

# 9 Synthetic Aperture Processing

## 9.1 Transmission of Power and Information by Sinusoidal Waves

Except for the discussion of image forming by tapped delay lines and by sampled storage circuits in Sections 2.1 and 2.3, we have based our discussion on periodic sinusoidal or pulsed sinusoidal waves. The reason was strictly that our current technology permits us to build equipment economically on the basis of sinusoidal functions. The price for this simpler implementation is reduced performance. We have seen in Sections 8.1 and 8.2 on range and Doppler images that signals with a small relative frequency bandwidth can produce effects and provide information that a pure sinusoidal wave cannot. This is in line with a basic rule of communications that a pure sinusoidal function transmits information at the rate zero, and that the transmitted information increases proportionately to the frequency bandwidth of a signal. Hence, we will investigate in this chapter what can potentially be accomplished with nonsinusoidal waves in acoustic imaging, although this theoretical investigation is well ahead of the available technology.

One of the most important features of sinusoidal waves is that they are not distorted when propagating through a linear, time-invariant medium. Hence, we need no distortion-correcting equipment if we operate our ac power transmission lines with sinusoidal currents rather than with ac currents having some other time variation. Communication lines, on the other hand, require *equalizers* to correct the distortions of the signals, which must be nonsinusoidal in order to transmit information at a rate larger than zero.

In first approximation, seawater is a linear, time-invariant transmission medium for sound waves. Hence, sinusoidal waves appear to be the waves of choice. However, this instant conclusion becomes already less convincing when we consider the efficiency of conversion of electric input power to acoustic power in the object plane.

The power source for any wave is almost always a dc power supply with a fixed voltage $V$. With two power supplies providing the voltages $+V$ and $-V$, we can produce a square wave with amplitude $V$. Let this square wave drive on electroacoustic transducer, and let the resulting wave propagate through seawater.[1] At some distance we will receive the fundamental

[1] See Roi (1970) about acoustic projectors for pulses.

sinusoidal component of the square wave with period $T$, and much more attenuated sinusoidal components with periods $T/3$, $T/5$, .... Consider now the transmission of a sinusoidal wave of period $T$. We must first produce a sinusoidal voltage from the dc voltages of the power supplies. The most efficient known way to do so is by[1] *class D conversion*. One produces the same square wave with amplitude $V$ as above, but suppresses all the harmonics with period $T/3$, $T/5$, ... in order to obtain the fundamental sinusoidal oscillation with period $T$ only. Using any reasonable acoustic transducer and letting the sinusoidal wave travel the same distance through seawater as the square wave, we will always receive less average power. This is, of course, a result of generating the sinusoidal wave by class D conversion from dc via a square wave. The result will not hold true if we find a more efficient way of generating a sine wave from dc. Let us calculate the efficiency of generating the sine wave via the square wave. The square wave shall have the amplitude $V$. A Fourier decomposition shows that the amplitude of its fundamental sine function is $4V/\pi$. The efficiency of power conversion becomes

$$\frac{\text{average power of fundamental sine function}}{\text{available dc power}} = \frac{1}{2}\frac{(4V/\pi)^2}{V^2} \doteqdot 0.81$$

Hence, the efficiency of class D power conversion is 81%. Only if we find a practical method of conversion with higher efficiency can the acoustic wave with sinusoidal time variation have a higher average power than the square wave at any distance. This is just the opposite of what one would obtain by assuming a sinusoidal and a square wave of equal average power at the electroacoustic converter, and ignoring that the power of both waves is coming from dc power supplies.

This point was somewhat theoretical. Let us now turn to a practical example of increasing power by not using sinusoidal waves. Consider a sonar that uses pulses of 1 $\mu$sec duration. Most of the energy of such pulses is concentrated in the band $0 < f < 1$ MHz. A sinusoidal carrier used for this pulse would have to have a frequency well above 10 MHz. Since the attenuation due to absorption, in decibels per unit distance, increases with the square of the frequency, one would need upward of 100 dB more signal power to compensate the higher absorption losses. Clearly, a sinusoidal signal giving the same range resolution is out of the question. Sonar with 1-$\mu$sec-long pulses may sound unrealistic for use underwater, but such sonar

[1] The efficiency of power conversion from dc to sinusoidal is 25% for class A amplifiers, 50% for class B amplifiers, and about 67% for class C amplifiers or oscillators if the impedance of the power source is matched to the load.

was built for medical diagnosis (von Ramm and Thurstone, 1975; Kisslo and von Ramm, 1975).

Let us turn from power to information transmission. As mentioned before, it is a fundamental rule of communications that a pure sinusoidal function transmits information at the rate zero. In beam forming, we receive information about the location of a source that emits a wave. In the one case, we deal with the time variable $t$ and in the other with the space variable $x$, but there must be a fundamental rule for space which is equivalent to the rule "the transmission rate of information is zero for sinusoidal functions."

For the derivation of this rule consider Fig. 9.1-1. A line array with

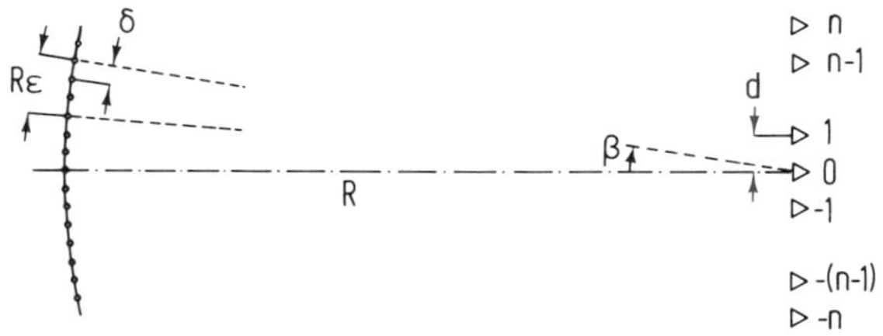


FIG. 9.1-1. The relative number of points resolved at a distance $R$ by an array of $2n + 1$ sensors using sinusoidal waves.

$2n + 1$ equally spaced sensors is shown on the right. This array can resolve $2n + 1$ plane waves with different angles of incidence, or $2n + 1$ "points." The resolution angle equals $\varepsilon = \lambda/(2n + 1)d$. The resolvable points at a distance $R$ have the distance $\Delta x = R\varepsilon$ from each other. It is usual to give them the locations $0, \pm\varepsilon R, \pm 2\varepsilon R, \ldots, \pm n\varepsilon R$. Note that we do not have to assume that $R$ is infinite. Our investigation is thus valid for focused beams that resolve points at a finite distance.

How many points are there on a circle of radius $R$ in the sector $-n\varepsilon < \beta < +n\varepsilon$? Figure 9.1-1 shows on the left points with a distance $\delta$, while only points with a distance $R\varepsilon$ can be resolved by the array. Since we are deriving a theoretical limit, we have three different choices for the distance $\delta$, or the number $N$ of points in the sector $-n\varepsilon < \beta < +n\varepsilon$:

(a) $\delta$ is zero in such a way that $N$ is nondenumerably infinite. This corresponds to our usual assumption of a continuous space.

(b) $\delta$ is zero in such a way that $N$ is denumerably infinite.

(c) $\delta$ is larger than zero and finite.

For the cases (a) and (b) the relative number of resolved points equals $(2n + 1)/N = 0$, regardless of the distance $R$. However, one is not easily impressed by the practical significance of such a result, since it is based on a mathematical abstraction. The case (c) with a finite value of $\delta$ is the only one with physical significance. At a distance $R$ there will be $(2n + 1)\varepsilon R/\delta$ points[1] with distance $\delta$ in the sector $-n\varepsilon < \beta < +n\varepsilon$. The relative number of resolvable points becomes $(2n + 1)/[(2n + 1)\varepsilon R/\delta] = \delta/\varepsilon R$. For $R \to \infty$ the ratio $\delta/\varepsilon R$ converges to zero. This is a physically significant result. Note that the physically nonsignificant cases (a) and (b) yielded a relative number of resolvable points that equaled zero for any value of $R$, while the significant case (c) only yields a convergence to zero for increasing values of $R$. Hence, for planar wavefronts, coming from points at infinity, we have the following theorem:

*The relative number of resolved points is zero for sinusoidal waves if the wavefronts are planar and if a line array with a finite number of equally spaced sensors is used.*

There are many ways to generalize this theorem, the most obvious ones being for beams focused to a finite distance $R$ or sets of beams focused to the distances $R$, $R + \Delta R$, $R + 2\Delta R$, .... We will not pursue this line, but concentrate on finding out how one could resolve more than $2n + 1$ points with $2n + 1$ sensors.

## 9.2 Introduction to Synthetic Aperture Theory

Synthetic aperture theory for radar was introduced by C. Wiley in 1951. A book by Harger (1970), a collection of papers in a book by Kovaly (1976), and a translation of a Russian book edited by Reutov (1970) provide a survey of the field. Although the theory of synthetic aperture radar is usually based on sinusoidal functions, the signals actually used have a bandwidth produced by a Doppler shift. The importance of this Doppler bandwidth was clearly stated by Kovaly[2] (1976):

"Each resolution function is inversely proportional to a bandwidth—in the range dimension, the RF bandwidth, in the azimuth dimension, the Doppler bandwidth. Fundamentally, then, resolution is obtained in both dimensions by generating a bandwidth."

The moving vehicle carrying the conventional sidelooking radar is a mechanical means for the modulation of a sinusoidal carrier to produce a signal with a finite bandwidth.[3]

We will investigate synthetic aperture techniques strictly from the stand-

[1] The number is actually either $2n\varepsilon R/\delta$ or $(2n + 1)\varepsilon R/\delta$, but this is of no consequence.

[2] Paragraph following Fig. 9 on the fifth page of the Introduction. Page is not numbered.

[3] The Doppler effect as used by the conventional sidelooking radar produces a signal with a finite bandwidth, not a frequency-shifted sinusoidal wave without a bandwidth, since the relative velocity between radar and target changes continuously.

point of nonsinusoidal waves and not assume a sinusoidal wave that is made mechanically nonsinusoidal by a moving vehicle. However, it will turn out that some of the results also apply if the nonsinusoidal signal is amplitude modulated onto a sinusoidal carrier.

For an explanation of this point, refer to Fig. 9.2-1. It shows a block

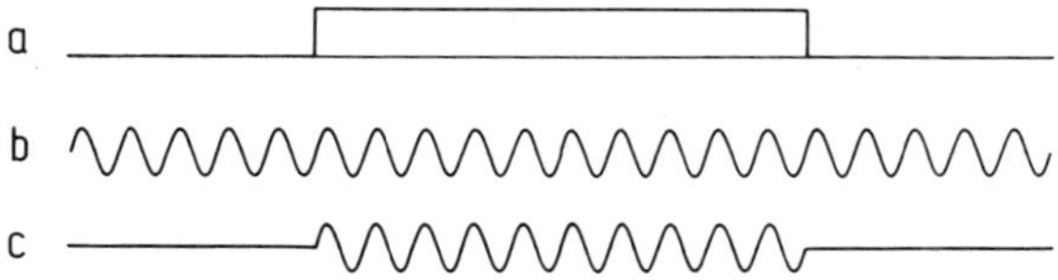


FIG. 9.2-1. Block pulse (a) multiplied with a sinusoidal function (b) yields the pulsed sinusoidal function (c).

pulse (a), which we generally assume as our signal. Consider now sonar or acoustic imaging based on sinusoidal waves. It uses the block pulse (a) as baseband signal, the periodic sinusoidal function (b) as carrier, and produces an amplitude modulated signal (c). Beam forming is always based on the sinusoidal carrier, even though the baseband signal and the carrier are contained in the radiated signal as equally important factors of a product. An investigation of synthetic aperture techniques for the block pulse (a) also applies to beam forming using the baseband signal in conventional radar and sonar.

Figure 9.2-2 shows two sonars $RA_1$ and $RA_2$, located at $-X$ and $+X$, measuring the two distances $d_{11}$ and $d_{21}$ to the target $TA_1$ with the coordinates $x_1$, $y$. These measurements require only the block pulse (a) in Fig. 9.2-1, since the sinusoidal carrier can be—and usually is—removed by demodulation. The distances $d_{11}$ and $d_{21}$ are expressed in cartesian coordinates:

$$d_{11} = [(x_1 + X)^2 + y^2]^{1/2} \tag{1}$$

$$\begin{aligned} d_{21} &= [(x_1 - X)^2 + y^2]^{1/2} \\ &= [(x_1 + X - 2X)^2 + y^2]^{1/2} = d_{11}(1 - 4Xx_1/d_{11}^2)^{1/2} \\ &\doteqdot d_{11}(1 - 2Xx_1/d_{11}^2) \qquad \text{for} \quad Xx_1 \ll d_{11}^2 \end{aligned} \tag{2}$$

Consider the *linear* difference between the two distances $d_{11}$ and $d_{21}$:

$$d_{11} - d_{21} \doteqdot 2Xx_1/d_{11} \tag{3}$$

It vanishes for large values of $d_{11}$:

$$\lim_{d_{11}\to\infty} (d_{11} - d_{21}) = 0 \tag{4}$$

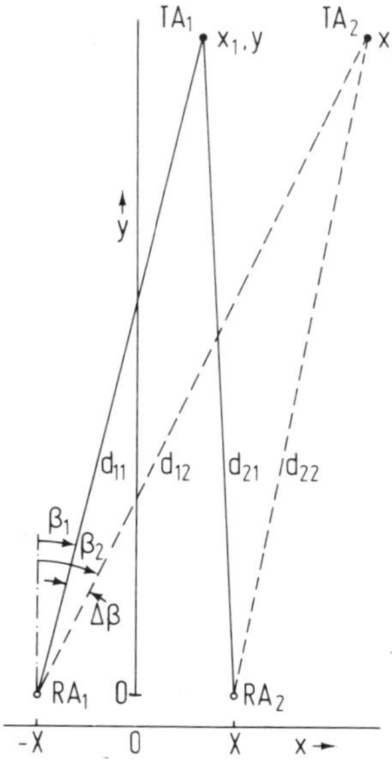


FIG. 9.2-2. Angular or polar coordinate resolution and cartesian coordinate resolution achievable with two sonars $RA_1$ and $RA_2$ located at $-X$ and $+X$.

Hence, for any finite value of $X$ and $x_1$ the linear difference $d_{11} - d_{21}$ will have the same value 0 if $d_{11}$ is large enough, and we will not be able to calculate $x_1$ from the linear difference $d_{11} - d_{21}$, the distance $d_{11}$ of the target, and the distance $2X$ between the two radars.

Let a sinusoidal wave be radiated, reflected, or backscattered by the target TA. The propagation times of a wavefront to the two sonar receivers equal $d_{11}/c$ and $d_{21}/c$. We cannot measure these times; we can only measure phase differences. However, if the distance $2X$ between the two sensors is no larger than half the wavelength $\lambda$, the measurement of the phase difference becomes equivalent to the measurement of the relative arrival time $t_{r,11}$ of the wavefront at sonar $RA_1$:

$$t_{r,11} = d_{11}/c - d_{21}/c, \qquad 2X \leq \lambda/2 \tag{5}$$

The right side of Eq. (5) equals the left side of Eq. (3), except for the constant $1/c$, and the relative time of arrival $t_{r,11}$ must thus vanish for large values of $d_{11}$ like the right side of Eq. (3). Hence, basing the resolution on the linear difference $d_{11} - d_{21}$ obtained from the absolute propagation time of two pulses or on the phase difference of two sinusoidal oscillations yields a very similar result. The two targets $TA_1$ and $TA_2$ in Fig. 9.2-2 must have an infinite distance $x_2 - x_1$ for $y \to \infty$, but the ratio $(x_2 - x_1)/y$ can have a finite value $\Delta\beta$:

$$\Delta\beta \doteq (x_2 - x_1)/y, \qquad x_2 - x_1 \to \infty, \qquad y \to \infty, \qquad \Delta\beta \ll 1 \tag{6}$$

Let us emphasize that Eqs. (3) and (5) do *not* yield the same resolution for a finite distance. The spacing $2X$ of the two sensors in Fig. 9.2-2 for a sinusoidal wave must not be larger than $\lambda/2$ according to Eq. (5), while there is no such restriction in Eq. (3). This is one reason why one should not ignore the baseband signal for beam forming.

The sinusoidal wave yields only the difference $d_{11} - d_{21}$, but the block pulse yields the two distances $d_{11}$ and $d_{12}$ separately. Hence, we can square $d_{11}$ and $d_{21}$ to obtain the *quadratic* difference from Eqs. (1) and (2):

$$d_{11}^2 - d_{21}^2 = 4Xx_1 \tag{7}$$

The quadratic difference does not vanish for large values of $d_{11}$ as does the linear difference of Eq. (3). Note that Eq. (7) is exact, while Eq. (3) is an approximation.

Let us calculate the minimum distance $x_2 - x_1$ between the targets $TA_1$ and $TA_2$ in Fig. 9.2-2, based on the quadratic difference of distances. In analogy to Eqs. (1), (2), and (7), one obtains:

$$\begin{gathered} d_{12} = [(x_2 + X)^2 + y^2]^{1/2}, \qquad d_{22} = [(x_2 - X)^2 + y^2]^{1/2} \\ d_{12}^2 - d_{22}^2 = 4Xx_2 \end{gathered} \tag{8}$$

The resolvable distance $x_2 - x_1$ follows from Eqs. (7) and (8):

$$x_2 - x_1 = [(d_{12}^2 - d_{22}^2) - (d_{11}^2 - d_{21}^2)]/4X \tag{9}$$

The smallest resolvable distance depends on the accuracy with which the distances $d_{11}$ to $d_{22}$ can be measured, that is, on the time resolution of the sonar, and it depends on the distance $2X$ between the two sonars. The coordinate $y$ is not contained in Eq. (9), and the resolvable distance $x_2 - x_1$ remains thus constant as $y$ increases toward infinity.[1] Such a result is too good to be true. Let us see where the limitation for the resolvable distance comes from.

Equation (3) is solved for $x_1$:

$$x_1 \doteq d_{11}(d_{11} - d_{21})/2X \tag{10}$$

A sonar cannot measure a distance $d_{ij}$ to a target with unlimited accuracy, but only with a certain error $\Delta d_{ij} \ll d_{ij}$. This error is partly due to features

[1] This is the same idealized result as known for the focused conventional synthetic aperture radar (Cutrona and Hall, 1962). The resolution angle is zero, and the minimum separation of two resolvable points does not depend on their distance from the radar.

of the equipment, such as jitter, and partly due to the features of the transmission path. We do not need to make any assumptions about the error $\Delta d_{ij}$ at this time beyond the condition that $\Delta d_{ij}$ be small compared with $d_{ij}$.

The error of $x_1$ due to inaccurate measurements of distance is denoted $\Delta x_1$. Introduction of the errors $\Delta d_{ij}$ into Eq. (10) yields the following result:

$$x_1 + \Delta x_1 \doteqdot (d_{11} + \Delta d_{11})(d_{11} + \Delta d_{11} - d_{21} - \Delta d_{21})/2X \tag{11}$$

Ignoring the error terms of higher than first order, we obtain:

$$\Delta x_1 = [d_{11}(2\Delta d_{11} - \Delta d_{21}) - d_{21}\Delta d_{11}]/2X \tag{12}$$

The corresponding error for the quadratic difference is calculated from Eq. (7):

$$x_1 + \Delta x = [(d_{11} + \Delta d_{11})^2 - (d_{21} + \Delta d_{21})^2]/4X \tag{13}$$

$$\Delta x_1 \doteqdot (d_{11}\Delta d_{11} - d_{21}\Delta d_{21})/2X \tag{14}$$

The comparison of Eqs. (12) and (14) shows that the error $\Delta x_1$ increases in both cases proportionately to the distances $d_{11}$ and $d_{21}$. The factors of $d_{21}$ are essentially the same in both equations, but the factors of $d_{11}$ are not. Since the errors $\Delta d_{ij}$ are not constants but random variables, the factor $2\Delta d_{11} - \Delta d_{21}$ in Eq. (12) is less desirable than the factor $\Delta d_{11}$ in Eq. (14). Hence, even though the quadratic difference $d_{11}^2 - d_{21}^2$ is not as good as it appears to be from Eq. (9), it is better than the linear difference $d_{11} - d_{12}$. Beyond the better resolution, the quadratic difference has a second advantage. The linear difference of Eq. (3) holds only for large values of $d_{11}$. For smaller values one must use more terms of the series expansion in Eq. (2). This is the process of focusing. The quadratic difference of Eq. (7) is correct for any distance, and the problem of focusing does not arise.

Let us now investigate the effect of an error $\Delta X$ in the distance $2X$ between the sonars in Fig. 9.2-2. For the linear difference we obtain from Eq. (3):

$$x_1 + \Delta x_1 = d_{11}(d_{11} - d_{21})/(2X + \Delta X) \tag{15}$$

$$\Delta x_1 \doteqdot -\Delta X d_{11}(d_{11} - d_{21})/4X^2$$

For the quadratic difference we obtain a corresponding formula from Eq. (7):

$$\Delta x_1 = -\Delta X(d_{11}^2 - d_{21}^2)/8X^2 \tag{16}$$

The error $\Delta x_1$ of Eq. (16) seems to be only half as large as the error $\Delta x_1$ in Eq. (15). However, one can readily verify that both equations yield the same error for a small difference between $d_{11}$ and $d_{21}$. Note that the error $\Delta X$ is multiplied in Eq. (15) and (16) with the squares of the distances to the target, while the errors $\Delta d_{ij}$ in Eqs. (12) and (14) are multiplied with the distances only.

We have now the principle of a synthetic aperture sonar based on pulses and the quadratic difference. The next step is to transform this principle into workable concepts. The most important limitation is the error $\Delta x_1$ defined by Eq. (14). We know that such random errors can be reduced by making many measurements and taking the average. These measurements can be made either as a sequence in time, using one or two sonars and repeating the measurements, or simultaneously by using many sonar receivers and making repeated measurements with all of them.

Consider a ship moving with velocity $v$ in the direction of the $x$-axis in Fig. 9.2-3. A sonar carried by it radiates pulses into a certain large sector

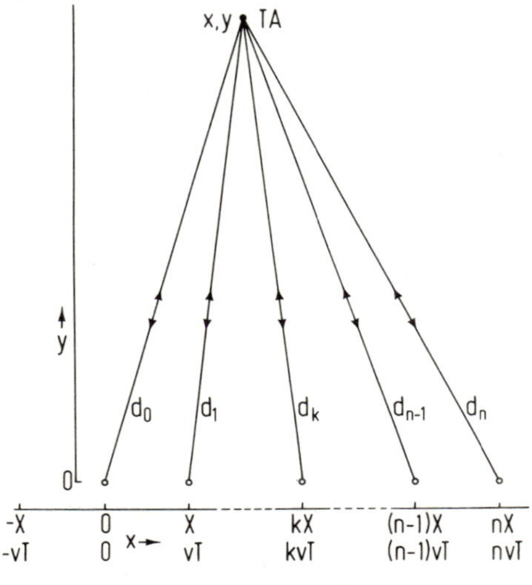


FIG. 9.2-3. Geometric relations for a sidelooking sonar moving along the $x$-axis with velocity $v$ and a target TA.

of the half-plane $y > 0$. The pulse radiated at the location $x = 0$ shall be *reflected* from a target with the coordinates $x$, $y$. The distance to the target is denoted $d_0$:

$$d_0 = (x^2 + y^2)^{1/2} \tag{17}$$

Further, pulses are radiated when the ship is at the locations $X, \ldots, kX, \ldots, nX$. The general distance $d_k$ is expressed with the help of the reference distance $d_0$:

$$d_k = [(x - kX)^2 + y^2]^{1/2} = d_0[1 - (2kXx - k^2X^2)/d_0^2]^{1/2} \tag{18}$$

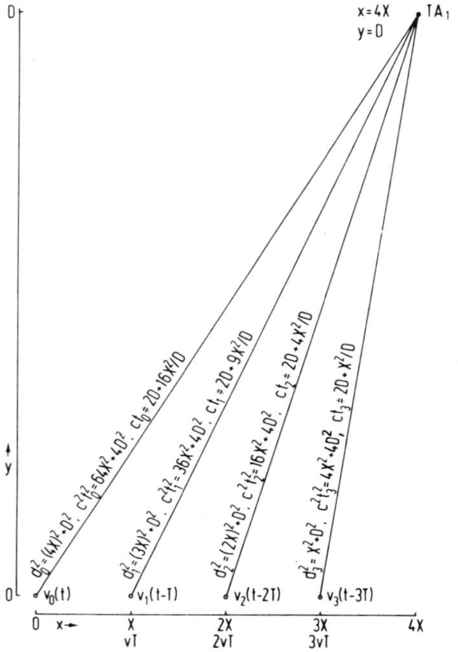


FIG. 9.2-4. Example of a sidelooking sonar measuring the distance of the target $TA_1$ from the locations $x = 0, X, 2X, 3X$ at the times $t = 0, T, 2T, 3T$. The sonar moves with velocity $v \ll c$ in the direction of the $x$-axis. The round trip travel time of a signal to the target $TA_1$ is negligibly small compared with $T$.

The round trip travel times of the two pulses radiated at $x = 0$ and $x = kX$ are $t_0$ and $t_k$:

$$t_0 = 2d_0/c, \qquad t_k = 2d_k/c \tag{19}$$

Since the sonar measures the times $t_0$ and $t_k$ rather than the distances $d_0$ and $d_k$, we will use the times. The square difference $t_0^2 - t_k^2$ follows from Eqs. (17) and (18):

$$t_0^2 - t_k^2 = 4(2kXx - k^2X^2)/c^2 \tag{20}$$

For the explanation of the required data processing, it is useful to rewrite this equation in two steps:

$$t_0^2 - [t_k^2 - 4(kX/c)^2] = 8kXx/c^2 \tag{21}$$

$$t_0^2 - [t_k^2 - 4(kX/c)^2 + 8(x/c^2)kX] = 0 \tag{22}$$

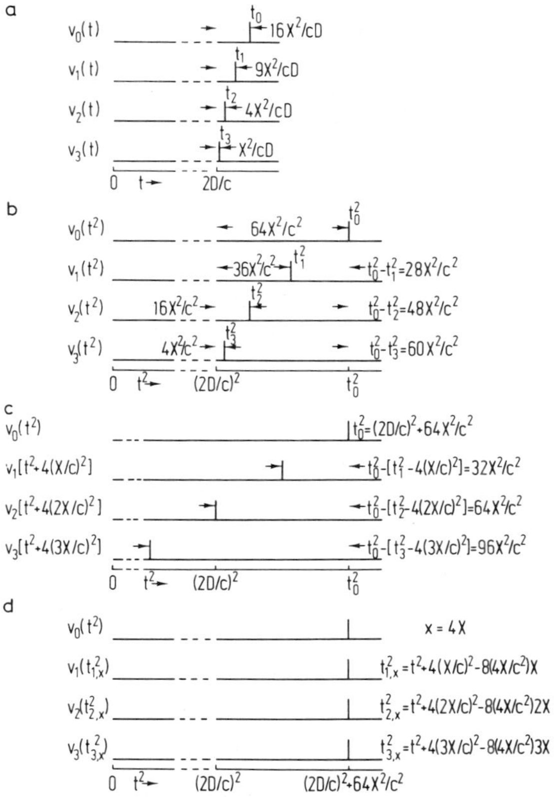

FIG. 9.2-5. Time diagram for the processing of the four voltages $v_0(t)$ to $v_3(t-3T)$ of Fig. 9.2-4 received by the sidelooking sonar.

In order to explain how the coordinate $x$ of the target TA in Fig. 9.2-3 is obtained from these equations, we will consider a specific numerical case. Figure 9.2-4 shows a sidelooking sonar moving along the $x$-axis and measuring the distance to the target $TA_1$ from the four locations 0, $X$, $2X$, $3X$. The coordinates of the target are $x = 4X$, $y = D$. The four squared distances $d_0^2, \ldots, d_3^2$ according to Eqs. (17) and (18) are shown. Also shown are the round trip travel times $c^2t_0^2, \ldots, c^2t_3^2$ according to Eq. (19), and the first-order approximations of $ct_0, \ldots, ct_3$. The output voltages of the sonar are denoted $v_0(t)$, $v_1(t-T)$, $v_2(t-2T)$ and $v_3(t-3T)$.

Let us now turn to Fig. 9.2-5. On top (a) are shown the four voltages $v_0(t)$ to $v_3(t)$ due to the target $TA_1$ in Fig. 9.2-4; the time shifts $-T$, $-2T$, and $-3T$ have been removed. The same voltages as functions of $t^2$ rather than $t$ are shown below (b). These voltages would be seen on a cathode ray tube if the deflection voltage for the time axis increased proportionately to $t^2$ rather than to $t$. In digital processing equipment, the transition from $t$ to $t^2$ means a relabeling of storage addresses from $t$ to $t^2$.

The square differences $t_0^2 - t_k^2$ of Eq. (20) are shown on the right side of Fig. 9.2-5b.

Figure 9.2-5c shows the transition from Eq. (20) to Eq. (21). A voltage $v_k(t^2)$ is time shifted to become $v_k[t^2 + 4(kX/c)^2]$. There is no difficulty in performing these shifts in digital processing equipment, while analog equipment would call for a sufficiently long delay to make sure that shifted voltages are always delayed and not advanced in time.

The final step from Eq. (21) to Eq. (22) requires the time shift $8kXx/c^2$ for a particular value of $x$; the value $x = 4X$ is chosen in Fig. 9.2-5d. The output voltages $v_0(t)$ to $v_3(t - 3T)$ are now properly lined up. The sum of the voltages in Fig. 9.2-5d yields four times—or generally $n$ times—the amplitude of the individual voltages. This increase in amplitude would not be achieved for a value $x \neq 4X$. We recognize here that the use of $n$ sonars in Fig. 9.2-3 instead of the two sonars in Fig. 9.2-2 increases the dynamic range or the contrast ratio.

Figure 9.2-5d yields the $x$-coordinate of the target $TA_1$ in Fig. 9.2-4. The $y$-coordinate follows from the time $t_0^2$ in Fig. 9.2-5d and Eqs. (17) or (18) and (19):

$$\begin{aligned} y^2 &= d_0^2 - x^2 = c^2t_0^2/4 - x^2 \\ y^2 &= d_k^2 - (x - kX)^2 = c^2t_k^2/4 - (x - kX)^2 \end{aligned} \tag{23}$$

The sign of $y$ remains undetermined, since the geometric relations in Fig. 9.2-4 would be the same for a target located at $y = -D$ rather than $y = +D$. There are a variety of ways to resolve this ambiguity. One is to use a sonar that does not radiate omnidirectionally but only into a sector that is no larger than the half-plane $y > 0$.

A sidelooking sonar is not normally used to locate a single target, but to make a map that contains many targets or points. We will show next how two targets—and by implication many targets with sufficient separation—are resolved, but we will have to return to the resolution of many points again at a later time.

Figure 9.2-6 shows the target $TA_1$ of Fig. 9.2-4 and in addition a second target $TA_2$ with the coordinates $x = 3X$, $y = D$. The values of $d_i^2$, $c^2t_i^2$, and $ct_i$ for $i = 0, \ldots, 3$ are shown for this second target, while the respective values for the first target $TA_1$ are the same as shown in Fig. 9.2-4. The four voltages $v_0(t)$, $v_1(t - T)$, $v_2(t - 2T)$, and $v_3(t - 3T)$—corrected for the delays $T$, $2T$, and $3T$—are shown in Fig. 9.2-7a. The pulses shown by solid lines are due to the target $TA_1$; they are the same pulses as shown in Fig. 9.2-5. The pulses shown by dashed lines are due to target $TA_2$. The

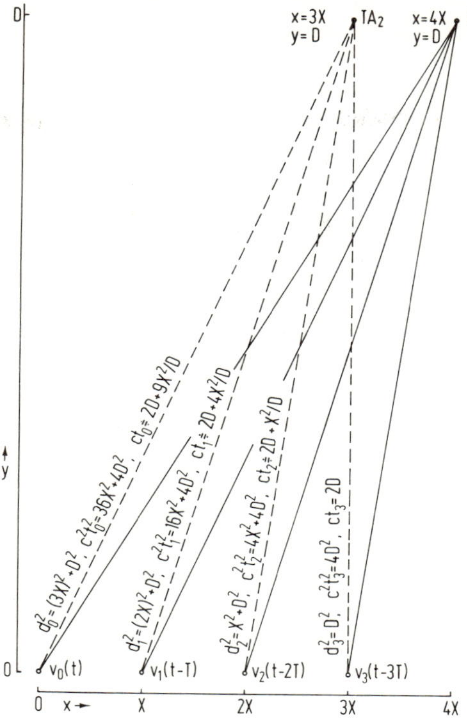


FIG. 9.2-6. Example of a sidelooking sonar measuring the distance to the targets $TA_1$ and $TA_2$ from the locations $x = 0, X, 2X, 3X$ at the times $t = 0, T, 2T, 3T$. The sonar moves with velocity $v \ll c$ in the direction of the $x$-axis. The round trip travel time of a signal to either target is negligibly small compared with $T$. The values of $d_i^2$, $c^2t_i^2$, and $ct_i$, $i = 0, \ldots, 3$ for target $TA_1$ are shown in Fig. 9.2-4.

round trip delay times $t_0 = 2D/c + 9X^2/cD$ to $t_3 = 2D/c$ are only shown for the dashed pulses; the values shown in Fig. 9.2-5 hold for the solid pulses.

The voltages $v_0(t^2)$ to $v_3(t^2)$ of the squared time are shown in Fig. 9.2-7b, and the shifted voltages $v_k[t^2 + 4(kX/c)^2]$, $k = 0, \ldots, 3$ in Fig. 9.2-7c. A further time shift corresponding to $x = 4X$ yields the voltages of Fig. 9.2-7d, while shifts corresponding to $x = 3X$ yield the voltages of Fig. 9.2-7e. Summation of the voltages of Fig. 9.2-7d produces one solid pulse with four times the amplitude of the original pulses, and four dashed pulses with the old amplitude; the contrast or the dynamic range is thus 4 : 1. An equivalent statement holds for Fig. 9.2-7e, except that the role of the solid and the dashed pulses is interchanged.

Equation (14) shows that the resolution error $\Delta x$ for the $x$-coordinate depends on the accuracy of the distance measurements and the spacing $2X$

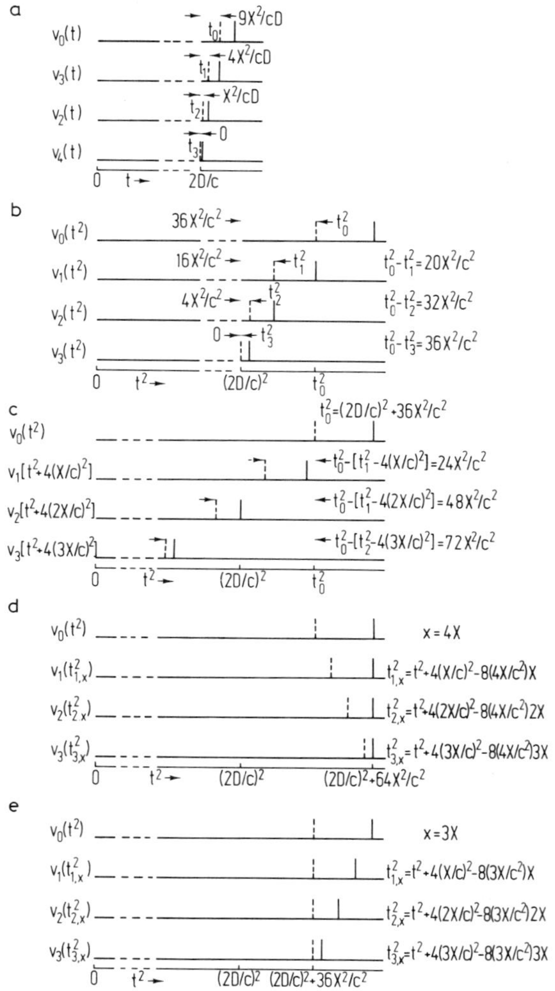


FIG. 9.2-7. Time diagram for the processing of the four voltages $v_0(t)$ to $v_3(t - 3T)$ of Fig. 9.2-6 received by the sidelooking sonar. The dashed pulses hold for target $TA_2$, the solid pulses for target $TA_1$; the numerical values for the solid pulses are shown in Fig. 9.2-5.

of the two sonars in Fig. 9.2-2. This spacing $2X$ is replaced by $nX$ or $nvT$ in Fig. 9.2-3, and by $3X$ in Figs. 9.2-4 and 9.2-6. The dynamic range or contrast is determined by the number of receivers, which is 2 in Fig. 9.2-2, $n$ in Fig. 9.2-3, and 4 in Figs. 9.2-4 and 9.2-6. Of course, the signal-to-noise ratio also increases with the number of receivers.

Let us note that the Doppler effect has not yet been used, even though the conventional sidelooking sonar would not work without it. There are two readily recognizable uses one can make of this fact. If the pulses that we have been using are transmitted with a sinusoidal carrier, one may use the Doppler shift of the carrier to distinguish between moving and nonmoving targets. On the other hand, since we do not need the Doppler shift, we can transmit the pulses without a carrier. This makes it possible to use pulses with a very short duration, but to avoid the high absorption losses that the transmission of these short pulses with a sinusoidal carrier would imply.

## 9.3 One- and Two-Dimensional Synthetic Apertures

Since we have not used the Doppler shift for the sidelooking sonar, one must be able to achieve the same resolution by using stationary sonar transmitters at the locations 0, $X$, ..., $nX$ in Fig. 9.2-3, or 0, $X$, $2X$, $3X$ in Figs. 9.2-4 and 9.2-6. This is not the same as using a phased array for beam forming based on sinusoidal waves. For instance, if we want to form a beam using 100 sensors for a wave with a frequency of 50 Hz, we must space the sensors at a distance of no more than $\lambda/2 = 15$ m according to Eq. (9.2-5); the length of the phased array is thus limited to $L = 1.5$ km, and the resolution angle to about $\varepsilon = \lambda/L = 0.02$. The only way to increase the aperture is to increase the number of sensors.

No restrictions to the spacing of the sensors is implied by Eqs. (9.2-3) or (9.2-7). One may space the sensors over an aperture of megameters rather than kilometers. The number of sensors will determine the dynamic range and the signal-to-noise ratio, but there will be no ambiguities due to grating lobes.

The possibility of generating a synthetic aperture without movement of the radiators and sensors makes synthetic aperture techniques applicable to acoustic imaging. We have to develop first the stationary synthetic aperture for sensors in one dimension, or a line array, then for sensors in two dimensions, or a two-dimensional array. Finally, we have to try to make this two-dimensional array smaller than it would be for sinusoidal waves without synthetic aperture techniques, since the reduction of the array size is the primary goal of synthetic aperture processing in acoustic imaging.

In the case of the sidelooking sonar according to Fig. 9.2-3, one must radiate a signal at each position 0, $X$, ..., $nX$. For a stationary sonar, we need to radiate only one signal rather than $n$, since all receivers are permanently at the positions 0, $X$, ..., $nX$. This implies an increase of the signal-to-noise ratio by a factor $n$, a reduction of the time required to produce an image by a factor $1/n$, and a reduction of the required radiators also by $1/n$.

Figure 9.3-1 shows a stationary sonar with a radiator at $x = 0$ and $n$ sensors at $x = 0, X, \ldots, nX$. The distances $d_0$ and $d_k$ are the same as given by Eqs. (9.2-17) and (9.2-18). The round trip travel time $t_0$ is the same as

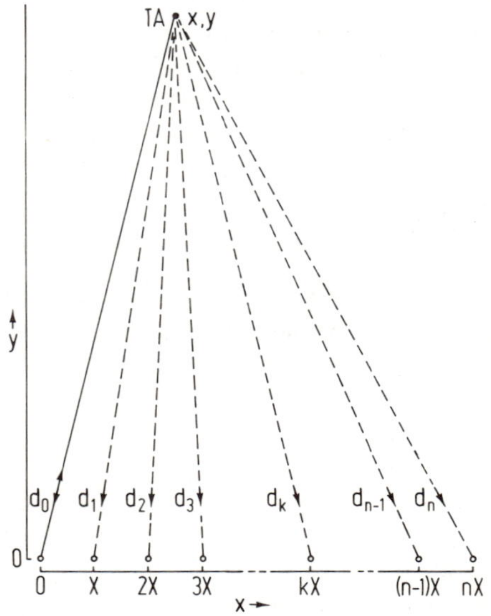


FIG. 9.3-1. Geometric relations for a stationary sonar with no restriction on the distance between the sensors at $x = 0, X, \ldots, nX$. The only radiator is located at $x = 0$, the target TA at $x, y$.

defined by Eq. (9.2-19), but $t_k$ has a different value, since the signal travels now from the radiator at $x = 0$ to the target TA and a sensor at $x = kX$:

$$t_0 = 2d_0/c, \qquad t_k = (d_0 + d_k)/c \tag{1}$$

With the help of Eq. (9.2-18) we obtain the following relation in place of Eq. (9.2-20):

$$\begin{aligned} t_0^2 - t_k^2 &= (3d_0^2 - 2d_0 d_k + d_k^2)/c^2 \\ &= [2d_0^2 + 2kXx - k^2X^2 - 2(d_0 - 2kXx + k^2X^2)^{1/2}]/c^2 \end{aligned} \tag{2}$$

In contrast to Eq. (9.2-20) we have now a square root. For large values of $d_0$ we can use a series expansion for the square root. The unfocused case is obtained if only the first two terms of the series expansion are used:

$$t_0^2 - t_k^2 \doteq 2(2kXx - k^2X^2)/c^2, \qquad d_0 \gg kX \tag{3}$$

The factor 2 occurs on the right side instead of the factor 4 in Eq. (9.2-20). This is in analogy to the well-known fact that the synthetic aperture of a conventional sidelooking sonar or radar needs to be only half as large as the real aperture of a stationary radar to obtain the same resolution. In Fig. 9.2-3 each signal travels twice over the respective distance $d_k, k = 0, \ldots, n$, but in Fig. 9.3-1 each signal travels only once over the distance $d_k$ on the way from the target TA to the sensor at $kX$, while the distance $d_0$ from the radiator to the target is the same for all signals.

Equation (3) is rewritten in analogy to Eqs. (9.2-21) and (9.2-22):

$$t_0^2 - [t_k^2 - 2(kX/c)^2] \doteq 4kXx/c^2 \quad (4)$$

$$t_0^2 - [t_k^2 - 2(kX/c)^2 + 4(x/c^2)kX] \doteq 0 \quad (5)$$

In order to explain the data processing required to obtain the coordinate $x$ of the target TA, we consider again a specific numerical case. Figure 9.3-2 shows four sensors at the locations $x = 0$, $X$, $2X$, $4X$ and one radiator at

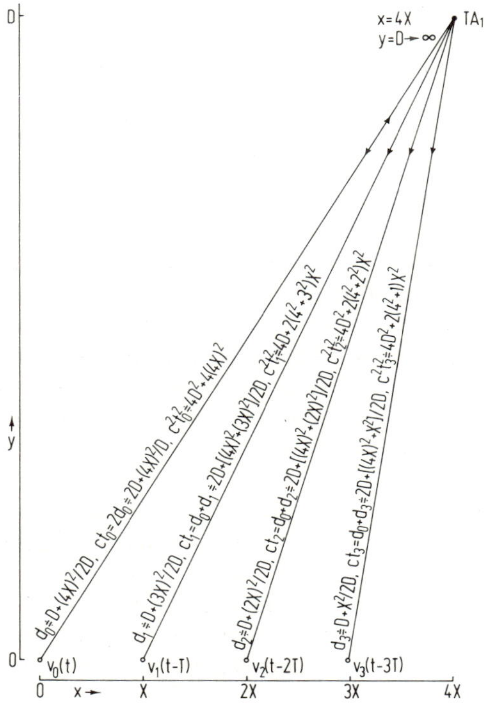


FIG. 9.3-2. Example of a stationary synthetic aperture sonar measuring the distances $2d_0$, $d_0 + d_1$, $d_0 + d_2$, and $d_0 + d_3$ to a target $TA_1$. One radiator at $x = 0$ and four sensors at $x = 0$, $X$, $2X$, $3X$ are used. The distances $d_0$ to $d_3$ are all large compared with the synthetic aperture $3X$.

$x = 0$. The distances $d_k$ and $d_0 + d_k$, as well as the squares $(d_0 + d_k)^2$ are shown for $k = 0, \ldots, 3$.

Figure 9.3-3 shows on top the received voltages $v_0(t)$ to $v_3(t)$ as functions of the time $t$, and below as functions of $t^2$. The implementation of Eq. (4) is shown by Fig. 9.3-3c, and the implementation of Eq. (5) by Fig. 9.3-3d.

Let us turn to the focused case. We start from Eq. (2), but do not use a series expansion for the square root. The distance $d_0^2$ is replaced by $x^2 + y^2$ according to Eq. (9.2-17):

$$t_0^2 - [t_k^2 - (kX/c)^2] = 2kXx/c^2 + 2(x^2 + y^2)\left[1 - 2\left(1 - \frac{2kXx - (kX)^2}{x^2 + y^2}\right)^{1/2}\right]\Big/c^2 \quad (6)$$

$$t_0^2 - [t_k^2 - (kX/c)^2 + 2(x/c^2)kX] = 2(x^2 + y^2)\left[1 - 2\left(1 - \frac{2kXx - (kX)^2}{x^2 + y^2}\right)^{1/2}\right]\Big/c^2 \quad (7)$$

$$t_0^2 - \Bigg\{t_k^2 - (kX/c)^2 + 2(x/c^2)kX + 2(x^2 + y^2)\left[1 - 2\left(1 - \frac{2kXx - (kX)^2}{x^2 + y^2}\right)^{1/2}\right]\Big/c^2\Bigg\} = 0 \quad (8)$$

Equations (6) and (7) are implemented like Eqs. (4) and (5), and the only difference being the reduction of the factors in front of the terms $(kX/c)^2$ and $(x/c^2)kX$ to one-half. Equation (8) calls for a different time shift for every value of $y$. Hence, this *focusing term* calls for a lot of computing, but creates no further problems; a set of voltages $v_k(t)$, $k = 0, \ldots, n$, permits the production of images of points at all coordinates $y$. This is in contrast to the usual focusing based on sinusoidal functions, which permits a focused image only of points on a circle with a *certain distance* from the center of the sensor array, but not for points on circles with *any distance* from the center of the sensor array.

Let us apply the results of the one-dimensional sensor array to a two-dimensional array. Figure 9.3-4 shows a two-dimensional array of $nm$ sensors that are located at the intersection of the $n$ columns and the $m$ rows. A radiator is located at $x = y = 0$. The distance between a sensor located at $x = kX$, $y = lY$, and the target TA is denoted $d_{kl}$, where $k = 0, \ldots, n$ and $l = 0, \ldots, m$. Equations (9.2-17) and (9.2-18) are replaced by the following generalized equations:

$$d_{00} = (x^2 + y^2 + z^2)^{1/2} \quad (9)$$

$$\begin{aligned} d_{kl} &= [(x - kX)^2 + (y - lY)^2 + z^2]^{1/2} \\ &= d_{00}[1 - (2kXx + k^2X^2 + 2lYy - l^2Y^2)/d_{00}^2]^{1/2} \end{aligned} \quad (10)$$

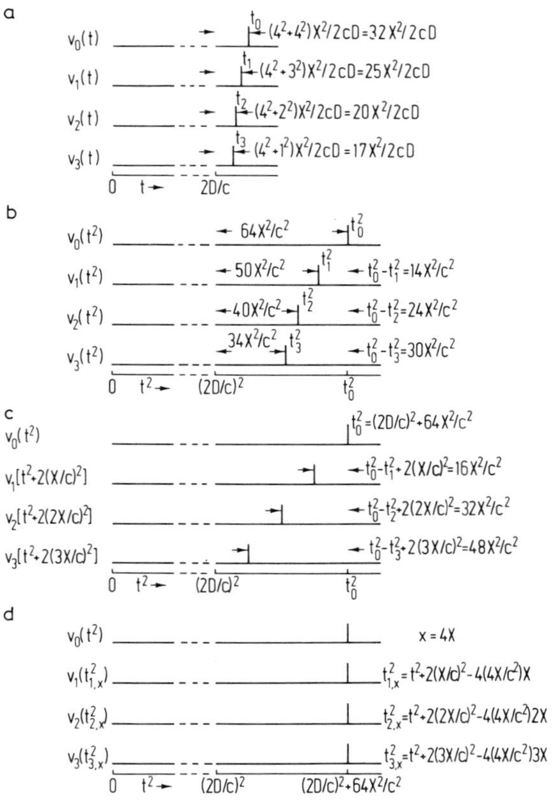


FIG. 9.3-3. Time diagram for the processing of the four voltages $v_0(t)$ to $v_3(t)$ of Fig. 9.3-2 received by the stationary synthetic aperture sonar.

The generalization of the travel times of Eq. (1) yields:

$$t_{00} = 2d_{00}/c, \qquad t_{kl} = (d_{00} + d_{kl})/c \tag{11}$$

One obtains the generalization of Eq. (2):

$$\begin{aligned} t_{00}^2 - t_{kl}^2 &= (3d_{00}^2 - 2d_{00}\,d_{kl} - d_{kl}^2)/c^2 \\ &= \{2d_{00}^2 + 2kXx - k^2X^2 + 2lYy - l^2Y^2 \\ &\quad - 2\,[d_{00} - 2kXx + k^2X^2 - 2lYy + l^2Y^2]^{1/2}\}/c^2 \end{aligned} \tag{12}$$

For large values of $d_{00}$ we use again a series expansion for the square root:

$$t_{00}^2 - t_{kl}^2 \doteqdot 2(2kXx - k^2X^2 + 2lYy - l^2Y^2)/c^2, \qquad d_{00} \gg kX,\ lY \tag{13}$$

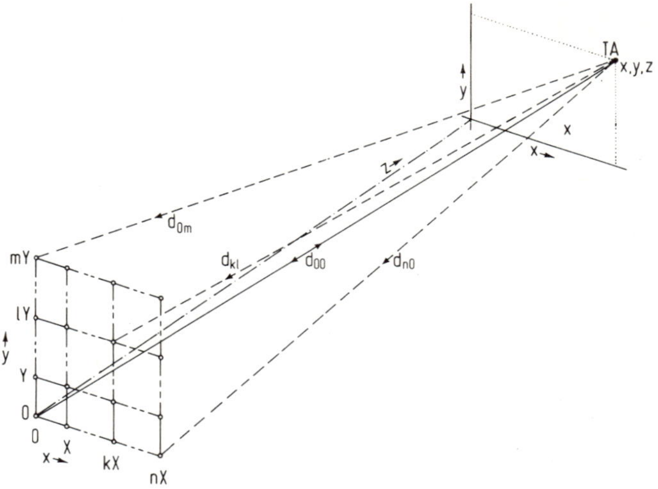


FIG. 9.3-4. Geometric relations for a stationary sonar with a two-dimensional sensor array. The sensors are located at the points $kX$, $lY$ with $k = 0, \ldots, n$ and $l = 0, \ldots, m$. The only radiator is located at $x = y = 0$, the target TA is located at $x$, $y$, $z$.

The two Eqs. (4) and (5) are replaced by four equations:

$$t_{00}^2 - [t_{kl}^2 - 2(kX/c)^2] \doteqdot 4kXx/c^2 + 2(2lY - l^2Y^2)/c^2 \qquad (14)$$

$$t_{00}^2 - [t_{kl}^2 - 2(kX/c)^2 - 2(lY/c)^2] \doteqdot 4kXx/c^2 + 4lYy/c^2 \qquad (15)$$

$$t_{00}^2 - [t_{kl}^2 - 2(kX/c)^2 - 2(lY/c)^2 + 4(x/c^2)kX] \doteqdot 4lYy/c^2 \qquad (16)$$

$$t_{00}^2 - [t_{kl}^2 - 2(kX/c)^2 - 2(lY/c)^2 + 4(x/c^2)kX + 4(y/c^2)lY] \doteqdot 0 \qquad (17)$$

Equation (14) corrects $t_{kl}^2$ for the column $kX$ in which the sensor is positioned, while Eq. (15) corrects for the row $lY$. For easier writing, we define the corrected time $t_{c,kl}$:

$$t_{c,kl}^2 = t_{kl}^2 - 2(kX/c)^2 - 2(lY/c)^2 \qquad (18)$$

Equation (16) shifts the received signals according to the $x$-coordinate of the target, and Eq. (17) shifts them according to the $y$-coordinate.

Let us again show the data processing by means of an example. Figure 9.3-5 shows an array with 4 × 4 sensors and a target located at $x = 4X$, $y = -Y$, $z = D \to \infty$. For simplicity, we assume that $X$ and $Y$ have the same magnitude. Since there is not enough space in Fig. 9.3-5 to show $d_{kl}$, $ct_{kl}$, and $c^2t_{kl}^2$ in analogy to Fig. 9.3-2, we list these quantities in Table 9.3-1 ordered by the voltages $v_{00}(t)$ to $v_{33}(t)$. Also shown are $t_{00}^2 - t_{kl}^2$, $t_{c,kl}^2$ and $t_{00}^2 - t_{c,kl}^2$.

Figure 9.3-6 is the time diagram for the two-dimensional array. It starts with the 16 voltages $v_{00}(t)$ to $v_{33}(t)$ of Fig. 9.3-6a. The pulses produced by the target TA in Fig. 9.3-5 and the respective travel times $t_{00}$ to $t_{33}$ are shown; also shown are the times $t_{00} - 2D/c = 34X^2/2cD$, $t_{10} - 2D/c = 27X^2/2cD, \ldots$.

The voltages $v_{00}(t^2)$ to $v_{33}(t^2)$ are shown in Fig. 9.3-6b. The differences $t_{00}^2 - (2D/c)^2 = 68(X/c)^2$, $t_{10}^2 - (2D/c)^2 = 54(X/c)^2, \ldots$ are listed as well as the differences $t_{00}^2 - t_{10}^2 = 14(X/c)^2$, $t_{00}^2 - t_{20}^2 = 24(X/c)^2, \ldots$. Figures 9.3-6a and b are in complete analogy to Figs. 9.3-3a and b.

Figure 9.3-6c shows the correction of the $x$-shift $2(kX/c)^2$ according to Eq. (14), which is analogous to Fig. 9.3-3c. The differences $t_{00}^2 - t_{10}^2 + 2(kX/c)^2 = 16(X/c)^2$, $t_{00}^2 - t_{20}^2 + 2(kX/c)^2 = 32(X/c)^2, \ldots$ are shown. Note that the first four voltages $v_{00}(t^2)$ to $v_{30}[t^2 + 2(3X/c)^2]$ are corrected relative to $v_{00}(t^2)$; the next four voltages $v_{01}(t^2)$ to $v_{31}[t^2 + 2(3X/c)^2]$ are corrected relative to $v_{01}(t^2)$; etc.

The correction of the $y$-shift according to Eq. (15) is shown in Fig. 9.3-6d. There is no analog for this diagram in Fig. 9.3-3. The differences $t_{00}^2 - t_{c,kl}^2$ are shown; for $l = 0$ they are identical to the shifts in Fig. 9.3-6c, but this is not so for $l = 1, 2, 3$. The additional shifts $4(Y/c)^2$ for $l = 1$, $8(Y/c)^2$ for $l = 2$, and $12(Y/c)^2$ for $l = 3$ are introduced.

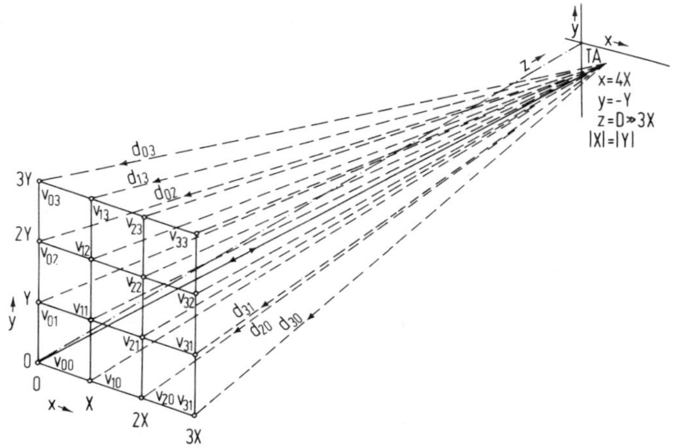


FIG. 9.3-5. Example of a stationary sonar with a two-dimensional synthetic aperture measuring the distances $2d_{00}, \ldots, d_{00} + d_{33}$ to a target TA. One radiator at $x = y = 0$ and 16 sensors at $x = kX$, $y = lY$ are used. The distances $d_{00}$ to $d_{33}$ are all large compared with the synthetic aperture $3X$ and $3Y$ in the $x$- and $y$-direction.

The time diagrams for the implementation of Eqs. (16) and (17) are not shown. In analogy to Fig. 9.3-3d one must introduce the shift $-4(x/c^2)kX$ for each coordinate $x$ for which one wants to produce an image, to

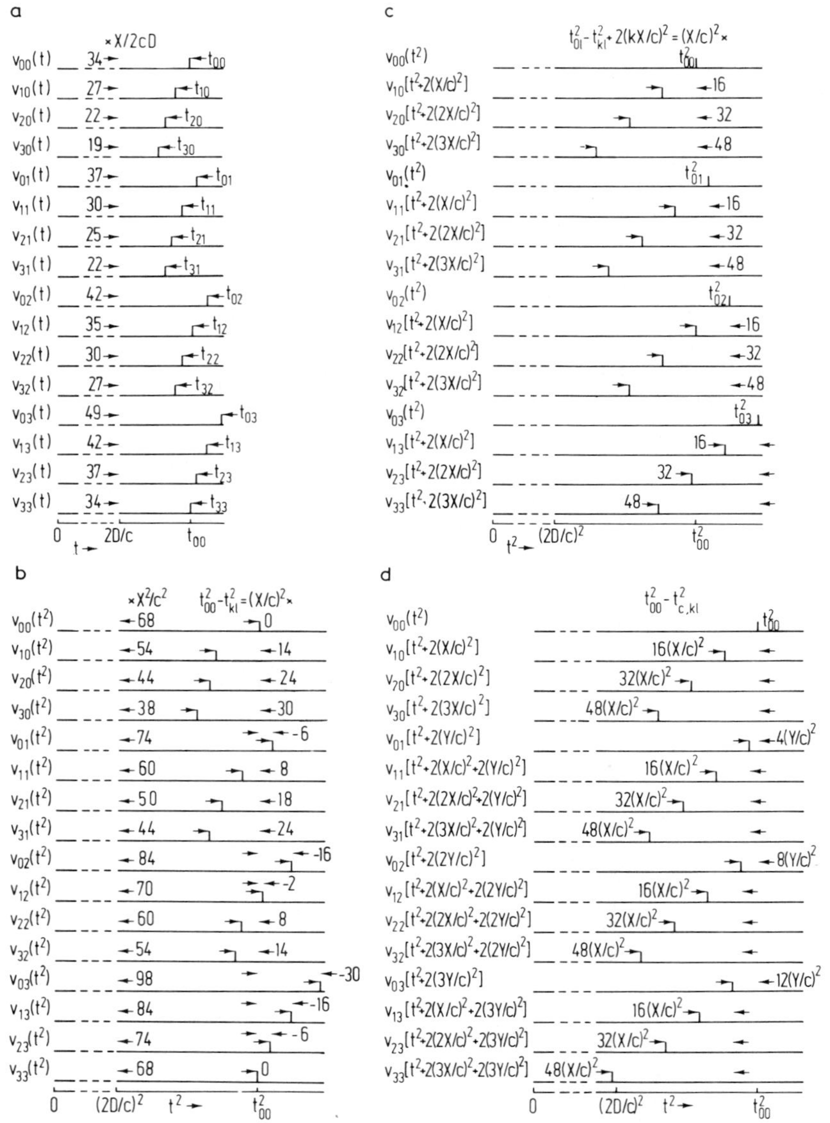


FIG. 9.3-6. Time diagram for the processing of the 16 voltages $v_{00}(t)$ to $v_{33}(t)$ of Fig. 9.3-5 received by the stationary sonar with two-dimensional synthetic aperture.

TABLE 9.3-1

THE VOLTAGES $v_{kl} = v_{kl}(t)$ OF FIG. 9.3-5 AND THE DISTANCES $d_{kl}$, TOGETHER WITH THE QUANTITIES $ct_{kl}$, $c^2t_{kl}^2$, $t_{00}^2 - t_{kl}^2$, $t_{c,kl}^2$, AND $t_{00}^2 - t_{c,kl}^2$ DERIVED FROM THEM

| | | | |
|---|---|---|---|
| $v_{03}(t)$<br>$d_{03} = D + [(4X)^2 + (4Y)^2]/2D$<br>$ct_{03} = d_{00} + d_{03} = 2D + (32X^2 + 17Y^2)/2D$<br>$c^2t_{03}^2 = 4D^2 + 2(32X^2 + 17Y^2)$<br>$t_{00}^2 - t_{03}^2 = 2(0X^2 - 15Y^2)/c^2$<br>$t_{c,03}^2 = t_{03}^2 - 2(0X/c)^2 - 2(3Y/c)^2$<br>$t_{00}^2 - t_{c,03}^2 = 0(X/c)^2 - 12(Y/c)^2$ | $v_{13}(t)$<br>$d_{13} = D + [(3X)^2 + (4Y)^2]/2D$<br>$ct_{13} = d_{00} + d_{13} = 2D + (25X^2 + 17Y^2)/2D$<br>$c^2t_{13}^2 = 4D^2 + 2(25X^2 + 17Y^2)$<br>$t_{00}^2 - t_{13}^2 = 2(7X^2 - 15Y^2)/c^2$<br>$t_{c,13}^2 = t_{13}^2 - 2(X/c)^2 - 2(3Y/c)^2$<br>$t_{00}^2 - t_{c,03}^2 = 16(X/c)^2 - 12(Y/c)^2$ | $v_{23}(t)$<br>$d_{23} = D + [(2X)^2 + (4Y)^2]/2D$<br>$ct_{23} = d_{00} + d_{23} = 2D + (20X^2 + 17Y^2)/2D$<br>$c^2t_{23}^2 = 4D^2 + 2(20X^2 + 17Y^2)$<br>$t_{00}^2 - t_{23}^2 = 2(12X^2 - 15Y^2)/c^2$<br>$t_{c,23}^2 = t_{23}^2 - 2(2X/c)^2 - 2(3Y/c)^2$<br>$t_{00}^2 - t_{c,23}^2 = 32(X/c)^2 - 12(Y/c)^2$ | $v_{33}(t)$<br>$d_{33} = D + [X^2 + (4Y)^2]/2D$<br>$ct_{33} = d_{00} + d_{33} = 2D + (17X^2 + 17Y^2)/2D$<br>$c^2t_{33}^2 = 4D^2 + 2(17X^2 + 17Y^2)$<br>$t_{00}^2 - t_{33}^2 = 2(15X^2 - 15Y^2)/c^2$<br>$t_{c,33}^2 = t_{23}^2 - 2(3X/c)^2 - 2(3Y/c)^2$<br>$t_{00}^2 - t_{c,33}^2 = 48(X/c)^2 - 12(Y/c)^2$ |
| $v_{02}(t)$<br>$d_{02} = D + [(4X)^2 + (3Y)^2]/2D$<br>$ct_{02} = d_{00} + d_{03} = 2D + (32X^2 + 10Y^2)/2D$<br>$c^2t_{02}^2 = 4D^2 + 2(32X^2 + 10Y^2)$<br>$t_{00}^2 - t_{02}^2 = 2(0X^2 - 8Y^2)/c^2$<br>$t_{c,02}^2 = t_{02}^2 - 2(0X/c)^2 - 2(2Y/c)^2$<br>$t_{00}^2 - t_{c,02}^2 = 0(X/c)^2 - 8(Y/c)^2$ | $v_{12}(t)$<br>$d_{12} = D + [(3X)^2 + (3Y)^2]/2D$<br>$ct_{12} = d_{00} + d_{12} = 2D + (25X^2 + 10Y^2)/2D$<br>$c^2t_{12}^2 = 4D^2 + 2(25X^2 + 10Y^2)$<br>$t_{00}^2 - t_{12}^2 = 2(7X^2 - 8Y^2)/c^2$<br>$t_{c,12}^2 = t_{12}^2 - 2(X/c)^2 - 2(2Y/c)^2$<br>$t_{00}^2 - t_{c,12}^2 = 16(X/c)^2 - 8(Y/c)^2$ | $v_{22}(t)$<br>$d_{22} = D + [(2X)^2 + (3Y)^2]/2D$<br>$ct_{22} = d_{00} + d_{22} = 2D + (20X^2 + 10Y^2)/2D$<br>$c^2t_{22}^2 = 4D^2 + 2(20X^2 + 10Y^2)$<br>$t_{00}^2 - t_{22}^2 = 2(12X^2 - 8Y^2)/c^2$<br>$t_{c,22}^2 = t_{22}^2 - 2(2X/c)^2 - 2(2Y/c)^2$<br>$t_{00}^2 - t_{c,22}^2 = 32(X/c)^2 - 8(Y/c)^2$ | $v_{32}(t)$<br>$d_{32} = D + [X^2 + (3Y)^2]/2D$<br>$ct_{32} = d_{00} + d_{32} = 2D + (17X^2 + 10Y^2)/2D$<br>$c^2t_{32}^2 = 4D^2 + 2(17X^2 + 10Y^2)$<br>$t_{00}^2 - t_{32}^2 = 2(15X^2 - 8Y^2)$<br>$t_{c,32}^2 = t_{32}^2 - 2(3X/c)^2 - 2(2Y/c)^2$<br>$t_{00}^2 - t_{c,32}^2 = 48(X/c)^2 - 8(Y/c)^2$ |
| $v_{01}(t)$<br>$d_{01} = D + [(4X)^2 + (2Y)^2]/2D$<br>$ct_{01} = d_{00} + d_{01} = 2D + (32X^2 + 5Y^2)/2D$<br>$c^2t_{01}^2 = 4D^2 + 2(32X^2 + 5Y^2)$<br>$t_{00}^2 - t_{01}^2 = 2(0X^2 - 3Y^2)/c^2$<br>$t_{c,01}^2 = t_{01}^2 - 2(0X/c)^2 - 2(Y/c)^2$<br>$t_{00}^2 - t_{c,01}^2 = 0(X/c)^2 - 4(Y/c)^2$ | $v_{11}(t)$<br>$d_{11} = D + [(3X)^2 + (2Y)^2]/2D$<br>$ct_{11} = d_{00} + d_{12} = 2D + (25X^2 + 5Y^2)/2D$<br>$c^2t_{11}^2 = 4D^2 + 2(25X^2 + 5Y^2)$<br>$t_{00}^2 - t_{11}^2 = 2(7X^2 - 3Y^2)/c^2$<br>$t_{c,11}^2 = t_{11}^2 - 2(X/c)^2 - 2(Y/c)^2$<br>$t_{00}^2 - t_{c,11}^2 = 16(X/c)^2 - 4(Y/c)^2$ | $v_{21}(t)$<br>$d_{21} = D + [(2X)^2 + (2Y)^2]/2D$<br>$ct_{21} = d_{00} + d_{21} = 2D + (20X^2 + 5Y^2)/2D$<br>$c^2t_{21}^2 = 4D^2 + 2(20X^2 + 5Y^2)$<br>$t_{00}^2 - t_{21}^2 = 2(12X^2 - 3Y^2)/c^2$<br>$t_{c,21}^2 = t_{21}^2 - 2(2X/c)^2 - 2(Y/c)^2$<br>$t_{00}^2 - t_{c,21}^2 = 32(X/c)^2 - 4(Y/c)^2$ | $v_{31}(t)$<br>$d_{31} = D + [X^2 + (2Y)^2]/2D$<br>$ct_{31} = d_{00} + d_{31} = 2D + (17X^2 + 5Y^2)/2D$<br>$c^2t_{31}^2 = 4D^2 + 2(17X^2 + 5Y^2)$<br>$t_{00}^2 - t_{31}^2 = 2(15X^2 - 3Y^2)/c^2$<br>$t_{c,31}^2 = t_{31}^2 - 2(3X/c)^2 - 2(Y/c)^2$<br>$t_{00}^2 - t_{c,31}^2 = 48(X/c)^2 - 4(Y/c)^2$ |
| $v_{00}(t)$<br>$d_{00} = D + [(4X)^2 + Y^2]/2D$<br>$ct_{00} = 2d_{00} = 2D + (32X^2 + 2Y^2)/2D$<br>$c^2t_{00}^2 = 4D^2 + 2(32X^2 + 2Y^2)$ | $v_{10}(t)$<br>$d_{10} = D + [(3X)^2 + Y^2]/2D$<br>$ct_{10} = d_{00} + d_{01} = 2D + (25X^2 + 2Y^2)/2D$<br>$c^2t_{10}^2 = 4D^2 + 2(25X^2 + 2Y^2)$<br>$t_{00}^2 - t_{10}^2 = 2(7X^2 + 0Y^2)/c^2$<br>$t_{c,10}^2 = t_{10}^2 - 2(X/c)^2 - 2(0Y/c)^2$<br>$t_{00}^2 - t_{c,10}^2 = 16(X/c)^2 - 0(Y/c)^2$ | $v_{20}(t)$<br>$d_{20} = D + [(2X)^2 + Y^2]/2D$<br>$ct_{20} = d_{00} + d_{20} = 2D + (20X^2 + 2Y^2)/2D$<br>$c^2t_{20}^2 = 4D^2 + 2(20X^2 + 2Y^2)$<br>$t_{00}^2 - t_{20}^2 = 2(12X^2 + 0Y^2)/c^2$<br>$t_{c,20}^2 = t_{20}^2 - 2(2X/c)^2 - 2(0Y/c)^2$<br>$t_{00}^2 - t_{c,20}^2 = 32(X/c)^2 - 0(Y/c)^2$ | $v_{30}(t)$<br>$d_{30} = D + (X^2 + Y^2)/2D$<br>$ct_{30} = d_{00} + d_{30} = 2D + (17X^2 + 2Y^2)/2D$<br>$c^2t_{30}^2 = 4D^2 + 2(17X^2 + 2Y^2)$<br>$t_{00}^2 - t_{30}^2 = 2(15X^2 + 0Y^2)$<br>$t_{c,30}^2 = t_{30}^2 - 2(3X/c)^2 - 2(0Y)^2$<br>$t_{00}^2 - t_{c,30}^2 = 48(X/c)^2 - 0(Y/c)^2$ |

implement Eq. (16). Then one has to introduce the additional shift $-4(y/c^2)lY$ for each coordinate $y$ for which one wants to produce an image, to implement Eq. (17).

Let us look at the effort required to produce images with a two-dimensional synthetic aperture sonar. If we want to resolve $n$ values of the $x$-coordinate and $m$ values of the $y$-coordinate, we have to produce a number of time shifts that is proportionate to $nm$. Hence, the data processing effort increases like $n^2$ for $n = m$, or exponentially. One would not expect to be able to do better than that. The advance in data processing equipment is one of the most conspicuous developments of our time, and an exponential increase in the data processing effort causes no concern. The situation is very different for the sensors. Figure 9.3-5 shows $4^2$ sensors, while Fig. 9.3-2 shows only 4 sensors. However, a comparison of Fig. 9.3-6d with Fig. 9.3-3c shows, that the dynamic range has increased from 4 : 1 to 16 : 1. A short reflection reveals that we need the same number of sensors for a one-dimensional and a two-dimensional array if we want the same dynamic range in both cases. A one-dimensional array with 100 sensors does not become a two-dimensional array with 10,000 sensors, but an array with 100 sensors positioned in a square pattern. The array with 16 × 16 hydrophones in Fig. 1.2-9 gave a very poor image when the usual beam forming based on sinusoidal functions was used. With synthetic aperture processing based on pulses, it would give a dynamic range of 256 : 1, which is much more than one usually will want.

## 9.4 Image Contrast, Pulse Compression

We have seen in Section 9.2 how returns from two targets received at four sensors can be separated by data processing. In imaging, the targets become scattering points, and there are always very many of them. Hence, we must investigate how the results of synthetic aperture processing are affected if we receive returns from many scattering points.

The voltages $v_0(t^2)$ to $v_0(t^2_{3,x})$ of Fig. 9.2-7d are shown again in Fig. 9.4-1a. The solid lines are pulses due to a target or scattering point $TA_1$ with the $x$-coordinate $x = 4X$, while the dashed lines are due to a scattering point $TA_2$ with the $x$-coordinate $x = 3X$. The voltages have been processed to emphasize all signals with the $x$-coordinate $x = 4X$. The sum of the four voltages in Fig. 9.4-1a shows that scattering point $TA_1$ produces a signal with four times the amplitude of the signal scattering point $TA_2$, even though the signals received by the four sensors had the same amplitude for both scattering points. We say that the contrast ratio or the dynamic range of the image is 4 : 1. Generally, with $n$ sensors instead of 4, one can achieve a contrast ratio of $n$ : 1.

Let us now look at Fig. 9.4-1b, which shows the voltages $v_0(t^2)$ to $v_3(t^2_{3,x})$

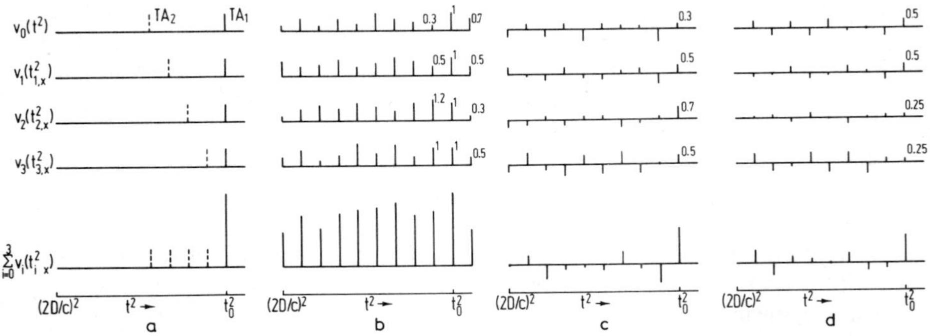

FIG. 9.4-1. Contrast of images produced by a synthetic aperture array using four receptors. (a) Pulse of Fig. 9.4-2, line a, two scattering points; (b) same pulse but many scattering points; (c) pulse of Fig. 9.4-2, line b, many scattering points; (d) pulse of Fig. 9.4-2, line c, many scattering points.

and their sum when many scattering points are present rather than only two. This will be the typical case if the synthetic aperture array is used to make a map. There is hardly any recognizable contrast ratio in Fig. 9.4-1b. Synthetic aperture imaging done in this way will show the contours of areas that return signals with significantly different amplitude, but there will be no halftones in terms of photography. How can we improve this image?

Figure 9.4-2a shows the block pulse on which our investigation has been based so far; this pulse may have been radiated directly or with the help of a sinusoidal carrier. A more complicated pulse is shown in Fig. 9.4-2b; the positive pulse of Fig. 9.4-2a is preceded by a negative pulse of equal magnitude and duration. The voltages $v_0(t^2)$ to $v_3(t^2_{3,x})$ in Fig. 9.4-1b are modified if this pulse is used. The amplitude of a pulse at a certain time $t^2$

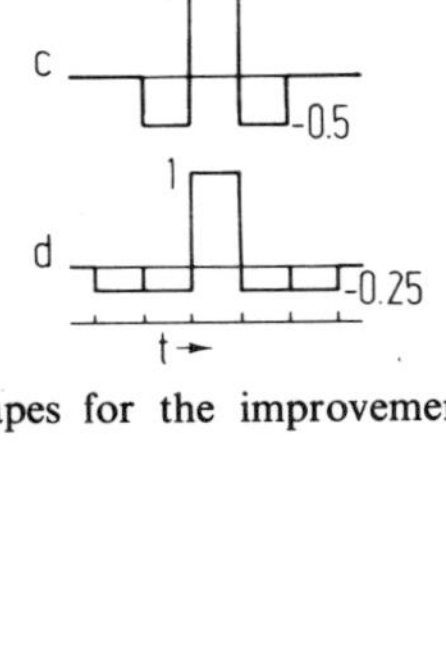

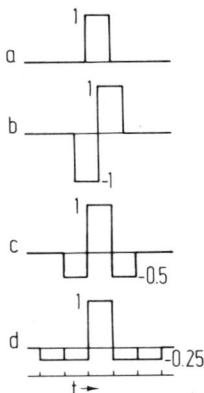

FIG. 9.4-2. Various pulse shapes for the improvement of the contrast of a synthetic aperture image.

is reduced by subtraction of the amplitude of the pulse immediately to its right. Consider the pulse at $t^2 = t_0^2$ in Fig. 9.4-1b with the amplitude 1. The pulses immediately to its right have the amplitudes 0.7, 0.5, 0.3, and 0.5. The differences $1 - 0.7 = 0.3$, $1 - 0.5 = 0.5$, etc. are shown for $t^2 = t_0^2$ in Fig. 9.4-1c. All the pulses for $v_0(t^2)$ to $v_3(t_{3,x}^2)$ in Fig. 9.4-1c have been obtained in this way from the pulses in Fig. 9.4-1b. The sum of the four voltages in Fig. 9.4-1c shows the pulse at $t^2 = t_0^2$ much more conspicuously than the sum in Fig. 9.4-1b.

The reason for the improvement is that pulses due to scattering points with the correct $x$-coordinate are summed algebraically, while pulses coming from scattering points with wrong $x$-coordinates will form a statistical sum of terms with positive or negative signs. The mean value of the sum is zero and its root-mean-square deviation increases proportionately to the square root $\sqrt{n}$ of the number of sensors. Since the algebraic sum increases proportionately to $n$ one can make the contrast ratio large by using many sensors.

The pulses of Figs. 9.4-2c and d are improvements over the pulse of Fig. 9.4-2b. The one large negative pulse is replaced by several but smaller negative pulses. The effect of the pulse of Fig. 9.4-2c is shown in Fig. 9.4-1d. The most important difference is that the magnitude of the large negative value of the sum $\sum v_i(t_{i,x}^2)$ just to the left of $t^2 = t_0^2$ in Fig. 9.4-1c has been reduced in Fig. 9.4-1d.

The use of short pulses yields images with good resolution, but the energy of the pulse is necessarily less than that of a longer pulse. It is not possible to simply increase the amplitude of the pulse to increase the energy, since practical equipment limits the peak power as well as the average power. Furthermore, seawater is not perfectly linear, which is a good reason to avoid high peak powers. The accepted way to increase the signal energy, and thus the signal-to-noise ratio, is by means of pulse compression.

Consider a sound wave with a time variation of the pressure according to Fig. 9.4-3a. There are six short pulses at the times $-20T$, $-12T$, ..., 0.

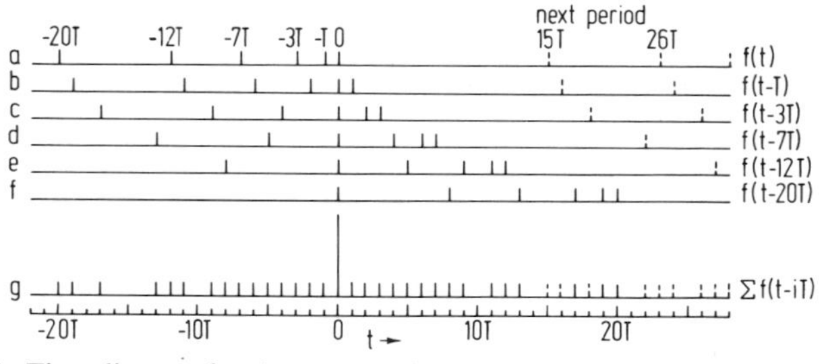


FIG. 9.4-3. Time diagram for the compression of six pulses into one. The pulses are short, but their time variation is otherwise not specified.

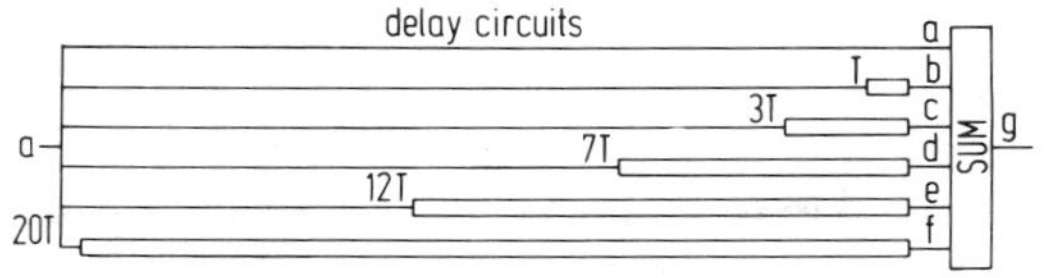


FIG. 9.4-4. Principle of a circuit for the pulse compression according to Fig. 9.4-3.

The shape of the pulses is not specified beyond the requirement that the pulses should be short compared with $T$. Let this sequence of pulses be fed to the input terminals of the delay circuits of Fig. 9.4-4. The outputs of the delay circuits at the points b–f are shown in the lines b–f of Fig. 9.4-3. The sum at the point g in Fig. 9.4-4 is shown in Fig. 9.4-3g. The six pulses of line a are compressed into one pulse with six times the amplitude in line g at the time $t = 0$. Without pulse compression, the peak power of the radiated wave would have had to be six times as high to produce the same signal-to-noise ratio for the pulse at $t = 0$.

Let the signal of Fig. 9.4-3 be repeated. The first positive pulse of the second period can occur at the earliest at the time $t = 15T$ in order to avoid an increase of the amplitudes of the sidelobes. The shortest possible period is thus $(20 + 15)T = 35T$. This is also the shortest possible ambiguity-free interval. The pulse density is $6/35T = 0.171$ pulses per time interval $T$. One may increase the ambiguity-free interval beyond $35T$ and pay for it with a reduced pulse density, which means with a lower signal-to-noise ratio. For instance, an increase of the ambiguity-free interval from $35T$ to $193T$ would reduce the pulse density to $6/193T = 0.031$ pulses per time interval $T$.

Can we increase the ambiguity-free interval without reduction of the signal-to-noise ratio? Figure 9.4-5 shows a sequence of 11 positive pulses in line a. The pulses at $t = 0, -T, \ldots, -20T$ are the same as those in line c of

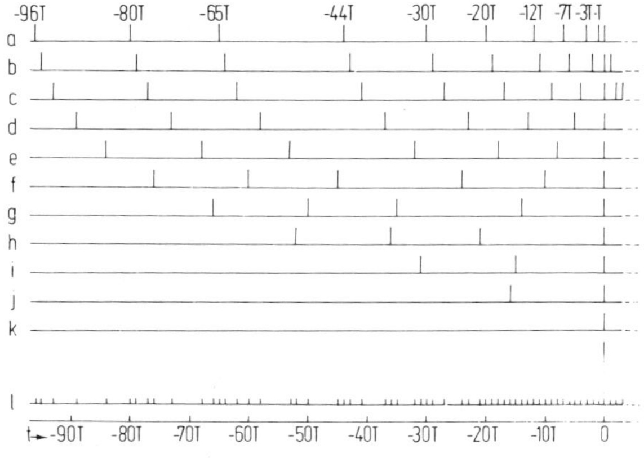


FIG. 9.4-5. Compression of eleven positive pulses into one.

Fig. 9.4-4. The pulses in lines b–k of Fig. 9.4-5 are delayed to yield the sum shown in line $l$ with a peak 11 times as large as that of the individual pulses. Line $l$ is symmetric for $t > 0$, and this section is therefore not shown. A second period of the pulses of line a may certainly start at $t = +97T$ without any increase in the sidelobes (a closer study shows that the smaller value $t = +71T$ is possible by interlacing the sidelobes of adjacent periods as in Fig. 9.4-3). This yields an ambiguity-free interval of $(96 + 97)T = 193T$, and a pulse density of $11/193T = 0.057$ pulses per time interval $T$.

Let us apply the technique of pulse compression to a pulse with the shape shown in Fig. 9.4-2b. A suitable time variation of the pressure of a sound wave is shown in Fig. 9.4-6, line a. This function has six positive and

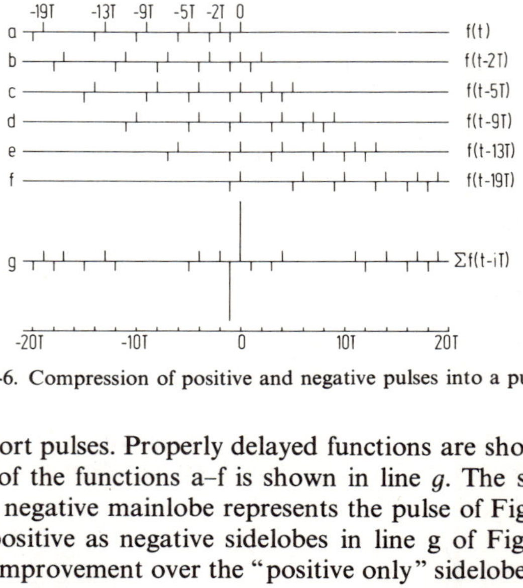


FIG. 9.4-6. Compression of positive and negative pulses into a pulse pair.

six negative short pulses. Properly delayed functions are shown in the lines b–f. The sum of the functions a–f is shown in line $g$. The succession of a positive and a negative mainlobe represents the pulse of Fig. 9.4-2b. There are as many positive as negative sidelobes in line g of Fig. 9.4-6. This is an important improvement over the "positive only" sidelobes in Figs. 9.4-3 and 9.4-5 when signals are returned from many scattering points.

## 9.5 RESOLUTION

We have developed enough theory to turn to the problem of primary interest in using synthetic aperture techniques in acoustic imaging. Can we reduce the array size without reducing the resolution by substituting pulses with a finite frequency bandwidth for sinusoidal waves with essentially no bandwidth? The concept of reducing significantly the classical limit of resolution, $\varepsilon = \lambda/A$, is usually referred to as superresolution, and it is thoroughly discredited. We know from communications that the transmittable information increases with the bandwidth $\Delta f$ and the logarithm of the average signal power $P$ if everything else remains unchanged, and that bandwidth can be interchanged with power according to the relation

$$\Delta f \log P = \text{constant} \tag{1}$$

Since the determination of the location of a scattering point means the transmission of information from the scattering point to the imaging equipment, we must assume that a law like Eq. (1) limits the resolution, rather than the classical law $\varepsilon = \lambda/A$, which contains neither a power nor a bandwidth. In essence, we have to decide whether the classical limit $\varepsilon = \lambda/A$ is a dogma, like Aristotle's dogma of heavier objects falling faster, or an axiom, like the laws of conservation of energy and momentum. Few would be willing to equate the classical limit of resolution with the conservation laws of physics, even if the importance of a finite frequency bandwidth for information transmission were not known. Hence, we will go ahead and investigate superresolution.

Equation (9.2-16) gives the error $\Delta x_1$ of the $x$-coordinate of a point $x_1$, $y_1$ as function of the error $\Delta X$ of the distance between two sensors:

$$\Delta x_1 \doteqdot -\Delta X(d_{11}^2 - d_{21}^2)/8X^2 \tag{2}$$

The error $\Delta x_1$ as function of the errors $\Delta d_{11}$ and $\Delta d_{21}$ of the measured distances is defined by Eq. (9.2-14):

$$\Delta x_1 \doteqdot (d_{11}\Delta d_{11} - d_{21}\Delta d_{21})/2X \tag{3}$$

The error $\Delta X$ in Eq. (2) can be made very small for a rigid array. The errors $\Delta d_{11}$ and $\Delta d_{12}$ of the distance measurement in Eq. (3) are the more interesting limitation on the achievable error $\Delta x_1$.

Let us assume that a block pulse according to Fig. 9.4-2a is used for the distance measurement; this may either be an inherently single pulse or a single pulse produced by pulse compression. The edges will not be infinitely steep, since the frequency bandwidth of the equipment will be finite. Let us assume that the frequency response of the processing equipment can be represented by an idealized low-pass filter with bandwidth $\Delta f$. A step function with amplitude $A$ and infinitely short transient time from 0 to $A$ will have the time variation of the sine integral[1] shown in Fig. 9.5-1 after

[1] The theoretical difficulties of using the sine integral and the practical way to overcome them are well known to any student of communications engineering (Close, 1966). It is usual to say that a function cannot be time and frequency limited; the assumption of a finite bandwidth $\Delta f$ creates thus the unrealistic infinitely long transition from 0 to $A$. Another point of view is that one needs a voltage source, a switch and a filter to produce a transient like the one in Fig. 9.5-1; the switch makes this a linear, time-variable circuit, while a description in terms of sinusoidal functions calls for a linear, time-invariant circuit. Either way, the use of the sine integral mathematically calls for an infinite delay, that can be approximated practically quite well by a finite delay.

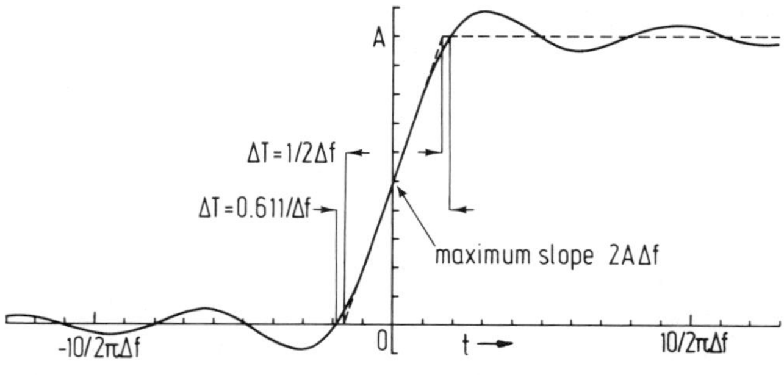


FIG. 9.5-1. Transient of a step function with amplitude $A$ after passing through a frequency low-pass filter with nominal bandwidth $\Delta f$, approximated by the sine integral (solid line) and by a linear transient with transient time $\Delta T = 1/2\Delta f$ (dashed line).

passing through such a filter. Following common practice, we use the simpler linear transient with transition time $\Delta T = 1/2\Delta f$ shown by the dashed line. The slope of the linear transient and the maximum slope of the sine integral have the same value $2A\Delta f$.

Let us use the approximation of the linear transient for the block pulse of Fig. 9.4-2a. If the pulse has the duration $\Delta T$ and the amplitude $A$, it will be transformed by an idealized low-pass filter with bandwidth $\Delta f = 1/2\Delta T$ into the triangular pulse with amplitude $A$ shown on top of Fig. 9.5-2a. A second pulse with amplitude $3A/4$, delayed by $\Delta T$, is shown below. The sum of both pulses is shown in the third row. Sampling at the

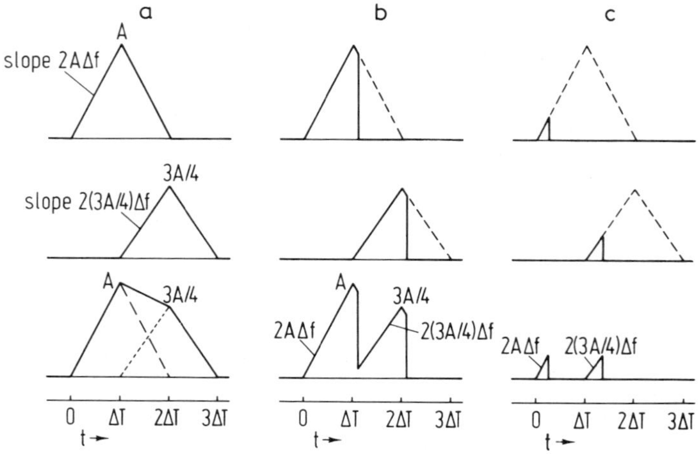


FIG. 9.5-2. Rectangular pulses according to Fig. 9.4-2a after passing through an idealized low-pass filter with transient time $\Delta T = 1/2\Delta f$ according to Fig. 9.5-1. The columns b and c show the suppression of the pulses after determination of their amplitude $A$.

times $\Delta T$ and $2\Delta T$ yields the amplitudes $A$ and $3A/4$. This is the basis for the usual statement that a bandwidth $\Delta f$ permits a time resolution $\Delta T = 1/2\Delta f$ and a distance resolution $\Delta d = c\Delta T/2 = c/4\Delta f$; the factor $\frac{1}{2}$ in the last equation is due to the fact that the signal has to travel twice the distance to the target or the scattering point.

Let us look at column b of Fig. 9.5-2. Once we know the position $t = \Delta T$ and the amplitude $A$ of the triangular pulse in the first row, we do not need the rest, or the trailing part, of the pulse. We can calculate that part of the pulse and subtract it from the received pulse. The resulting truncated pulse is shown by the solid line. The pulse in the second row with amplitude $3A/4$ is truncated in the same way. It would be difficult to do this truncation with analog circuits. However, synthetic aperture *radar* images are routinely produced by computer processing, usually not directly but from data stored on magnetic tape, and the pulse truncation discussed here is not a difficult process for a computer.

The sum of the two truncated pulses is shown in the third row of Fig. 9.5-2b. This sum looks better than the one in column a, but one actually gains nothing. The amplitude measured at the time $t = \Delta T$ would be the amplitude $A$ of the first pulse plus a certain fraction of the amplitude $3A/4$ of the second pulse if the pulses were not at least the time $\Delta T$ apart. Hence, a wrong amplitude would be sampled for a spacing between the pulses shorter than $\Delta T$, and the use of this wrong amplitude for the truncation of the first pulse would also modify the amplitude of the second pulse.

In order to achieve a better resolution than $\Delta T = 1/2\Delta f$, we must obtain the amplitudes $A$ and $3A/4$ of the pulses before they have reached their peak values. Since the slope of a pulse with amplitude $A$ equals $2A\Delta f$, and the bandwidth $\Delta f$ is known, we can derive $A$ from the slope. In the absence of noise one obtains the slope from the difference of two samples taken closely together. The knowledge of the slope $2A\Delta f$ gives us the amplitude $A$ of the pulse, while the knowledge of the time and the amplitudes of the samples taken gives us the time $t = \Delta T$ where the peak of the triangular pulse is located. With this knowledge we can calculate the instantaneous amplitudes of the triangular pulse at any time $t$ and subtract them from the received instantaneous amplitudes. The results are the truncated pulses in the first two rows of Fig. 9.5-2c. Their sum in the third row shows that a resolution time $\Delta T'$ smaller than $\Delta T = 1/2\Delta f$ can be achieved. The minimum value of $\Delta T'$ is no longer determined by the bandwidth $\Delta f$, but by statistical disturbances of the signals and the imperfect working of the equipment.

Let us assume a sample with amplitude $A_1$ is obtained at the time $t_1$ and a sample with amplitude $A_1 + \Delta A$ at the time $t_1 + \Delta T'$. The slope has the value

$$[(A_1 + \Delta A) - A_1]/\Delta T' = \Delta A/\Delta T' = 2A\Delta f \qquad (4)$$

and the amplitude $A$ of the received pulse equals

$$A = \Delta A/\Delta T'\Delta f \tag{5}$$

The values of $\Delta T'$ and $\Delta f$ are fixed and not subject to statistical disturbances. Hence, the error of $A$ caused by statistical disturbances is strictly due to $\Delta A$. Let the statistical disturbances be additive thermal noise. The signal-to-noise ratio of $\Delta A$ is $20 \log(\Delta A/A)$ decibels lower than that of $A$. In order to obtain $A$ with the same statistical error from Eq. (5) as from sampling the peak values of the triangular pulses in Fig. 9.5-2a, one must increase the signal power by $20 \log(A/\Delta A)$ decibels.

Let us next investigate the precision with which the time of arrival of a pulse can be determined. Figure 9.5-2a only shows that we can measure the amplitudes $A$ and $3A/4$ independently if we know that $A$ occurs at $t = \Delta T$ and $3A/4$ at $t = 2\Delta T$, but nothing is said about where this knowledge of the sampling times came from.

Figure 9.5-3a (top) shows how the time $t_p$ of the peak of a triangular

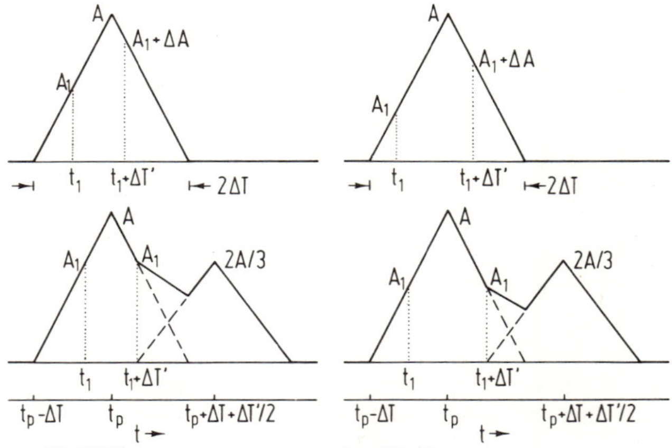


FIG. 9.5-3. Determination of the time $t_p$ of the peak amplitude of a triangular pulse by a vanishing difference $\Delta A$ of two sampled amplitudes $A_1$ and $A_1 + \Delta A$.

pulse can be determined. A sample $A_1$ is taken at the time $t_1$ and another sample $A_1 + \Delta A$ at the time $t_1 + \Delta T'$. We obtain the arrival time $t_p$ of the peak when $\Delta A$ equals zero:

$$t_p = t_1 + \Delta T'/2 \qquad \text{for} \quad (A_1 + \Delta A) - A_1 = \Delta A = 0 \tag{6}$$

Instead of using the difference $(A_1 + \Delta A) - A_1$, one may also use the quotient $(A_1 + \Delta A)/A_1$:

$$(A_1 + \Delta A)/A_1 = 1 \qquad \text{or} \quad \Delta A/A_1 = 0 \tag{7}$$

In either case, the minimum distance between two adjacent pulses is no longer $\Delta T$ as in Fig. 9.5-2a but $\Delta T + \Delta T'/2$, as one can readily see from Fig. 9.5-3a (bottom). In the absence of noise one may make $\Delta T'$ very small, but one will hesitate to do so when the signals are disturbed. To see what determines the choice of $\Delta T'$, we plot $\Delta A$ as function of the differences between sampling time $t_1$ and pulse peak time $t_p$ for various values of $\Delta T'$ in Fig. 9.5-4; the plotting is done for $(t_1 - \Delta T') - (t_p - \Delta T)$ rather than for $t_1 - t_p$ in order to let all curves start at zero.

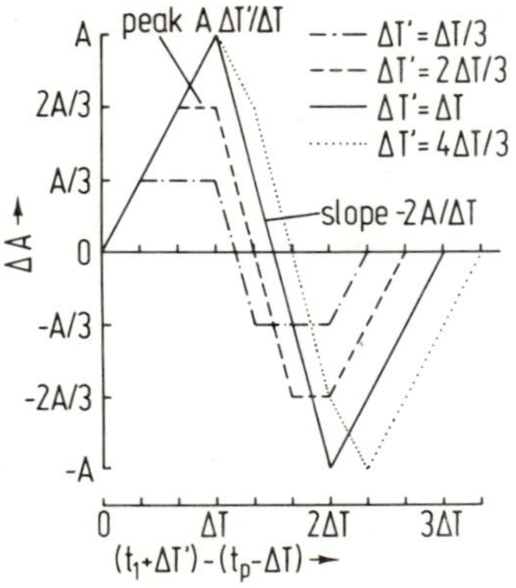


FIG. 9.5-4. Variation of $\Delta A$ in Fig. 9.5-3. as a function of the sampling times $t_1$ and $t_1 + \Delta T'$ relative to the time $t_p - \Delta T$ of the beginning of the pulse for various values of $\Delta T'$.

The rising slope of all four curves shown equals $A/\Delta T$; it becomes zero at $\Delta A = A\Delta T'/\Delta T$ and stays so until the time $\Delta T$ is reached. Then follows a falling slope $-2A/\Delta T$, equal for all curves with $\Delta T' \leq \Delta T$; the more complicated curve for $\Delta T' > \Delta T$ is not of interest here.

In the presence of additive thermal noise we are interested in a long and steep negative slope. For instance, a noise amplitude within the limits $+A/3$ and $-A/3$ superimposed on the sampled amplitude $A_1 + \Delta A$ can make the solid curve ($\Delta T' = \Delta T$) in Fig. 9.5-4 zero anywhere in the interval

$$4\Delta T/3 \leq (t_1 + \Delta T') - (t_p - \Delta T) \leq 5\Delta T/3,$$

but the dashed–dotted curve ($\Delta T' = \Delta T/3$) can be made zero anywhere in the much larger interval

$$0 \leq (t_1 + \Delta T') - (t_p - \Delta T) \leq 7\Delta T/3$$

A value $\Delta T' > \Delta T$ does not yield either a steeper slope nor a longer one in Fig. 9.5-4; it increases the required distance between two resolvable pulses in Fig. 9.5-3 without giving any advantage in return. Hence, $\Delta T'$ should vary from almost zero at large signal-to-noise ratios to $\Delta T$ at small signal-to-noise ratios.

In a practical case, one generally tries to use $\Delta T' = \Delta T$. This choice implies that the middle of the rising and falling edges of a pulse are used to measure its time of arrival. According to Fig. 9.5-1 this point ($t = 0$) is least affected by the ringing at the beginning and end of the transient; experience shows that this holds generally, not only for the idealized sine integral transient.

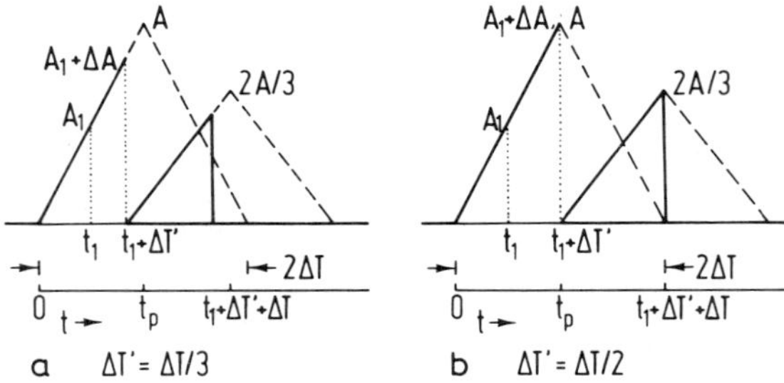


FIG. 9.5-5. Determination of the time $t_p$ of the peak amplitude of a triangular pulse by a vanishing difference $\Delta A - A_1$ of two sampled amplitudes $A_1$ and $A_1 + \Delta A$.

Let us turn to the determination of the arrival time $t_p$ of the pulse peak if the pulse is truncated according to Fig. 9.5-2. From Fig. 9.5-5 and Eq. (4) one can readily recognize the following relation:

$$(A - A_1)/(t_p - t_1) = 2A\Delta f = \Delta A/\Delta T' \tag{8}$$

Reordering of the terms yields $t_p$:

$$t_p = t_1 + 1/\Delta f + A_1/(\Delta A/\Delta T') \tag{9}$$

The values of $t_1$, $\Delta f$, and $\Delta T'$ are not subject to statistical disturbances, but the amplitude $A_1$ and the slope $\Delta A/\Delta T'$ are. With the notation $A_1/\Delta A = q$ we may rewrite Eq. (9) in the form of Eqs. (6) and (7):

$$t_p = t_1 + 1/\Delta f + q\Delta T' \qquad \text{for} \quad A_1/\Delta A = q \quad \text{or} \quad q\Delta A - A_1 = 0 \tag{10}$$

We choose $q = 1$ to simplify the investigation. The calculation of the best value of $q$ for a certain model of noise is rather difficult.[1]

[1] Let the noise samples $a_1$ and $a_2$ be added to the sampled signal amplitudes $A_1$ and $A_1 + \Delta A$ to yield $A_1 + a_1$ and $A_1 + \Delta A + a_2$. The difference $q\Delta A - A_1$ in Eq. (10) yields $q(\Delta A - a_1 + a_2) - (A_1 + a_1)$, which equals $\Delta A - A_1 + a_2 - 2a_1$ for $q = 1$. The difference $\Delta A$ in Eq. (6) becomes $(A_1 + \Delta A + a_2) - (A_1 + a_1) = \Delta A + a_2 - a_1$. If $a_1$ and $a_2$ are statistically independent, a decision based on the equation $\Delta A + a_2 - a_1 = 0$ is less affected by noise than a decision based on the equation $\Delta A - A_1 + a_2 - 2a_1 = 0$. However, the assumption for Eq. (10) is that $\Delta T'$ is smaller than $\Delta T$, and this implies that $a_1$ and $a_2$ are not statistically independent. Hence, the best choice of $q$ and the effect of noise on the decisions based on Eqs. (6) and (10) depends not only on the noise model and the filter, but also on $\Delta T'$.

The minimum distance between two adjacent pulses equals $t_1 + \Delta T'$ according to Fig. 9.5-5. For $q = 1$ one obtains $t_1 = \Delta T'$, and the minimum distance between two pulses becomes $2\Delta T'$. Since $2\Delta T'$ can be made arbitrarily small in the absence of noise, we must determine how noise influences the choice of $\Delta T'$. We plot in Fig. 9.5-6 the difference $\Delta A - A_1$ as a function of the difference between sampling time $t_1$ and pulse peak time $t_p$ for various values of $\Delta T'$; as before, the plotting is done for $(t_1 - \Delta T') - (t_p - \Delta T)$ rather than for $t_1 - t_p$ in order to let all curves start at zero.

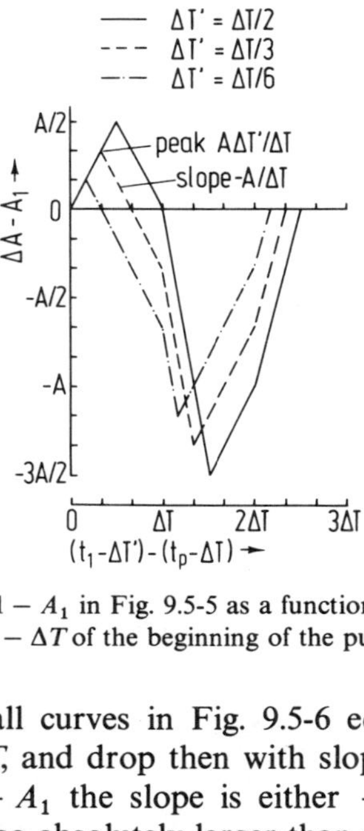


FIG. 9.5-6. Variation of $\Delta A - A_1$ in Fig. 9.5-5 as a function of the sampling times $t_1$ and $t_1 + \Delta T'$ relative to the time $t_p - \Delta T$ of the beginning of the pulse for various values of $\Delta T'$.

The rising slope of all curves in Fig. 9.5-6 equals $A/\Delta T$; the curves reach the peak $A\Delta T'/\Delta T$, and drop then with slope $-A/\Delta T$ to zero. For negative values of $\Delta A - A_1$ the slope is either $-A/\Delta T$ or steeper. The negative peak value is also absolutely larger than the positive one.

Let us compare the curves of Fig. 9.5-6 with the curve for $\Delta T' = \Delta T$ in Fig. 9.5-4, which had been found to be the best in the presence of thermal noise. The minimum time between two resolvable pulses in Fig. 9.5-4—determined by $\Delta A = 0$—equals $\Delta T + \Delta T'/2 = 3\Delta T/2$ for $\Delta T' = \Delta T$, while this time in Fig. 9.5-6—determined by $\Delta A - A_1 = 0$—equals $2\Delta T' \leq \Delta T$. The ratio of the minimum distance of resolvable pulses is denoted $\xi$:

$$\xi = 2\Delta T'/(3\Delta T/2) = 4\Delta T'/3\Delta T \tag{11}$$

To determine the price to be paid in terms of signal power for this improved time resolution, we consider two cases. For a large signal-to-noise ratio the slope around $\Delta A = 0$ in Fig. 9.5-4 and around $\Delta A - A_1 = 0$ in Fig. 9.5-6 will be the important quantity; note that we do not assume more about the noise than that the signal-to-noise ratio be large. The slope in Fig. 9.5-6 equals $-A/\Delta T$ while the slope in Fig. 9.5-4 equals $-2A/\Delta T$. Doubling the pulse amplitude $A$ and thus increasing the signal power by 6 dB makes the slope in Fig. 9.5-6 as large[1] as in Fig. 9.5-4.

If the signal-to-noise ratio is not large, we will want to make the peak $A\Delta T'/\Delta T$ in Fig. 9.5-6 equal to the peak $A$ of the curve for $\Delta T' = \Delta T$ in Fig. 9.5-4. This requires that the amplitude $A$ of the pulses in Fig. 9.5-5 be increased to $A\Delta T/\Delta T'$. The slope $-A/\Delta T$ in Fig. 9.5-6 becomes in this case $-A/\Delta T'$, which means it is steeper than in Fig. 9.5-4 for $\Delta T' < \Delta T/2$; hence, the timing error will actually be smaller for the curves of Fig. 9.5-6 than for the curve $\Delta T' = \Delta T$ in Fig. 9.5-4. We introduce the amplitude ratio $\eta$,

$$\eta = (A\Delta T/\Delta T')/A = \Delta T/\Delta T' \tag{12}$$

and the power ratio $\eta^2$:

$$\eta^2 = (\Delta T/\Delta T')^2 \tag{13}$$

The product of the ratio $\xi$ of the minimum distance of resolvable pulses and the amplitude ratio $\eta$ is constant:

$$\xi\eta = (4\Delta T'/3\Delta T)(\Delta T/\Delta T') = \tfrac{4}{3} \tag{14}$$

In readily understandable terminology, we have a linear interchange between resolution time and pulse amplitude, and a quadratic interchange between resolution time and "pulse power," where pulse power stands either for peak power or average power of the pulse.

Let us return to Eq. (3), which gives the error $\Delta x_1$ of the $x$-coordinate as a function of the distance errors $\Delta d_{11}$ and $\Delta d_{21}$. The minimum time distance $2\Delta T'$ of two resolved pulses yields the bounds $\pm\Delta d = \pm c\Delta T'/2$ of the range error. The bounds for the error of the $x$-coordinate follow from Eq. (3):

$$-c\Delta T'(d_{11} + d_{12})/4X \leq \Delta x_1 \leq c\Delta T'(d_{11} + d_{12})/4X \tag{15}$$

If one makes many measurements, one obtains an average error that is zero and an rms error that is generally substantially smaller than the bounds of Eq. (15). We do not discuss this improvement of the resolution by

[1] The equality of the slope only implies that the influence of noise is approximately the same. The noise samples $a_1$ and $a_2$ in the equation $\Delta A + a_2 - a_1 = 0$ in footnote on p. 212 are statistically independent if we assume thermal noise, due to the assumption $\Delta T' = \Delta T$ for this case. On the other hand, the samples $a_1$ and $a_2$ neither occur in the form $a_2 - a_1$ in the equation $\Delta A - A_1 + a_2 - 2a_1 = 0$ nor are they statistically independent due to the relation $\Delta T' \leq \Delta T/2$.

statistical methods, since a very sophisticated model of the causes of the errors is required to make such an investigation worthwhile.

Equation (14) may be rewritten in terms of bandwidth $\Delta f$ and average signal power $P$. Let $\Delta T = 1/2\Delta f$ be substituted from Fig. 9.5-1 into Eq. (11):

$$\xi = 4\Delta T'/3(1/2\Delta f) = \tfrac{8}{3}\Delta T'\,\Delta f \tag{16}$$

Let $P_0$ denote the average power of a rectangular pulse of duration $\Delta T$, which produces the triangular pulses with amplitude $A$ and duration $2\Delta T$ in the top rows of Figs. 9.5-2 and 9.5-3; the average power of a pulse of duration $\Delta T'$ is denoted $P$. The amplitudes of these two rectangular pulses are $A = K\sqrt{P_0}$ and $A\Delta T/\Delta T' = K\sqrt{P}$, where $K$ is a constant. We obtain from Eq. (12)

$$\eta = (P/P_0)^{1/2} \tag{17}$$

Substitution of Eqs. (16) and (17) into Eq. (14) yields

$$\Delta f\sqrt{P} = \tfrac{3}{8}\sqrt{P_0}/\Delta T' = \text{constant} \tag{18}$$

This equation states, that a certain resolution time $\Delta T'$ and a certain signal-to-noise ratio represented by the reference signal power[1] $P_0$ can be achieved for various values of $\Delta f$ and $P$, as long as the product $\Delta f\sqrt{P}$ remains unchanged.

Let us compare Eqs. (1) and (18). The structure is the same, but $\log P$ is replaced by $\sqrt{P}$. Since $\sqrt{P}$ increases faster than $\log P$, one will expect that one still can improve upon Eq. (18), but $\Delta f\sqrt{P} = \text{constant}$ is a remarkably good exchange of power for bandwidth.

In connection with Fig. 9.1-1 we had derived the distance $\Delta x$ of two resolvable points at a distance $R$ with $\Delta x = R\varepsilon = R\lambda/A$. Neither signal power nor bandwidth enter this equation. The signal power is not in it, since the resolution is limited by the diffraction pattern rather than by noise, and an increase of the signal power does not change the diffraction pattern. The bandwidth is not in the equation because it is zero.

Let $\Delta T'$ be expressed from Eq. (18),

$$\Delta T' = \tfrac{3}{8}\sqrt{P_0}/\Delta f\sqrt{P} \tag{19}$$

and inserted into Eq. (15):

$$-\tfrac{3}{8}c(\sqrt{P_0}/\Delta f\sqrt{P})(d_{11} + d_{12})/4X \le \Delta x_1 \le \tfrac{3}{8}c(\sqrt{P_0}/\Delta f\sqrt{P})(d_{11} + d_{12})/4X \tag{20}$$

[1] We have chosen $P_0$ arbitrarily, but it would actually be determined by the signal-to-noise ratio one wants to achieve.

The error $\Delta x_1$ now takes the place of the distance $\Delta x$ of two resolvable points. The error $\Delta x_1$ can be reduced by using a large "aperture" $X$, but also by increasing the bandwidth $\Delta f$ or the signal power $P$. The frequency bandwidth $\Delta f$, or rather the wavelength bandwidth $c/\Delta f$, may readily be viewed as a generalization of the wavelength $\lambda$ in the classical formula $\Delta x = R\lambda/A$; furthermore, $d_{11}$ and $d_{12}$ are obvious generalizations of $R$. The most important term in Eq. (20) is $\sqrt{P}$, since it has no equivalent in the classical formula.

Let us return to the theorem derived at the end of Section 9.1, which said that the relative number of resolved points is zero for sinusoidal waves if the wavefronts are planar and if a line array with a finite number of equally spaced sensors is used. The basis for this theorem was that one could produce only $2n + 1$ beams with $2n + 1$ sensors. The situation is now completely different. If $\Delta x_1$ is as large as $\delta$ in Fig. 9.1-1, we can maintain the ratio $\Delta x_1/\delta = 1$ for any distance $R$ by keeping the ratio $(d_{11} + d_{12})/\sqrt{P}$ in Eq. (20) constant. This completes the analogy for imaging or beam forming of the theorem in communications, which says that information can only be transmitted at a rate larger than zero with signals that have a bandwidth larger than zero.

Let us end with a short proof that the determination of the amplitude of a sample from its slope does not lead to an ever increasing error if the processing is not done with unlimited precision. Assume that an amplitude determined to have the value $A_0$ actually has the value $(1 + q)A_0$, where $1 < q < -1$. The error $qA_0$ is thus introduced. At the next determination of the amplitude, which should have the value $A_1$, we will determine[1] $A_1 + qA_0$, and this value should actually be $(1 + q)(A_1 + qA_0)$. Hence, the error is now $qA_1 + q^2A_0$. After $n$ samples we will have the total error $qA_n + q^2A_{n-1} + , \ldots, + q^{n+1}A_0$. Let all amplitudes $A_0, \ldots, A_n$ have the same value. We may then use the sum of the geometric series, $qA_0/(1 - q)$, for the total error. For an error of $\pm 10\%$ or less, $q = \pm 0.1$, we obtain for the total error $qA_0(1 + q)$. The single error $qA_0$ is about 10%, and the total error $qA_0(1 + q)$ about 11%. Hence, imperfect processing does not lead to a significant error accumulation.

This brief argument about error accumulation will not satisfy everybody, but the principle is evident and a more thorough investigation will require several scientific papers.

[1] Generally, the error $qA_0$ will be somewhat smaller when the second, third, . . . sample is determined. We are calculating here a worse than typical case.

# References


Alais, P. (1974). Real time acoustical imaging with a 256 × 256 matrix of electrostatic transducers, *Acoust. Hologr.* **5**, 671–684.

Albers, V. M. (1965). "Underwater Acoustics Handbook." Penn. State Univ. Press, University Park, Pennsylvania.

Aldridge, E. E. (1969). Ultrasonic holography, *Phys. Bull.* **20**/1, 10–17.

Anderson, R. (1974). Potential medical applications for ultrasonic holography, *Acoust. Hologr.* **5**, 505–513.

Aoki, Y. (1968). Three-dimensional information storage in acoustical hologram, *Appl. Opt.* **7**, 1402–1403.

Aoki, Y. (1969a). Space division multiplexing holography, *Proc. IEEE* **57**, 353–359.

Aoki, Y. (1969b). High-order images reconstructed from a sampled sound wave hologram, *J. Appl. Phys.* **40**, 2294–2295.

Aoki, Y. (1974). Image reconstruction by computer in acoustical holography, *Acoust. Hologr.* **5**, 551–572.

Aoki, Y., Ioshida, N., Tzukamoto, N., and Suzuki, M. (1967). Sound wave hologram and optical reconstruction, *Proc. IEEE* **55**, 1622–1623.

Auld, A., Gilbert, R., Hyllested, K., Roberts, C., and Webb, D. C. (1972). A. 1.1 GHz scanned acoustic microscope, *Acoust. Hologr.* **4**, 73–96.

Auld, B. A., Addison, R. C., and Webb, D. C. (1970). Focusing and scanning of acoustic waves in solids, *Acoust. Hologr.* **2**, 117–132.

Babin, L. V., and Gurevich, S. B. (1972). Acoustic holography, *Sov. Phys.—Acoust.* **17**/4, 419–437 [*Akust. Zh.* **17**, 489–512 (1971)].

Beaver, W. L., Maginness, M. G., Plummer, J. D., and Meindl, J. D. (1976). Ultrasonic imaging using two-dimensional transducer arrays, *in* "Cardiovascular Imaging and Image Processing, Theory and Practice" (D. C. Harrison, H. Sander, and H. A. Miller, eds.), Vol. 72, pp. 17–23. Soc. Photo-Opt. Instrum. Eng., Palos Verdes Estates, California.

Beaver, W. L., Dameron, D. H., and Macovski, A. (1977). Ultrasonic imaging with an acoustic lens, *IEEE Trans. Sonics Ultrason.* **SU-24**, 235–243.

Berger, H. (1969). Ultrasonic imaging system for nondestructive testing, *J. Acoust. Soc. Am.* **45**, 859–867.

Booth, N. O., and Sutton, J. L. (1974). Holographic Acoustic Imaging, Rep. NUC-IP-424. Nav. Undersea Cent., San Diego, California.

Boutin, H. and Mueller, R. K. (1967). Real-time display of sound holograms by KD P modulation of a coherent light source, *J. Acoust. Soc. Am.* **42**, 1169.

Brenden, B. B. (1972). Real time acoustical imaging by means of liquid surface holography, *Acoust. Hologr.* **4**, 1–9.

Brenden, B. B. (1975). History and present status of liquid surface acoustical holography, *J. Acoust. Soc. Am.* **58**, 951–955.

Brown, P., and Galloway, P. (1976). Recent developments in underwater imaging using the ultrasonic image converter tube, *Ultrasonics* **14**, 273–277.

Bruneel, C., Nongaillard, B., Torguet, R., Bridoux, E., and Rouvaen, J. M. (1977). Reconstruction of an acoustical image using an "acousto-electronic" lens device, *Ultrasonics* **15**, 263–264.

Burckhardt, C. B. (1978). Speckle in ultrasound B-mode scans, *IEEE Trans. Son. Ultrason.* **SU-25**, 1–6.

Carter, W. H. (1968). Aliasing in sampled holograms, *Proc. IEEE* **56**, 96–98.

Clark, J. A., and Stone, K. L. (1974). A solid plate acoustical viewer for underwater diving, *Acoust. Hologr.* **5**, 701–714.

Close, C. M. (1966). "The Analysis of Linear Circuits," p. 479. Harcourt, New York.

Coello-Vera, A. E., Schlussler, L., Fontana, J. R., and Wade, G. (1978). Motion effects in scanned acoustic holography, *IEEE Trans. Son. Ultrason.* **SU-25**, 167–176.

Collier, R. J., Burckhardt, C. B., and Lin, L. H. (1971). "Optical Holography." Academic Press, New York.

Cutrona, L. J., and Hall, G. O. (1962). A comparison of techniques for achieving fine azimuth resolution, *IEEE Trans. Mil. Electron.* **MIL-6**, 119–121.

Deschamps, J., and Kleehammer, R. J. (1977). Scan converter for ultrasonic medical imaging, *Acoust. Hologr.* **7**, 283–289.

Dick, M. L., Dick, D. E., McLeod, F. D., and Kindig, N. B. (1977). Ultrasonic synthetic aperture imaging, *Acoust. Hologr.* **7**, 327–346.

Domarkas, V. I. (1976). Design aspects of ultrasonic instruments operating in the megahertz range with electrically controlled directivity patterns, *Sov. Phys. Acoust.* **22**, 81 [*Akust. Zh.* **22**, 144–145 (1976)].

Duck, F., Johnson, S., Greenleaf, J., and Samayoa, W. (1977). Digital image focusing in the near field of a sampled acoustic aperture, *Ultrasonics* **15**, 83–88.

Farnow, S. A., and Auld, A. (1975). An acoustic phase plate imaging device, *Acoust. Hologr.* **6**, 259–274.

Farrah, H. R., Marom, E., and Mueller, R. K. (1970). An underwater viewing system using sound holography, *Acoust. Hologr.* **2**, 173–183.

Feleppa, E. V. (1969). Biomedical applications of holography, *Phys. Today* **22**/7, 25–32.

Fenner, W. R., and Stewart, G. E. (1974). An ultrasonic holographic imaging system for medical applications, *Acoust. Hologr.* **5**, 481–492.

Folds, D. L. (1975). Focusing properties of a solid four-element ultrasonic lens, *J. Acoust. Soc. Am.* **58**, 72–77.

Fraser, J., Havlice, J., Kino, G., Leung, H., Shaw, H., Toda, K., Waugh, T., Winston, D., and Zitelli, L. (1975). An electronically focused two-dimensional acoustic imaging system, *Acoust. Hologr.* **6**, 275–304.

Fritzler, D., Marom, E., and Mueller, R. K. (1969). Ultrasonic holography via the ultrasonic camera, *Acoust. Hologr.* **1**, 249–255.

Gabor, D. (1949). Microscopy by reconstructed wave-fronts, *Proc. R. Soc. London, Ser. A* **197**, 454–487.

Gabor, D. (1951). Microscopy by reconstructed wave fronts, II, *Proc. Phys. Soc., London, Sect. B* **64**/6, 449–464.

Gaubatz, D. A. (1976). FFT based analog beamforming processor, *1976 Ultrason. Symp. Proc.* (IEEE 76CH1120-5SU), pp. 676–681.

Gaubatz, D. A. (1977). Fast beamforming processor, *Acoust. Hologr.* **7**, 495–507.

Goetz, G. G. (1970). Real time holographic reconstruction by electro-optic modulation, *Appl. Phys. Lett.* **17**, 63–66.

Graeme, J. G. (1973). "Applications of Operational Amplifiers." McGraw-Hill, New York.

Green, P. S., Bellin, J. L., and Knollman, G. C. (1968). Acoustic imaging in a turbid underwater environment, *J. Acoust. Soc. Am.* **44**, 1719–1730.

Green, P. S., Schaefer, L., and Macovski, L. (1972). Considerations for diagnostic ultrasonic imaging, *Acoust. Hologr.* **4**, 97–111.

Green, P. S., Schaefer, L., Jones, E., and Suarez, J. (1974). A new high-performance ultrasonic camera system, *Acoust. Hologr.* **5**, 493–513.

Greene, D. C. (1969). Use of acoustical holography for the imaging of sources of radiated acoustic energy, *J. Acoust. Soc. Am.* **46**, 44–45.

Halstead, J. (1968). Ultrasound holography, *Ultrasonics* **6**/2, 79–87.

Hanafy, A., and Zambuto, M. (1977). Acoustic image converter for three-step acoustic holography, *Acoust. Hologr.* **7**, 117–132.

Harger, R. O. (1970). "Synthetic Aperture Radar Systems: Theory and Design." Academic Press, New York.

Harmuth, H. F. (1976). Generation of images by means of two-dimensional spatial electric filters, *Adv. Electron. Electron Phys.* **41**, 167–248.

Harmuth, H. F. (1977). "Sequency Theory—Foundations and Applications." Academic Press, New York.

Harmuth, H. F., Kamal, J., and Murthy, S. S. R. (1974). Two-dimensional spatial hardware filters for acoustic imaging *in* "Applications of Walsh Functions and Sequency Theory" (H. Schreiber and G. F. Sandy, eds.), pp. 94–126 (74CH0861-5EMC). IEEE, New York.

Havlice, J. F. (1969). Visualization of acoustic beams using liquid crystals, *Electron. Lett.* **5**/20, 477–478.

Hildebrand, B. P., and Brenden, B. B. (1972). "An Introduction to Acoustical Holography." Plenum, New York.

Hildebrand, B. P., and Haines, K. A. (1969). Holography by scanning, *J. Opt. Soc. Am.* **59**, 1–6.

Holt, D., and Coldrick, J. R. (1969). Techniques of acoustical holography; an outline of principles and possible applications, *Wireless World* **75**, 425–428.

Jacobs, J. E. (1968). Ultrasound image converter system utilizing electron-scanning techniques, *IEEE Trans. Son. Ultrason.* **SU-15**, 146–152.

Jones, C. H., and Gilmour, G. A. (1974). Sonic Cameras, Rep. No. 74-1M7-DPSUB-P1, 46 pp. (Available from Research Laboratories Library, Westinghouse Corp., Pittsburg, Pennsylvania 15235.)

Jones, C. H., and Gilmour, G. A. (1976). Sonic cameras, *J. Acoust. Soc. Am.* **59**, 74–85.

Jones, H. W., and Williams, C. J. (1977). Lenses and ultrasonic imaging, *Acoust. Hologr.* **7**, 133–153.

Kanevskii, I. N., and Surikov, B. S. (1971). Homogeneous spherical lenses for the focusing of sound waves, *Sov. Phys.—Acoust.* **17**/1, 43–47 [*Akust. Zh.* **17**, 55–60 (1971)].

Kanevskii, I. N., and Surikov, B. S. (1973). Calculation of the sound field in the focal region of a closed homogeneous cylindrical lens, *Sov. Phys.—Acoust.* **19**/1, 28–31 [*Akust. Zh.* **19**, 42–46 (1973)].

Kanevskii, I. N., and Surikov, B. S. (1976). Two-layer acoustic lenses, *Sov. Phys.—Acoust.* **22**/4, 292–294 [*Akust. Zh.* **22**, 526–530 (1976)].

Kapustina, O. A., and Lupanov, V. N. (1977). Experimental investigation of a liquid crystal acousto-optical image converter, *Sov. Phys.—Acoust.* **23**/3, 218–221 [*Akust. Zh.* **23**, 390–396 (1977)].

Keating, P. N., Koppelmann, R. F., Mueller, R. K., and Steinberg, R. F. (1974). Complex on-axis holograms and reconstruction without conjugate images, *Acoust. Hologr.* **5**, 515–526.

Kessler, L. W., Palermo, P. R., and Korpel, A. (1972). Practical high resolution microscopy, *Acoust. Hologr.* **4**, 51–71.

Kheifets, E. I. (1973). Quantitative analysis of sound fields visualized by holographic techniques, *Sov. Phys.—Acoust.* **19**/3, 279–283 [*Akust. Zh.* **19**, 434–443 (1973)].

Kingslake, R. (1967). "Applied Optics and Optical Engineering," Vol. 4, "Optical Instruments, Part 1." Academic Press, New York.

Kisslo, J., and von Ramm, O. T. (1975). Cardiac imaging using a phased array ultrasound system, II. Clinical technique and application, *Circulation* **53**, 262–267.

Kleshev, A. A. (1973). Synthesis of an acoustic array having a curvilinear (spheroidal) surface over a wide range of wave dimensions, *Sov. Phys.—Acoust.* **18**/3, 347–351 [*Akust. Zh.* **18**, 413–420 (1972)].

Knollman, G. C., Weaver, J. L., Hartog, J. J., and Bellin, J. L. (1975). Real-time ultrasonic imaging methodology in nondestructive testing, *J. Acoust. Soc. Am.* **58**, 455–470.

Kock, W. E. (1967). Use of lens arrays in holography, *Proc. IEEE* **56**, 1103–1104.

Kock, W. E. (1968). Stationary coherent (hologram) radar and sonar, *Proc. IEEE* **56**, 2180–2181.

Kock, W. E. (1973). "Radar, Sonar and Holography." Academic Press, New York.

Korpel, A. (1968). Acoustic imaging and holography, *IEEE Spectrum* **5**/10, 45–52.

Korpel, A., and Desmares, P. (1969). Rapid sampling of acoustic holograms by laser scanning techniques, *J. Acoust. Soc. Am.* **45**, 881–884.

Korpel, A., and Whitman, R. L. (1969). Probing of acoustic surface perturbation by coherent light, *Appl. Opt.* **8**, 1567–1576.

Kostanty, R. G. (1972). Doubling op amp summing power, *Electronics* Feb. 14, 73–75.

Kovaly, J. J. (1976). "Synthetic Aperture Radar." Artech House, Dedham, Massachusetts.

Landry, J., Powers, J., and Wade, G. (1969). Ultrasonic imaging of internal structure by Bragg-diffraction, *Appl. Phys. Lett.* **15**/6, 186–188.

Landry, J., Keyani, H., and Wade, G. (1972). Bragg-diffraction imaging: A potential technique for medical diagnosis and material inspection, *Acoust. Hologr.* **4**, 127–146.

Lee, C. H., Heidbreder, G., Wade, G., and Coello-Vera, A. (1977). On-line interactive computer processing of acoustic images, *Acoust. Hologr.* **7**, 227–223.

Levi, L. (1968). "Applied Optics." Wiley, New York.

Liberman, M. Y. (1976). Recording of ultrasonic holograms by the suspension method, *Sov. Phys.—Acoust.* **22**, 143–145 [*Akust. Zh.* **22**, 257–260 (1976)].

Maginness, M. G., Plummer, J. D., and Meindl, J. D. (1974). An acoustic image sensor using a transmit–receive array, *Acoust. Hologr.* **5**, 619–631.

Maginness, M. G., Plummer, J. D., Beaver, W. L., and Meindl, J. D. (1976). State-of-the-art two-dimensional ultrasonic transducer array technology, *Med. Phys.* **3**/5, 312–318.

Marom, E., Fritzler, D., and Mueller, R. K. (1968). Ultrasonic holography by electronic scanning of piezoelectric crystals, *Appl. Phys. Lett.* **12**/2, 26–28.

Marom, E., Mueller, R. K., Koppelman, R. F., and Zilinskas, G. (1971). Design and preliminary test of an underwater viewing system using sound holography, *Acoust. Hologr.* **3**, 191–209.

Massey, G. A. (1968). Acoustic imaging by holography, *IEEE Trans. Son. Ultrason.* **SU-15**, 141–146.

Matzuk, T., and Skolnik, M. C. (1978). Novel ultrasonic real-time scanner featuring servo controlled transducers displaying a sector image, *Ultrasonics* **16**/4, 171–178.

Meindl, J. D. (1976a). Integrated electronics in medicine, *Adv. Biomed. Eng.* **6**, 45–98.

Meindl, J. D. (1976b). Integrated electronics for acoustic imaging arrays, *Acoust. Hologr.* **7**, 127–188.

Meindl, J. D., Beaver, W. L., Maginness, M. G., Plummer, J. D., and Macovski, A. (1976). A high resolution three-dimensional real-time ultrasonic imaging system, *L. H. Gray Conf., 7th: Med. Images* (G. A. Hay, ed.), pp. 99–114. Inst. Phys., New York.

Metherel, A. F., Spinak, S., and Pisa, E. J. (1969). Temporal-reference acoustical holography, *Appl. Opt.* **8**, 1543–1550.

Metherel, A. F., Erikson, K. R., Wreede, J. E., Norton, R. E., and Watts, R. M. (1974). A medical imaging acoustical holography system using linearized subfringe holographic interferometry, *Acoust. Hologr.* **5**, 453–470.

Mezrich, R. S. (1977). High resolution, high sensitivity ultrasonic C-scan imaging system, *Acoust. Hologr.* **7**, 51–64.

Mezrich, R. S., Etxold, K. F., and Vilkomerson, D. (1975). System for visualizing and measuring ultrasonic wavefronts, *Acoust. Hologr.* **6**, 165–182.
Miller, E. B., and Thurstone, F. L. (1977). Linear ultrasonic array design for echosonography, *J. Acoust. Soc. Am.* **61**, 1481–1491.
Mohamed, N. J. (1976). Generation of Images by Means of Sound Waves Using a Two-Dimensional Hydrophone Array and a Sampling Filter Processor. Ph.D. Thesis, Sch. Eng., Catholic Univ., Washington, D.C.
Mueller, R. K. (1971). Acoustic holography, *Proc. IEEE* **59**, 1319–1335.
Mueller, R. K., and Keating, P. N. (1969). The liquid–gas interface as a recording medium for acoustical holography, *Acoust. Hologr.* **1**, 49–55.
Mueller, R. K., and Sheridan, N. K. (1966). Sound holograms and optical reconstruction, *Appl. Phys. Lett.* **9**, 328–329.
Mueller, R. K., Marom, E., and Fritzler, D. (1969). Some problems associated with optical image formation from acoustical holograms, *Appl. Opt.* **8**, 1537–1542.
Penttinen, A., and Luukkala, M. (1977). A high resolution ultrasonic microscope, *Ultrasonics* **15**, 205–210.
Phillips, D., Smith, S., von Ramm, O., and Thurstone, F. (1975). Sampled aperture techniques applied to B-mode echo encephalography, *Acoust. Hologr.* **6**, 103–120.
Plonus, M. A. (1968). Ultrasonic camera system for viewing acoustical holograms, *Proc. IEEE* **56**, 1135–1136.
Preston, K. J., and Kreuser, J. L. (1967). Ultrasonic imaging using a synthetic holographic technique, *Appl. Phys. Lett.* **10**, 150–152.
Prokhorov, V. G. (1972). Piezoelectric matrices for the reception of acoustic images and holograms, *Sov. Phys.—Acoust.* **18**, 408–410 [*Akust. Zh.* **18**, 482–484 (1972)].
Redman, J. D., Walton, W. P., Fleming, J. E., and Hall, A. M. (1969). Holography display of data from ultrasonic scanning, *Ultrasonics* **7**, 26–29.
Reutov, A. P. (1970). "Sidelooking Radar." Natl. Tech. Inf. Serv., Springfield, Virginia (AD 787 070). Russian original, "Radiolokatsionnye Stantsii Bokovogo Obzora." *Sov. Radio*, Moscow, 1970.
Roi, N. A. (1970). Pulsed electrodynamic radiators, *Sov. Phys.—Acoust.* **16**(1), 94–100 [*Akust. Zh.* **16**, 121–128 (1970)].
Ross, E. S. (1965). Birds that "see" in the dark with their ears, *Natl. Geogr.* **127**/2, 282–290.
Sabet, C., and Turner, C. W. (1976). Parametric transducer for high-speed real-time acoustic imaging, *Electron. Lett.* **12**/2, 44–45.
Sasaki, K., Sato, T., and Nakamura, Y. (1978). An effective utilization of spectral spread in holographic passive imaging systems, *IEEE Trans. Son. Ultrason.* **SU-25**, 177–184.
Sato, T., and Ikeda, O. (1977). Super-resolution ultrasonic imaging by combined spectral and aperture synthesis, *J. Acoust. Soc. Am.* **62**, 341–345.
Sato, T., Sasaki, K., Nakamura, Y., and Nonaka, M. (1977). Bispectral passive holographic imaging system. *Acoust. Hologr.* **7**, 179–205.
Sato, T., and Wadaka, S. (1975). Incoherent ultrasonic imaging systems, *J. Acoust. Soc. Am.* **58**, 1013–1017.
Sato, T., Wadaka, S., and Ishii, J. (1977). New ultrasonic imaging system using a moving random phase mask and a stationary-point receiver, *J. Acoust. Soc. Am.* **62**, 102–107.
Sato, T., Wadaka, S., Ishii, J., and Sunada, T. (1977). A new ultrasonic imaging system by using a rotating random phase disk and power spectral and third order correlation analysis, *Acoust. Hologr.* **7**, 167–178.
Shaw, A., Paton, J. S., Gregory, N. L., and Wheatley, D. J. (1976). A real time two-dimensional ultrasonic scanner for clinical use, *Ultrasonics* **14**, 35–40.
Shibata, S., Koda, T., and Yamaga, J. (1978). C-mode ultrasonic imaging by an electronically scanned coaxial circular spherical receiving array, *Ultrasonics* **16**/2, 65–68.

Skattebol, L. V. (1968). Acoustic holograms, *Electron. Lett.* **4**, 583–584.

Skudryzk, E. (1971). "The Foundations of Acoustics." Springer-Verlag, Vienna and New York.

Smaryshev, M. D. (1973). "Directivity of Underwater Acoustic Arrays" (in Russian). Sudostroenie, Leningrad.

Smith, R. B., and Wade, G. (1971). Noise characteristics of Bragg imaging, *Acoust. Hologr.* **3**, 93–128.

Solov'ev, D. K. (1975). Spatial dependence of the frequency spectrum of a scanning acoustic array, *Sov. Phys.—Acoust.* **21**/2, 166–170 [*Akust. Zh.* **21**, 268–275 (1975)].

Stephanis, C. G., and Hampsas, G. D. (1978). Holographic acoustical lens, *J. Acoust. Soc. Am.* **63**, 860–862.

Svet, V. D. (1977). Application of wideband signals in acoustic holography, *Sov. Phys.-Acoust.* **23**/6, 531–533 [*Akust. Zh.* **23**, 929–923].

Szilard, J., and Kidger, M. (1976). A new ultrasonic lens, *Ultrasonics* **14**, 268–272.

Thorn, J. V., Booth, N. O., Sutton, J. L., and Saltzer, B. A. (1974). Test and Evaluation of an Experimental Holographic Acoustic Imaging System, Rep. NUC-TP-398. Nav. Undersea Cent., San Diego, California.

Thurstone, F. L. (1967). Holographic imaging with ultrasound, *J. Acoust. Soc. Am.* **45**, 895–899.

Thurstone, F. L. (1968). On holographic imaging with long-wavelength fields, *Proc. IEEE* **56**, 768–769.

Thurstone, F. L., and von Ramm, O. (1974). A new ultrasound imaging technique employing two-dimensional electronic beam steering, *Acoust. Hologr.* **5**, 248–259.

Turner, W. R. (1977). Experiments in ultrasonic imaging for weld inspection, *J. Acoust. Soc. Am.* **62**, 361–369.

Urick, R. J. (1967). "Principles of Underwater Sound for Engineers." McGraw-Hill, New York.

Valasek, J. (1969). "Introduction to Theoretical and Experimental Optics." Wiley, New York.

Vasil'ev, S. A. (1975). Exact solution of the problem of the formation of a real holographic image by a scalar field, *Sov. Phys.—Acoust.* **21**/1, 14–18 [*Akust. Zh.* **21**, 24–32 (1975)].

Vilkomerson, D. (1972). Analysis of various ultrasonic holographic imaging methods for medical diagnosis, *Acoust. Hologr.* **4**, 11–32.

von Ramm, O. T., and Thurstone, F. L. (1975). Cardiac imaging using a phased array ultrasound system, I. System design, *Circulation* **53**, 258–262.

Wade, G. (1975). Acoustic imaging with holography and lenses, *IEEE Trans. Son. Ultrason.* **SU-22**, 385–394.

Wang, K. Y. (1974). Comparison of ideal performance of some real-time acoustic imaging systems, *J. Acoust. Soc. Am.* **56**, 922–928.

Wang, K. Y., and Wade, G. (1972). Threshold contrast for various acoustic imaging systems, *Acoust. Hologr.* **4**, 431–462.

Wang, K. Y., and Wade, G. (1974). A high-sensitivity real-time acoustic imaging system for biomedical diagnosis, *Proc. IEEE* **62**, 650–651.

Wang, K. Y., and Wade, G. (1975). A scanning-focused-beam system for real-time diagnostic imaging, *Acoust. Hologr.* **6**, 213–228.

Wang, K. Y., Burns, V., Wade, G., and Elliott, S. (1977a). Opto-acoustic transducers for potentially sensitive ultrasonic imaging, *Opt. Eng.* **16**, 432–439.

Wang, K. Y., Chang, H., Shen, H., Wade, G., Su, K., Lo, M., and Elliott, S. (1977b). A transmitter for diagnostic ultrasonic imaging, *Acoust. Hologr.* **7**, 359–374.

Waugh, T. M., and Kino, G. S. (1977). Real time imaging with shear waves and surface waves, *Acoust. Hologr.* **7**, 103–115.

Waugh, T. M., Kino, G. S., DeSilets, C. S., and Fraser, J. D. (1976). Acoustic imaging techniques for nondestructive testing, *IEEE Trans. Son. Ultrason.* **SU-23**, 313–317.

Wells, P. N. T. (1977). "Biomedical Ultrasonics." Academic Press, New York.

White, D., and Brown, R. E. (eds.) (1977). "Ultrasound in Medicine," vols. 3A and 3B. Plenum Press, New York.

Whitman, R. L., Ahmed, M., and Korpel, A. (1972a). A progress report on the laser scanned acoustic camera, *Acoust. Hologr.* **4**, 11–32.

Whitman, R. L., Korpel, A., and Ahmed, N. (1972b). A novel technique for real time depth-gated acoustic image holography, *Appl. Phys. Lett.* **20**/7, 370.

Winter, D. (1973). Noise reduction in acousto-optic imaging systems by holographic recording, *Appl. Phys. Lett.* **22**/4, 151–152.

Wollman, M., and Wade, G. (1974). Experimental results from an underwater acoustical holographic system, *Acoust. Hologr.* **5**, 159–174.

Young, J. W. (1977). Electrically scanned and focussed receiving array, *Acoust. Hologr.* **7**, 387–403.

# Index*



* This index contains the names cited on pp. 1–216 as well as those of coauthors and editors in the References on pp. 217–223. The names of the alphabetically listed authors in the References are not repeated.

## F



## G



## H



## I

## N



## O



## P



## Q



## R



## S

**T**